Photovoltaic Materials

SERIES ON PROPERTIES OF SEMICONDUCTOR MATERIALS

Series Editor: Ron C. Newman

Forthcoming

Semiconductor Defect Modelling *ab initio* Methods
R. Jones and P. R. Briddon

Muon Spectroscopy of Semiconductors
S. F. J. Cox and R. Lichti

Complex Semiconductor Defects
M. Spaeth and F. Koschnick

Series on Properties of Semiconductor Materials – Vol. 1

Photovoltaic Materials

Richard H. Bube

Stanford University, USA

Imperial College Press

ICP

Published by

Imperial College Press
203 Electrical Engineering Building
Imperial College
London SW7 2BT

Distributed by

World Scientific Publishing Co. Pte. Ltd.

P O Box 128, Farrer Road, Singapore 912805

USA office: Suite 1B, 1060 Main Street, River Edge, NJ 07661

UK office: 57 Shelton Street, Covent Garden, London WC2H 9HE

Library of Congress Cataloging-in-Publication Data
Bube, Richard H., 1927–
    Photovoltaic materials / Richard H. Bube.
       p.   cm. -- (Series on properties of semiconductor materials ; vol. 1)
    Includes bibliographical references and index.
    ISBN 1-86094-065-X
    1. Photovoltaic cells -- Materials.   2. Semiconductors.   I. Title.
II. Series.
TK8322.B83   1998
621.3815'42--dc21   97-36334
                                                          CIP

British Library Cataloguing-in-Publication Data
A catalogue record for this book is available from the British Library.

This book is printed on acid-free paper.

Printed in Singapore by Uto-Print

The heavens are telling the glory of God;
and the firmament proclaims His handiwork. ...
In [the heavens] He has set a tent for the sun ...
Its rising is from the end of the heavens ...
and there is nothing hid from its heat.

Psalm 19:1,4,6

This book is dedicated to Betty Jane Meeker Bube,
my beloved wife of 48 years,
who went home to be with the Lord
when this book was in preparation.

PREFACE

Interest in photovoltaics has grown rapidly during the last half of the current century. The importance of alternative energy sources has increased in significance both for energy supply and ecological conservation reasons. In spite of limitations due to short-term economic considerations, research and development of photovoltaic solar cells has increased and is playing an increasingly practical role all over the world.

The performance of photovoltaic solar cells is intimately related to the properties of the materials from which they are made, and many materials science problems are encountered in the understanding of existing solar cells and the development of more efficient, less costly and more stable cells.

Every semiconducting material in a suitable electronic environment is capable of exhibiting properties that might properly be called "photovoltaic", i.e. the generation of an electric current and potential difference under absorbed illumination. What may be considered surprising, therefore, is that so few materials are known that are able to form photovoltaic devices with sufficient efficiency to make them of potential interest for practical applications. Allowing for the additional existence of a number of solid solution compounds based on this small number, it is still noteworthy that only the following materials have exhibited a solar efficiency in excess of ten percent: silicon, gallium arsenide, indium phosphide, cadmium telluride, copper indium diselenide, and cuprous sulfide. Of these the most versatile is silicon, which can be used to produce efficient cells in single crystal, polycrystalline or amorphous form.

It is the purpose of this book to describe the properties of these materials that play an important role in their photovoltaic applications, and to discuss the experimental and theoretical developments that have led to the leading contenders among photovoltaic cells for practical applications today. In conjunction with this is included (where appropriate) a briefer discussion of other systems that have been tested but have been found to exhibit major problems primarily in the areas of low efficiency or poor stability under operating conditions, as well as systems that are considered exploratory today but may develop into devices of interest in the future. It is not the purpose of the book to attempt an exhaustive history of the rapidly expanding field of photovoltaic materials, but to concentrate instead, primarily on developments in the last decade.

The book starts with a summary of the major properties of the semiconductor junctions that are involved in conventional photovoltaic devices, together with the typical quantitative relationships that describe electrical transport in these junctions. The major purpose of this summary is to provide a common framework for the discussions to follow, and it is assumed that the reader has a basic grounding in semiconductor electronics. It is anticipated that the book will be used in a teaching or research environment.

The focus of the book is on the materials properties important for photovoltaic applications, and does not deal in detail with device design or systems considerations, except where these are essential to the discussion. It is based on over twenty-five years of research experience, which has touched at one time or another on properties relevant to photovoltaic applications of all of the major materials.

The book covers research done by a wide variety of investigators in many countries. I am grateful for the assistance that has been given to me by many colleagues from all over the world in acquiring reprints of key research papers.

In my own program at Stanford University, thirty-one students have completed their Ph.D. research on photovoltaic materials and have added many significant memories in the process. I gratefully acknowledge support for that research from the Air Force Materials Laboratory at Wright–Patterson Air Force Base, the NASA Lewis Research Center, the National Science Foundation — Research Aimed at National Needs (NSF–RANN), ERDA, the Division of Basic Energy Science (BES) of the Department of Energy, the Solar Energy Research Institute (SERI), the Electric Power Research Institute (EPRI), ARCO Solar, and the National Renewable Energy Laboratory (NREL). The research benefited throughout from facilities made available to Stanford University by the National Science Foundation through the Center for Materials Research at Stanford University.

I am indebted to the many Visiting Scholars from all over the world who have participated in this research, and particularly to my immediate colleagues at Stanford: Dr. Alan L. Fahrenbruch, Dr. David Redfield, and Dr. Adolfo Lopez-Otero. I am also most thankful for the sun that has shone in my own life throughout these years because of the constant love and support of my wife Betty.

Stanford 1997   Richard H. Bube

CONTENTS

CHAPTER 1

PHOTOVOLTAIC CELLS AND PHENOMENA

1.1. Overview

A material (or device) is said to be "photovoltaic" when exposure of the material to light that can be absorbed by the material is able to transform the energy of the light photons into electrical energy in the form of a current and voltage. The concept is simple and the number of materials that are able to exhibit photovoltaic characteristics is large. What is not large, however, is the number of such materials or devices that are able to make the transformation of solar radiation to electrical energy with high efficiency, of the order of 20%, at low cost, and with high stability under operation.

The development of photovoltaic cells can be dated originally to Becquerel's discovery in 1839 of a photovoltage produced by the action of light on an electrode in an electrolyte solution (Becquerel, 1839). About 40 years later, Adams and Day observed the photovoltaic effect in selenium (Adams and Day, 1877). Solar efficiencies of about 1% characterized selenium and copper oxide cells by about 1914. The modern era of semiconductor photovoltaics started in 1954 when Chapin, Fuller and Pearson obtained a solar efficiency of 6% for a silicon junction cell (Chapin et al., 1954), a value that was increased to 14% by 1958, and to 28% by 1988 (Verlinden et al., 1988). The year 1954 also dates the announcement of the first all-thin-film cell composed of a Cu_xS/CdS junction (Reynolds et al., 1954) with an efficiency of 6%, later increased to over 9% (Bragagnolo et al., 1980), which unfortunately had the stability problems described below. The first mention of a cell based on GaAs was for a 4% p–n homojunction in 1956 (Jenny et al., 1956); later developments were to produce cells with efficiencies greater than 30%, as discussed in Chapter 4. Concern about energy resources motivated a strong surge of interest in solar cells for terrestrial applications in the early 1970's. In spite of the fact that the 1980's and '90s have been a period in which public and government support for photovoltaics have been underemphasized, considerable activity and progress in research and development of solar cells has continued.

1

The Photovoltaic Process

The basic concept of the photovoltaic process is also simple. When light with photon energy greater than the band gap is absorbed by a semiconductor material, free electrons and free holes are formed by optical excitation in the semiconductor. The crucial characteristic needed for the photovoltaic effect is the presence of some kind of internal electric field (from any of a variety of causes: e.g. different doping in different regions, contacts, surfaces, etc.) that is able to separate the freed electrons and holes so that they can pass out of the material into the external circuit before they recombine with one another. The flow of carriers into the external circuit constitutes a reverse electrical current density, J amp cm^{-2}, which, under short-circuit conditions, is known as the short-circuit current density, J_{sc}. (For consistency, in this book the term current will always mean current density J unless specifically stated otherwise. The total current $I = JA$, where A is the area through which the current flows.) At the same time, the separation of the charges sets up a forward

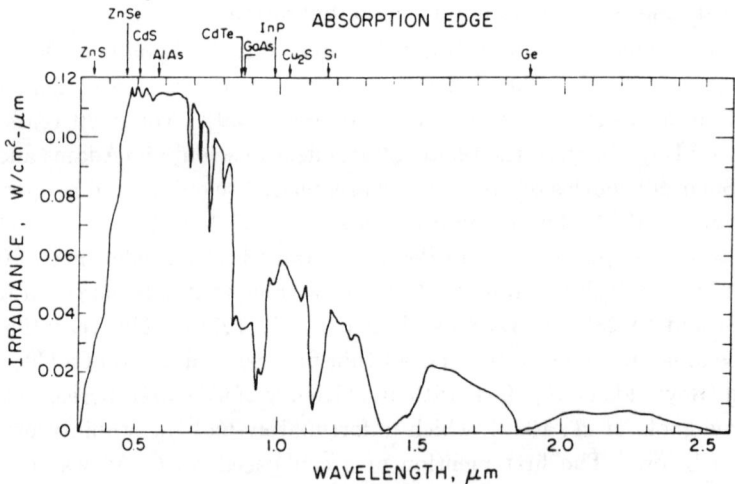

Fig. 1.1. Average solar spectrum at the surface of the earth, with the band gaps of various semiconductors indicated. As discussed in this chapter, a homojunction or a front-wall heterojunction will absorb all the solar spectrum to the left of the indicated absorption edge for a particular material. A back-wall heterojunction will absorb the solar spectrum between the two band gaps of the materials making up the heterojunction. (Reprinted from R. H. Bube, "Solar Cells" in *Handbook on Semiconductors. Device Physics.* Vol. 4C. C. Hilsum, ed., 1993, p. 797, with kind permission from Elsevier Science — NL, Sara Burgerhartstraat 25, 1055 KV Amsterdam, The Netherlands.)

potential difference between the two ends of the material, ϕ, which under open-circuit conditions is known as the open-circuit voltage, ϕ_{oc}. The polarity of the open-circuit voltage, therefore, is such as to drive electrons (holes) in the opposite direction (the forward-bias direction) to that of the electron (hole) motion in the short-circuit current. Specific examples of photovoltaic systems are described in more detail in following sections of this chapter.

Maximizing Photovoltaic Performance

It is desirable to maximize both J_{sc} and ϕ_{oc}. In order to maximize J_{sc}, it is desirable (1) to absorb as much of the incident light as possible, i.e. to have a small band gap with high absorption over a wide energy range, and (2) to

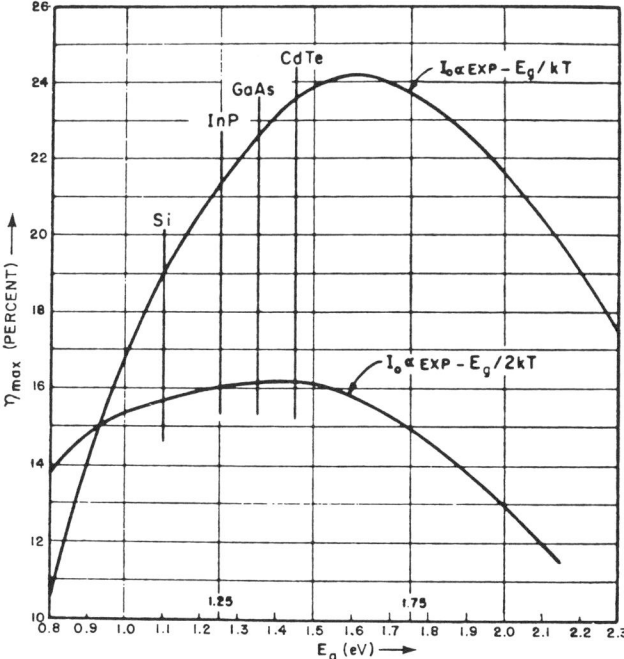

Fig. 1.2. An early calculation of the theoretical solar efficiency versus semiconductor band gap for ideal p–n homojunction cells with no surface recombination loss. Curves are shown for two different junction transport mechanisms: (top) $A = 1$ for injection dominated current and (bottom) $A = 2$ for recombination in the depletion layer. (Reprinted with permission from J. J. Loferski, *J. Appl. Phys.* **27**, 777 (1956). Copyright 1956, American Institute of Physics.)

have material properties such that the photoexcited electrons and holes are able to be collected by the internal electric field and pass into the external circuit before they recombine, i.e. a material with a high minority carrier lifetime and mobility. In order to maximize ϕ_{oc}, it is also preferred to have the forward current driven by the photo-induced potential difference be as small as possible since this current will reduce the potential difference set up by light. The details of this forward current depend critically on the actual mechanisms of transport involved, but in general the forward current varies inversely as the band gap of the material.

From these two purely qualitative considerations, therefore, it may be concluded that there will be an optimum band gap or band gap range for

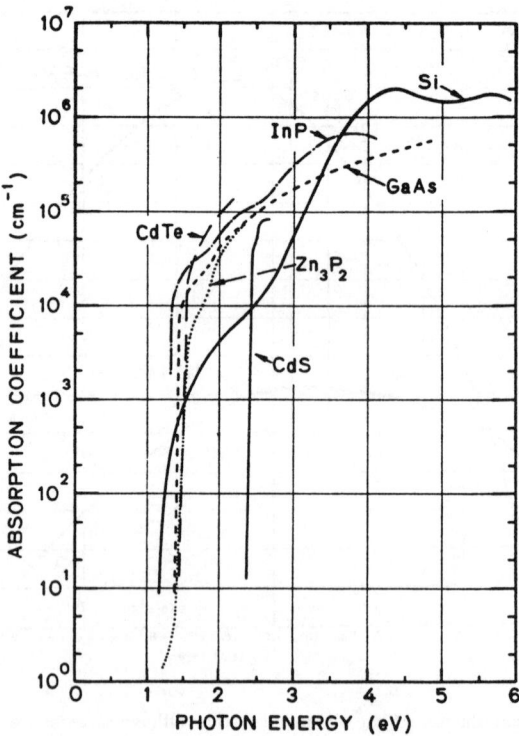

Fig. 1.3. Optical absorption coefficients of various single-crystal semiconductors commonly used in photovoltaic devices. (Reprinted with permission from A. L. Fahrenbruch and R. H. Bube, *Fundamentals of Solar Cells: Photovoltaic Solar Energy Conversion*, Copyright 1983, Academic Press, Orlando, FL.)

maximizing both J_{sc} and ϕ_{oc}, and hence the efficiency of the photovoltaic device itself. For interaction with the solar spectrum (see Fig. 1.1), this optimum band gap range lies approximately between 1.2 and 1.8 eV. The results of one of the first calculations of the dependence of efficiency on band gap for two different types of junction current are pictured in Fig. 1.2 (Loferski, 1956). Details of the concepts described in the figure caption are discussed later in this chapter. For comparison, Fig. 1.3 shows the dependence of optical absorption coefficients on photon energy for various single-crystal semiconductors used in photovoltaic devices (Fahrenbruch and Bube, 1983).

Description of Solar Spectrum

As shown in Fig. 1.1, the sun is a complex radiator with a spectrum that can be approximated by the spectrum of a 6050 K black body. This spectrum is then modified by temperature variations across the sun's disk, the effects of the solar atmosphere, Fraunhofer absorption lines, and the path length of the radiation through the earth's atmosphere, where the primary effects are due to water content, turbidity, ozone, cloudiness, and ground reflection.

For terrestrial applications, the path length through the atmosphere is of fundamental importance. This path length can be conveniently described in terms of an equivalent "air mass", m_r. If a zenith angle z is defined as the angle from the normal to the plane containing the horizon circle, which describes the declination of the sun, the path length for a zenith angle z is just sec z times the path length for $z = 0$, and the air mass m_r, is defined as $m_r = \sec z$. Specific solar spectra are labeled AMm_r (read: air mass m_r). AM0 corresponds to the solar spectrum in outer space; AM1 to the solar spectrum at the earth's surface for the sun overhead. AM2 corresponds to an approximately average solar spectrum at the earth's surface.

Simplest Photovoltaic Device

In order to make these introductory concepts more specific, let us consider the simplest photovoltaic device and the mathematical description of the electrical properties of such a device. In subsequent sections of this chapter, we summarize some of the possible variations and complications in more detail. Here we use the simple equivalent circuit for a photovoltaic cell as shown in Fig. 1.4, including a current generator corresponding to photoexcitation, a diode containing the internal electric field necessary for driving photoexcited carriers to the external circuit, a series resistance, R_s, and a parallel resistance, R_p. We

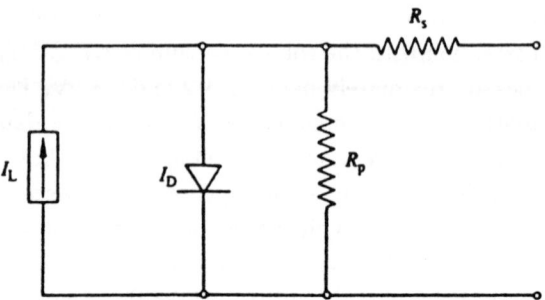

Fig. 1.4. Simple equivalent circuit for a photovoltaic cell, including a current generator with total light current I_L, a diode with total dark current I_D, a series resistance R_s, and a parallel resistance R_p. (Reprinted with permission from R. H. Bube, *Photoelectronic Properties of Semiconductors*, Copyright 1992, Cambridge University Press.)

make the model even simpler by considering at this point an "ideal device" with $R_s = 0$ and $R_p = \infty$. If we make the simple assumption that the current generated by light can simply be added to the current flowing in the dark ("superposition"), then the current density J flowing in the device in the presence of photoexcitation can be expressed as

$$J = J_o[\exp(q\phi/AkT) - 1] - J_L \tag{1.1}$$

Here the first term on the right of Eq. (1.1) is the forward current driven by the voltage ϕ, and the second term is the reverse current associated with photoexcitation. J_o is often called the "reverse saturation current" of the diode, the value of J in the dark for large negative values of ϕ in ideal junctions, which depends on the actual transport mechanism for the diode current (J_o is often called simply the "pre-exponential coefficient" in practical devices for which reverse saturation may not occur), and A is the so-called "ideality factor" that has a value depending on the mechanism of the junction transport (e.g., $A = 1$ if the transport process is diffusion, $A \approx 2$ if the transport process is recombination in the depletion region). Typical variations of total current I in both the dark and the light as a function of ϕ are given in Fig. 1.5. If the voltage is zero (short-circuit condition), then of course there is zero current in the dark, but in the light we have

$$J_{sc} = -J_L \tag{1.2}$$

and the short-circuit current is controlled only by the photoinduced current generation and the recombination processes. If the total current under

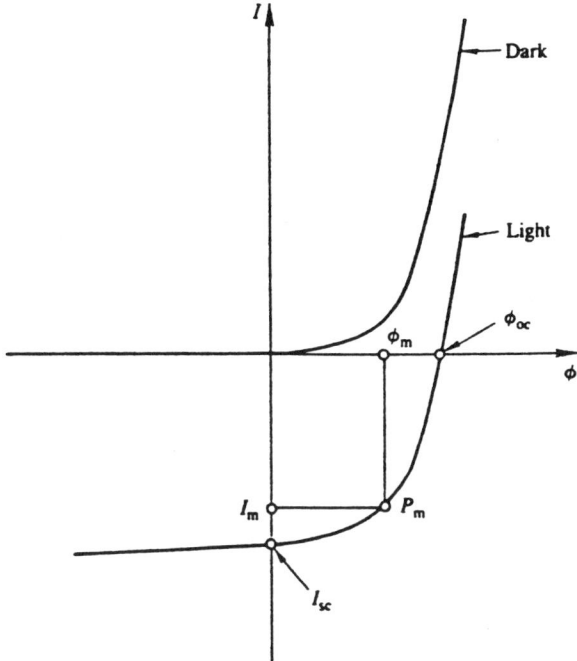

Fig. 1.5. Typical idealized total light and dark current versus voltage curves for a photovoltaic cell in which the principal cell parameters do not depend on photoexcitation, showing the open-circuit voltage ϕ_{oc}, the short-circuit current I_{sc}, and the maximum power point P_m. (Reprinted with permission from A. L. Fahrenbruch and R. H. Bube, *Fundamentals of Solar Cells: Photovoltaic Solar Energy Conversion*, Copyright 1983, Academic Press, Orlando, FL.)

illumination is zero (open-circuit condition), then solution of Eq. (1.1) for $J = 0$ gives

$$\phi_{oc} = (AkT/q) \ln[(J_L/J_o) + 1] \tag{1.3}$$

Thus the open-circuit voltage is controlled by the current generation and recombination processes, but also by the nature of the junction transport currents depending on A and J_o. Combination of Eqs. (1.2) and (1.3) shows that the relation between J_{sc} and ϕ_{oc} is given by

$$J_{sc} = J_o[\exp(q\phi_{oc}/AkT) - 1] \tag{1.4}$$

which is exactly similar to Eq. (1.1) for J versus ϕ in the dark for this ideal junction device. The equivalence of these two dependences is a basic test for the absence of any light-related changes in the parameters entering the equations.

Photovoltaic Efficiency

Since photovoltaic efficiency is of central importance, we see next how this efficiency is expressed in terms of this simple model. The efficiency of a photovoltaic solar cell is a maximum when the product of the current density J and the voltage ϕ is a maximum. The efficiency η itself can be expressed as

$$\eta = P_m/P_{\text{rad}} = J_m\phi_m/P_{\text{rad}} = J_{sc}\phi_{oc}ff/P_{\text{rad}} \qquad (1.5)$$

where P_{rad} is the total radiation power incident on the cell, and ff is called the fill factor; ff is a measure of the "squareness" of the light J–ϕ curve, as shown in Fig. 1.5. J_m and ϕ_m are respectively the values of current density and voltage at the condition corresponding to maximum power. The definition of the fill factor ff can be obtained from Eq. (1.5),

$$ff = J_m\phi_m/J_{sc}\phi_{oc} \qquad (1.6)$$

The value of ϕ_m can be obtained by multiplying Eq. (1.1) by ϕ and maximizing the power with respect to ϕ.

$$\phi_m = \phi_{oc} - (AkT/q)\ln[(q\phi_m/AkT) + 1] \qquad (1.7)$$

which can be solved iteratively for ϕ_m. Once a value for ϕ_m has been obtained, the value of J_m, and hence the maximum power $P_m = J_m\phi_m$, can be obtained from Eq. (1.1) by substituting $\phi = \phi_m$. The fill factor for a junction describable by Eq. (1.4) is a function of A and ϕ_{oc}, increasing with decreasing A and increasing ϕ_{oc} (Lindmayer, 1972).

Figure 1.6 gives general insights into the factors that determine the efficiency in an actual photovoltaic solar cell. The figure shows the various components to the loss of efficiency in a standard single-crystal silicon cell for irradiation by sunlight. Specific numbers depend, of course, on the details of the material and the solar cell, and those given here are intended to be primarily illustrative. The analysis is similar to that carried out by Wolf (1971). Starting from the top of the diagram down:

(1) Some of the cell area is obscured by the current-collecting grid, thus reducing the effective cell area: 4% loss;
(2) some of the incident photons are reflected and not absorbed: 2% loss;
(3) some of the photons are absorbed in spurious absorption processes such as at antireflection coatings, at defects, etc., which do not lead to free carriers: 1% loss;

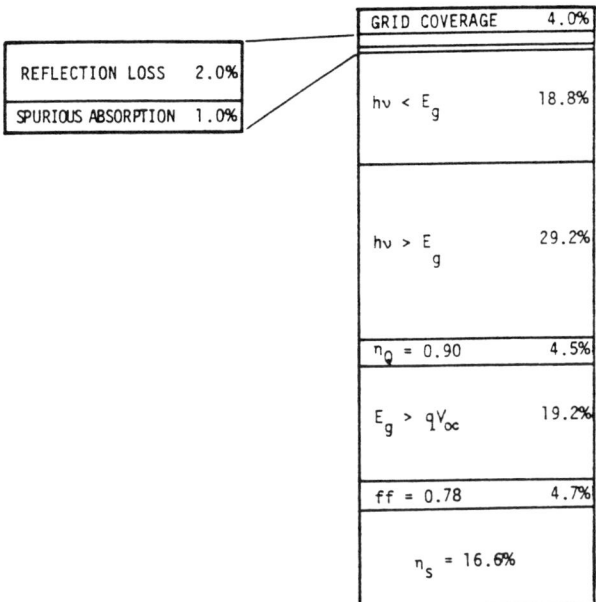

Fig. 1.6. Power loss chart for a silicon cell showing the percentage of total input power P_{rad} lost to each of the loss mechanisms. Values of $J_o = 5 \times 10^{-12}$ A cm^{-2} and $A = 1$ are assumed. (Reprinted with permission from A. L. Fahrenbruch and R. H. Bube, *Fundamentals of Solar Cells: Photovoltaic Solar Energy Conversion*, Copyright 1983, Academic Press, Orlando, FL.)

(4) some of the incident photons have an energy less than the band gap of the semiconductor ($h\nu < E_g$), and hence have insufficient energy to produce free carriers in the absorbing material: 18.8% loss;

(5) some of the absorbed photons have an energy larger than the band gap E_g of the absorbing material; the excess energy ($h\nu - E_g$) is non-usefully dissipated in producing hot carriers which transform the light energy to heat as the carriers thermalize to near the band edge: 29.2% loss;

(6) not every photoexcited free carrier is collected by the internal electric field; this is expressed by the quantum efficiency $\eta_Q < 1$: if $\eta_Q = 0.90$, this amounts to a 4.5% loss;

(7) in general the band gap $E_g > q\phi_{oc}$, i.e., the energy used to create the free carriers ($h\nu \geq E_g$) is greater than the energy associated with the open-circuit voltage, and there is therefore an energy loss of 19.2%;

(8) finally in a real solar cell forward-biased to ϕ_m, with finite R_s and R_p, the fill factor $ff < 1$: if $ff = 0.78$, this amounts to a 4.7% loss.

Given the various estimates listed here, the cell efficiency is finally 16.6%. Losses (1) to (3), and (6) to (8), can be reduced by careful material control and cell design. Losses (4) and (5) are difficult to avoid with simple solar cells, but more complicated structures (graded gap or multijunction structures) can be designed to help here as well. Photovoltaic cells with efficiency over 30% have been designed and produced.

Summary of Important Materials Properties

We now summarize the most significant materials properties that are important for the preparation of high efficiency photovoltaic solar cells. Rothwarf (1987) has listed some twenty to thirty materials and processing related parameters that must be controlled and optimized for maximizing solar-cell performance.

(1) Band gap of the absorbing material. The band gap of the absorbing material must be small enough to allow absorption of an appreciable portion of the solar spectrum, and at the same time large enough to minimize the reverse saturation junction current density J_o.

(2) Diffusion length of minority carriers. The diffusion length of minority carriers must be as large as possible so that carriers excited by light some distance from the actual semiconductor junction will be able to diffuse to the junction and be collected before they recombine with carriers of the opposite sign. The diffusion length of minority carriers L_{min} is given by $L_{min} = (D_{min}\tau_{min})^{1/2} = [(kT/q)(\mu\tau)_{min}]^{1/2}$, where D_{min} is the diffusion constant, τ_{min} is the lifetime, and μ_{min} is the mobility, for minority carriers. It is desired, therefore, to have a material in which the minority carriers have as large a mobility and lifetime as can be obtained. The value of the mobility is more or less determined by the choice of material and does not vary over a wide range. The value of the lifetime, however, is very sensitive to a variety of phenomena in the bulk and at the surface that contribute to recombination of photoexcited carriers. Optimization of efficiency must include solar cell growth and deposition conditions that maximize the minority carrier lifetime. Cell design must include photoexcitation of minority carriers within a diffusion length of the collecting junction.

(3) Desirable junction properties. The actual junction structure and composition determines the magnitude of the junction transport current density J_o and of the ideality factor A. Formation of the semiconductor junction must be carefully controlled, therefore, to produce junctions with as low a junction current as possible. Various typical mechanisms for the junction current are summarized in Sec. 1.4.

(4) In the simple analysis of this section, we have assumed that $R_s = 0$ and $R_p = \infty$ in Fig. 1.4. In real solar cells, however, finite values of these two resistances will be present and can be a major factor, particularly in determining the effective value of the fill factor ff. Contributions to the series resistance R_s can arise both from the resistance of the semiconductor bulk and from the contact resistance to the semiconductor to complete the circuit. Problems involving semiconductor doping and control of contact resistance can play a significant role in some cases. The parallel resistance R_p can be reduced by grain boundaries or other defects that enhance forward junction current and contribute to an increase in J_o and a decrease in ϕ_{oc}. In polycrystalline thin film solar cells, grain boundaries at the junction interface can critically affect junction transport properties.

(5) Solar cells are intended for use in exposed areas for long periods of time without failure. This means that a variety of phenomena that might lead to a decrease in cell efficiency with time of exposure must be carefully considered. In some cases, as we shall see, these instability problems may play a dominant role in determining cell efficiency and utility.

All five of these areas of materials properties must be carefully designed and controlled to maximize the efficiency of an actual solar cell. One of the purposes of this book is to describe the ways in which practical considerations of this type have been considered and dealt with.

Reference Information

This book provides three kinds of reference information at the end of the volume: (1) a representative bibliography of books that have been written on the subject of photovoltaics, with entries listed chronologically; (2) a representative list of review papers on aspects of photovoltaics, with entries also listed chronologically; and (3) a traditional list of references to specific research referred to throughout the book by name of author and date of publication, listed alphabetically by first author's name.

In addition, general attention should be called to the published papers presented at a number of meetings dedicated to the subject of photovoltaics. The proceedings of the IEEE Photovoltaic Specialists Conference, with meetings held at 18 month intervals and its 25th meeting in May 1996, are a rich source of information on progress in the field. At this 25th meeting, for example, 370 papers were presented involving almost 1000 authors. The first World Conference on Photovoltaic Energy Conversion met in Hawaii as part of the 24th IEEE Photovoltaic Specialists Conference in 1994, and the Second World

Conference is being held in Vienna in July 1998. There are comparable proceedings of the Photovoltaic Solar Energy Conferences of the Commission of the European Communities; the 14th Conference was held in Barcelona in July 1997.

Another valuable source of continuing input is available through the Program Review Proceedings and the Annual Reports published through the years by NSF-RANN (1973–1975), ERDA (1975–1978), and DOE, and the Solar Energy Research Institute (SERI), now the National Renewable Energy Laboratory (NREL) (since 1978). The Office of Scientific and Technical Information of DOE also published a bimonthly collection of Current Abstracts in Photovoltaic Energy. In 1996 this service was replaced by a listing of abstracts on photovoltaic technology on the World Wide Web at: http://www.doe.gov/phv/phvhome.html. Beginning in October 1996, it also became possible to receive a listing of the citations to photovoltaic reports processed by the Office of Scientific and Technical Information (OSTI) during the preceding two months, by regular mail, e-mail, or fax.

Some of the background material included in the book, particularly in this first chapter, is adapted with minor revision from the chapter by the author on "Solar Cells", in *Handbook on Semiconductors, Device Physics*, Vol. 4, ed. C. Hilsum, 797–839 (1993), with kind permission from Elsevier Science, NL, Sara Burgerhartstraat 25, 1055 KV Amsterdam, The Netherlands. Considerable use has also been made of two other books involving the author for background material and insights into earlier years of research on photovoltaics: *Fundamentals of Solar Cells: Photovoltaic Solar Energy Conversion* by A. L. Fahrenbruch and R. H. Bube, Academic Press, N.Y. (1983), and a chapter on "Photovoltaic Effects" in *Photoelectronic Properties of Semiconductors* by R. H. Bube, Cambridge Univ. Press, Cambridge (1992).

1.2. Applications of Photovoltaics

Photovoltaics are primarily devices to provide alternative sources of electrical energy using the radiant energy that comes to us from the sun. Most of these attempts are motivated by the realization of the limited supply of fossil fuels, the undesirable consequences of nuclear fission, and the stress that energy supply places on the environment in a world with a growing population, a relatively small fraction of which uses far more than its proportional share of energy at the expense of the rest. If the future of the earth is considered in terms of hundreds, rather than tens of years, traditional sources of energy, such as coal, oil and natural gas, provide only a limited resource. Furthermore, growing usage of these fossil fuels contributes to a variety of environmental problems: air pollution, acid rain, and the greenhouse effect, to name only a few. In principle nuclear energy might be considered as a way to overcome these problems, but nuclear fission is an uninviting prospect in view of the dangers in plant safety, waste disposal, unfavorable economics, and potential for misuse, and nuclear fusion is still an uncertain possibility.

The utility of photovoltaics is especially significant with respect to remote applications, and one of the chief advantages of photovoltaics in the future is decentralization of the power grid, a consideration also related to remote energy use.

A Fundamental Limitation

It should be remembered that there is a fundamental thermodynamic constraint on our ability to convert energy on earth, considering the accompanying heat, without catastrophic effects. If we were to release on earth, through the use of either fossil fuels or nuclear energy, heat corresponding to 1% of the solar power density falling on the earth averaged over a year, the average temperature of the earth would be increased by about 1°C, with major effects on climatic conditions all over the world. Today the average level of energy density consumption in the United States is already about 0.2% of the incident solar power density, with considerably higher values in urban areas. Only renewable energy sources such as solar and wind energy enable us to avoid these constraints.

What Photovoltaic Energy is Available?

Is photovoltaic use of solar energy adequate to the task? Sunlight falls on the earth with an intensity of about 1 kW/m^2. If this energy could be converted to electricity with a relatively modest 10% overall efficiency, 1 kW of electricity would be generated for every 10 m^2 of active area as long as the sun was shining near its peak intensity. This is of the order of the average electricity usage per residence. Because the sun does not shine at its peak intensity for 24 hours on any day, and because the periods of sunny days are interspersed with periods of cloudy days, both a greater area than this 10 m^2 (by a factor of about ten) and a means of electrical storage are required to distribute the use over non-sunny periods.

Main Areas of Concern for Photovoltaics

The basic concerns for long-range use of photovoltaics for terrestrial solar energy conversion lie in three areas: (1) efficiency, (2) cost, and (3) operating lifetime. These three areas are interrelated in a complex way. Already for a number of years photovoltaics have been applied to a variety of localized tasks, such as pumping water or telephone communications, in developing nations or elsewhere in the world where a power grid does not currently exist. It is the major use of photovoltaics to supply an appreciable fraction of the electrical power on a general basis that raises critical issues.

(1) The achievable efficiency with a particular photovoltaic system is highly important and it has become generally agreed that cell efficiencies of about 20% (increased from an earlier estimate of 10%) is a lower limit for large-scale power applications. Even if cells with appreciably lower efficiency are very inexpensive, the cost of the cell then becomes insignificant compared to the costs of the installation required for its use. Contemporary efficiencies range from about 14% for single-junction thin-film polycrystalline cells, to close to 30% for single-junction single-crystal cells to be used with solar concentration (up to a factor of 1000 or more). Efficiencies greater than 30% have been achieved with multijunction cells, devices in which the sunlight is passed sequentially through two or more single-junction cells to maximize the match between sunlight absorbed and semiconductor band gap. Most of these cells are made with either silicon (Si) or III–V compounds related to gallium arsenide (GaAs). Only three materials used in single-junction thin-film cells without solar concentration have been demonstrated to provide an efficiency greater than 10%: hydrogenated amorphous silicon (a-Si:H), cadmium telluride (CdTe), and copper indium diselenide ($CuInSe_2$). Each of these major materials has a chapter devoted to it in this book.

(2) The basic cost of photovoltaic systems decreased dramatically by a factor of twenty between 1956 and 1976, and in the last decade it has decreased from fifteen dollars to 30 cents per kilowatt-hour; the cost goal by the end of the next decade is six cents per kilowatt-hour. Photovoltaic cells for terrestrial applications will take one of two major forms: a large-area, thin-film, flat-plate form, probably installed at a fixed angle with respect to the earth, suitable for large-area dispersed applications on rooftops etc., or a single-crystal form to be used with a concentrator system built to maintain maximum sunlight on the cell throughout the day through tracking of the concentrator, for centralized power production. Reduction of cost for the thin-film cells is achieved by minimization of the amount of material used, the possibility of inexpensive materials processing methods, and the use of inexpensive mountings. The single crystal cells are themselves generally

highly complex and expensive, but the small amount of material needed with intense concentration of sunlight again effects a cost reduction.

(3) To be effective as an alternative energy source, a solar cell must generate at least enough energy in its operating lifetime to pay back both the financial and energy costs required to produce the cell in the first place, and hopefully several times this. It is estimated that an operating lifetime of about 20 years would be a workable value. This lifetime may be determined by a variety of external factors such as physical damage, corrosion, deterioration of cell support structures etc., or by what is of more direct concern to us in this book, by a variety of internal materials-related factors such as diffusion, photogeneration of defects etc. For cells in space, as compared to terrestrial-use cells, radiation damage is also a major factor in degrading performance.

The Illustrative Case of Cu_xS/CdS Solar Cells

Stability for long-term operation has proven to be a real problem for several thin-film technologies. The original all-thin-film, solar cell system consisting of a junction between Cu_xS and CdS showed promise of reasonable efficiency and ease of preparation (Reynolds *et al.*, 1954; Boer and Meakin, 1975; Stanley, 1975; Rothwarf and Barnett, 1977; Barnett, 1977). For a period of almost 20 years, dating from the late '50s, this photovoltaic system was the only all-film cell available. Solar cells with 10% efficiency were also made with $Cu_xS/Cd_{1-y}Zn_yS$ thin films ($y = 0.10, 0.16$) with open-circuit voltages greater than that in Cu_xS/CdS cells (Hall *et al.*, 1981). A review is given in Fahrenbruch and Bube (1983) and by Hill and Meakin (1985). Although commonly prepared by the procedure of simply dipping CdS into a warm solution of cuprous ions to effect a replacement reaction in which Cu_xS forms topotaxially on the CdS, other methods of depositing the Cu_xS were also investigated, such as the vacuum evaporation of CuCl followed by heat treatment, or the deposition of Cu_xS by reactive sputtering.

The complexity of this system arises from the variety of Cu_xS phases that may exist at room temperature with quite different photovoltaic properties, and from the ability to change from one phase to another during cell operation due to interaction with the atmosphere or diffusion of Cu into the CdS. Cu_xS can exist in the chalcocite phase ($x = 1.995$–2.000), which has superior photovoltaic properties, the djurleite phase ($x = 1.96$), and the digenite phase ($x = 1.8$) Reduction in the value of x can occur by diffusion of Cu into the CdS or by oxidation of Cu at the free surface of Cu_xS, and results in a degradation of photovoltaic properties. Any heat treatment of the cell causes appreciable Cu diffusion into the CdS, which forms deep acceptor states in the CdS that widen the depletion layer and lead to greater carrier recombination before collection, giving rise to *enhanced* (narrow depletion layer), and *quenched* (wide depletion layer) conditions. In addition to these effects, an optical degradation process occurs through a reversible photoinduced defect reaction (Redfield and Bube, 1996), caused by photoexcitation after Cu diffusion into the CdS has occurred (Kanev *et al.*, 1969, 1971). This leads to a state where lifetime-killing defects have been formed that can be thermally annealed away, giving rise to *degraded* (lifetime killer defects formed), and *restored* (lifetime-killer defects annealed away) conditions. In an appropriately heat-treated junction of this type, the value of J_{sc} for the enhanced and restored state (narrow depletion layer and photoinduced defects annealed away), can be over 10^3 times that for the quenched and degraded state (wide depletion layer and photoinduced defects active). The effects associated with the diffusion of Cu into the CdS upon heat treatment used in cell preparation, have been investigated for Cu_xS formed on single crystal CdS (Gill and Bube, 1970; Lindquist and Bube, 1972a,b). Detailed summaries of the results have been described (Fahrenbruch and Bube, 1974, 1983; Bube, 1992).

Many attempts to produce a stable cell in spite of these problems looked promising but proved unsuccessful and further work on the system was finally abandoned, and we will not discuss the details of this particular system further in this book. We refer the interested reader to the above discussions in the literature for what is scientifically a fascinating problem. Among today's thin-film cell materials, stability is a special problem for a-Si:H cells because light-induced formation of defects decreases the lifetime and diffusion length of free carriers; this particular problem remains a live issue and is discussed in more detail in Chapter 3.

1.3. Types of Semiconductor Junctions

There are six different types of semiconductor junctions that have possible application in photovoltaic solar cells, each junction having the fundamental role of supplying the internal electric field needed to separate the photoexcited carriers and to cause them to flow as a current in the external circuit: (1) homojunctions, (2) heterojunctions, (3) heteroface junctions or buried homojunctions, (4) metal-semiconductor junctions (Schottky barriers), (5) *p–i–n* junctions, and (6) semiconductor-electrolyte junctions. We describe the characteristic energy-band structure for each of these junctions in this section. Discussion of the use of different materials in these possible junction configurations for photovoltaic solar energy conversion makes up the rest of this book.

Homojunctions

A typical energy-band diagram for a homojunction is given in Fig. 1.7. A homojunction consists of a junction between two portions of the same semiconductor, one doped p-type and the other doped n-type, hence the name, p–n junction. Typical details of such an energy-band diagram are shown in the figure: the vacuum level E_{vac}, the conduction band edge E_c, the Fermi level E_F, the valence band edge E_v, the band gap E_G, the electron affinity χ_s, and the diffusion potential $q\phi_D$. The work function $q\phi_W$ of a semiconductor is defined as the energy difference $(E_{vac} - E_F)$. Since the work function of the p-type portion of the material is greater than that of the n-type portion in Fig. 1.7,

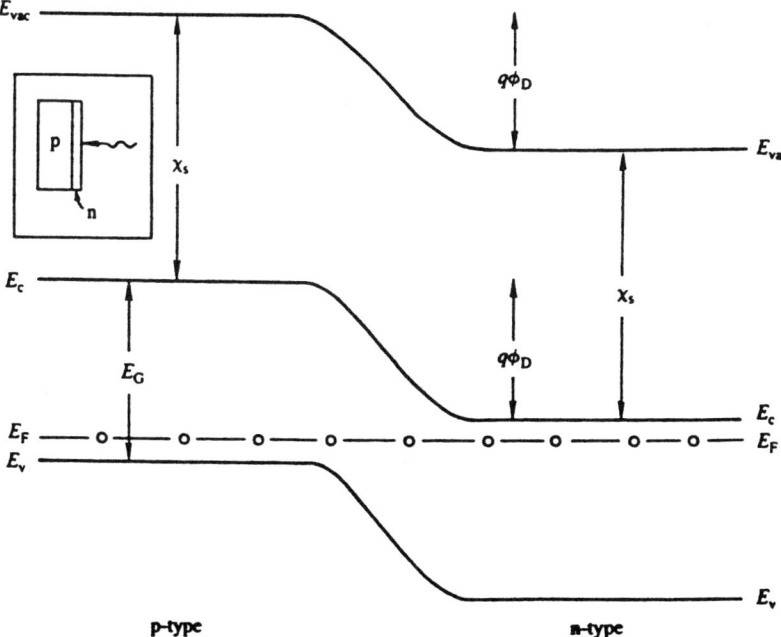

Fig. 1.7. Energy-band diagram for a p–n homojunction with equal densities of doping in the p- and n-type portions.

$(q\phi_{W_p} - q\phi_{W_n}) = q\phi_D$, the energy bands between p- and n-type portions are curved, indicating the presence of an internal electric field. Physically one can think of the diffusion potential $q\phi_D$ as resulting from a transfer of electrons from n-type to p-type material in order to equalize the Fermi energy on both sides when the junction was formed, giving rise to a positive charge of ionized donors in an electron-depleted layer near the junction interface on the n-type side and a negative charge of ionized acceptors in a hole-depleted layer near the junction interface on the p-type side, resulting in a flat Fermi level across the whole structure. The diagram shown in Fig. 1.7 is specifically for the situation where the doping of the p- and n-type regions is the same, resulting in equal widths for the depletion layers on both sides of the junction.

Photoexcitation produces free minority carrier electrons in the p-type region, and free minority carrier holes in the n-type region. Each of these then diffuse toward the junction, and if they reach the junction without being removed by recombination, they pass over the junction, are collected by the junction field, diffuse through the other portion of the semiconductor and pass into the external circuit. Since carriers must be excited within by about a diffusion length of the junction in order to be able to diffuse to the junction before recombination occurs, the geometry shown in the insert in Fig. 1.7 is frequently used, with the portion of the semiconductor on the illuminated side being much thinner than that on the opposite side. If the semiconductor has the high optical absorption characteristic of a direct band gap, then the illuminated side of the junction must be very thin (≈ 0.1 μm) to allow light to penetrate to within a diffusion length of the junction, but this has the disadvantage of causing many carriers to be generated near the surface where the probability of recombination due to surface defects is greater than that in the bulk. On the other hand if the semiconductor has a lower optical absorption characteristic of an indirect band gap, then the portion of the semiconductor on the opposite side to that illuminated must be quite thick to allow most of the light to be absorbed, but then much of this absorption occurs more than a diffusion length away from the junction unless the diffusion length of the material is large. These considerations show immediately the need for a high-quality, well-engineered material to serve in an efficient p–n homojunction type of solar cell. High-efficiency, single crystal Si solar cells are usually of the p–n homojunction type.

Heterojunctions

A p–n heterojunction is a p–n junction formed between two different semiconductors with different band gaps and electron affinities. A typical band

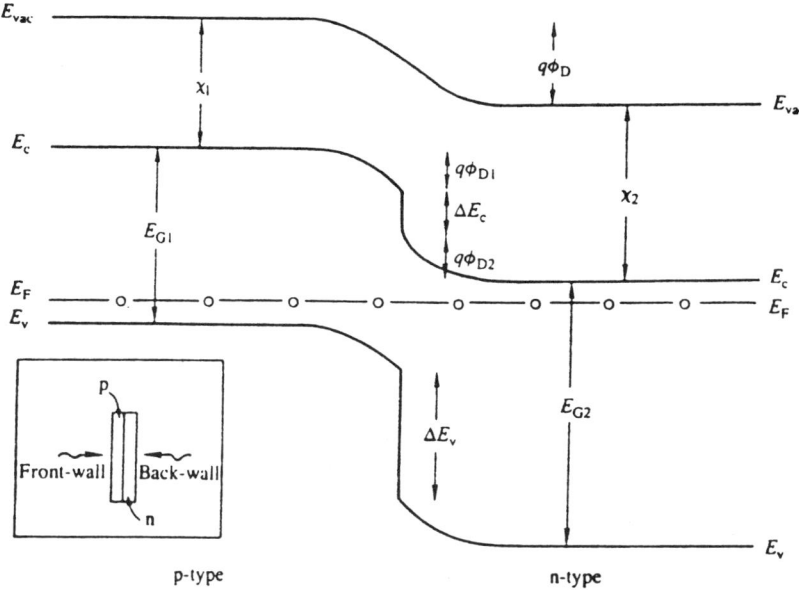

Fig. 1.8. Energy-band diagram for a p–n heterojunction with equal densities of doping in the p- and n-type portions, and a choice of material parameters ($\chi_1 < \chi_2$, $E_{G1} < E_{G2}$) such that there are no energy spikes at the junction interface.

diagram for a p–n heterojunction is given in Fig. 1.8, where the p-type material is assumed to have a smaller band gap E_{G1} than that of the n-type material E_{G2}. As in Fig. 1.7, the assumption is made that the doping is of the same magnitude for both p- and n-type materials, giving equal-width depletion layers on both sides of the junction.

A variety of complex phenomena can occur at the junction interface in such a heterojunction, which we have simplified in Fig. 1.8, following the Anderson abrupt-junction model (Anderson, 1962). In this approach we neglect any effects of interface dipoles or interface states, and recognize that differences between the electron affinities and band gaps of the two material give rise to discontinuities ΔE_c in the conduction band and ΔE_v in the valence band, which can in principle be either positive or negative. Figure 1.8 has been drawn with the desirable assumption that $\chi_1 < \chi_2$, so that an energy spike does not occur in the conduction band impeding electron transport from the p- to the n-type material.

As indicated in the insert in Fig. 1.8, photoexcitation could be either on the n-type material (back-wall) or on the p-type material (front-wall).

Back-wall excitation profits from the larger band gap of the n-type material, which acts essentially like a window even for light that is highly absorbed in the p-type region, and allows the light to penetrate through to the junction with minimum loss. On the other hand, other problems are introduced which are related to the likelihood of lattice mismatch at the junction between the two semiconductors, and which may in itself produce localized interface states that facilitate carrier loss through recombination at what is now a kind of "internal surface". Such localized interface states may also play a role in increasing the reverse saturation current density J_o and hence reducing ϕ_{oc}.

A typical summary of design choices for materials to be used in a heterojunction is given in Table 1.1.

Table 1.1. Material considerations for use in a heterojunction solar cell.

Property	Criteria
Band gap of smaller band-gap material	Direct gap near 1.4 eV
Band gap of larger band-gap material	As large as possible while maintaining low series resistance
Conductivity type	Smaller band-gap material should be p-type because of longer electron diffusion lengths
Electron affinities	Such that no potential spike occurs at the junction for minority carriers
Diffusion voltage, ϕ_D	Large, since maximum $\phi_{oc} \propto \phi_D$
Diffusion length	Long electron diffusion length in p-type material
Lattice mismatch	Small as possible to avoid interface states at the junction
Electrical contacts	Low-resistance contacts to both n- and p-type materials
Material availability	Good supply of material available
Material cost	Material costs must be competitive
Material toxicity	Materials should be non-toxic, or possible effective control of toxicity
Cell stability	Materials should be free of interactions leading to changes in properties with time or operation

Sometimes a variation on the p–n heterojunction between two semiconductors (sometimes called an SS junction, where 'S' stands for 'semiconductor') is made by including a thin layer of an insulating material between the two semiconductors (to form an SIS junction) to help reduce the junction currents that decrease ϕ_{oc}.

Buried Homojunctions or Heteroface Junctions

The energy band diagram of Fig. 1.9 pictures what is called a "buried homojunction" or a "heteroface junction". It is an effort to benefit from the best properties of a homojunction and a heterojunction, while minimizing their problems. The structure shown consists of a large band gap p^+-type material which forms a heterojunction to a smaller band gap p-type material, which in turn forms a homojunction with an n-type material with the same band gap. The primary purpose of the structure is to decrease losses due to surface recombination at the illuminated surface in the homojunction configuration by providing the p^+–p junction with good lattice matching. One of the most successful developments of this type is the p-GaAlAs/p-GaAs/n-GaAs cell (Hovel *et al.*, 1972; MacMillan *et al.*, 1988), for which the lattice constant of AlAs is 0.5661 nm and that of GaAs 0.5654 nm; the proportion of Al in the GaAlAs

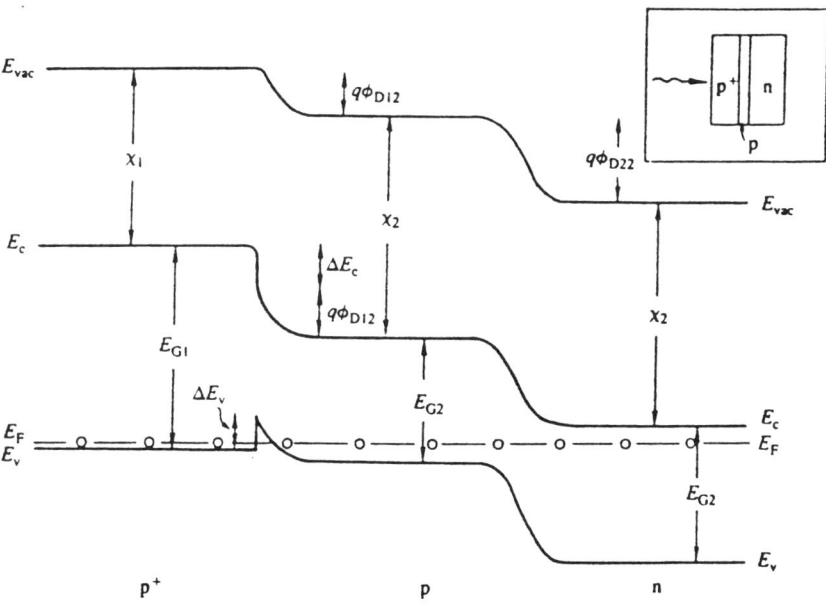

Fig. 1.9. Energy-band diagram for a p^+–p–n buried homojunction or heteroface junction in which the p^+ material acts as a large band gap window and an ohmic contact to the p-type material. Inset shows the typical direction of illumination for use as a solar cell.

solid solution compound is chosen to provide (a) a large enough band gap to act as a window layer for solar energy, and (b) a good lattice match to the GaAs, hence providing a low interface-recombination velocity at the p^+–p interface.

Schottky Barriers

The previous three types of photovoltaic junctions all involve junctions between two semiconductors. A conceptually simpler junction can be obtained from a Schottky barrier metal contact to a semiconductor (an MS junction).

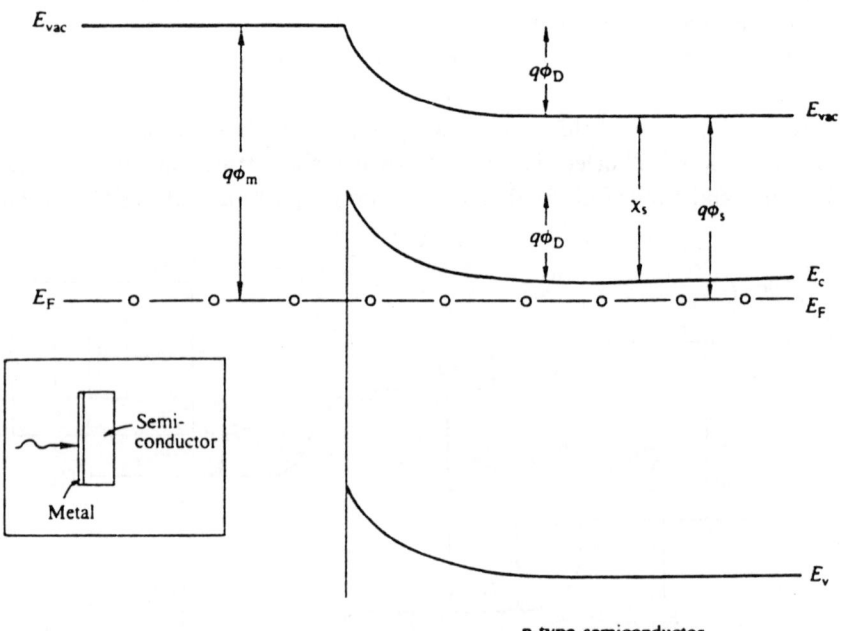

n-type semiconductor

Fig. 1.10. Energy-band diagram for a Schottky barrier metal contact to an n-type semiconductor, based on the energy parameters of the materials without including interface effects.

A typical energy band diagram for a Schottky barrier on an n-type material is given in Fig. 1.10. In the case shown, $q\phi_M > q\phi_S$, and upon contact electrons flow from the n-type semiconductor to the metal, causing a depletion layer in the n-type semiconductor and an internal electric field next to the junction, which is able to collect photoexcited carriers and enable them to contribute to electric current in an external circuit. [For a p-type semiconductor, a Schottky barrier is formed if $q\phi_M < q\phi_S$, and holes flow from the semiconductor to the metal upon contact, causing a depletion layer for holes next to the junction.] Although a simple approach equates the diffusion potential of the Schottky barrier to the difference between the work functions of the metal and the semiconductor ($q\phi_D = q\phi_M - q\phi_S$), many more complex interactions can occur at the interface that cause variations from such a prediction (e.g. see Fahrenbruch and Bube, 1983). The actual height of a specific Schottky barrier must often be determined experimentally.

Analogous to the case of the SS and SIS junctions, it is frequently found desirable to insert a thin layer of insulator in a Schottky barrier junction to reduce J_o, forming an MIS junction, sometimes referred to as an MOS junction if the insulator is an oxide (Anderson *et al.*, 1977; Stirn and Yeh, 1977). Suitable energy band structures can exist so that the insulator layer reduces the leakage currents of the Schottky barrier, and, as long as it is thin enough not to impede carrier collection, can contribute to improved cell performance.

p–i–n Junctions

What we refer to here as a *p–i–n* junction (see Fig. 3.15 for the energy band diagram of a *p–i–n* junction for a-Si:H) differs from the SIS junction mentioned above in two major ways: (1) the insulator in an SIS or MIS junction is a thin layer confined to the junction region, whereas the insulator in a *p–i–n* junction is usually a thick undoped layer of the same semiconductor where the principal absorption of light occurs; and (2) transport of carriers in an SIS or MIS junction occurs by diffusion between the depletion layer and the contacts, whereas transport in a *p–i–n* junction occurs by drift under an electric field that exists throughout the insulator. Since a collecting electric field exists throughout the region where free carriers are being formed by photoexcitation, there are definite advantages to the *p–i–n* structure. It becomes possible, for example, to use a semiconductor with desirable properties of other kinds but with a diffusion length too small to be useful in a conventional *p–n* structure.

One of the most efficient developments in crystalline silicon solar cells is the *p–i–n* back point-contact (BPC) cell (Swanson *et al.*, 1984) which is discussed in Sec. 2.4. a-Si:H solar cells are usually fabricated in the form of a *p–i–n* junction (Carlson, 1977). An example of a heterojunction *p–i–n* structure is the *n*-CdS/*i*-CdTe/*p*-ZnTe solar cell (Meyers, 1989). Here an *n*-CdS layer is deposited on SnO_2-coated glass, an *i*-CdTe layer is deposited by electrodeposition on the CdS, and a *p*-ZnTe layer is deposited on the CdTe.

Photoelectrochemical Cells

The sixth type of junction goes back all the way to the first discovery of the photovoltaic effect by observing what happens when a semiconductor-electrolyte interface is illuminated. Two types of phenomena are of interest: photoelectrolysis (Bocarsly *et al.*, 1977) and photoelectrochemical cell performance (Gerischer, 1975; Chai *et al.*, 1977; Lewis, 1995). A typical band diagram for photoelectrolysis is given in Fig. 1.11(a), corresponding to the experimental arrangement shown in Fig. 1.11(b), in which electron-hole pairs are excited in a large band-gap semiconductor, which can then be used to dissociate water. A photoelectrochemical arrangement is shown in Fig. 1.11(c), involving an oxidation reaction associated with photoabsorption at one surface and a reduction reaction at another surface, providing a flow of electrons to the external circuit.

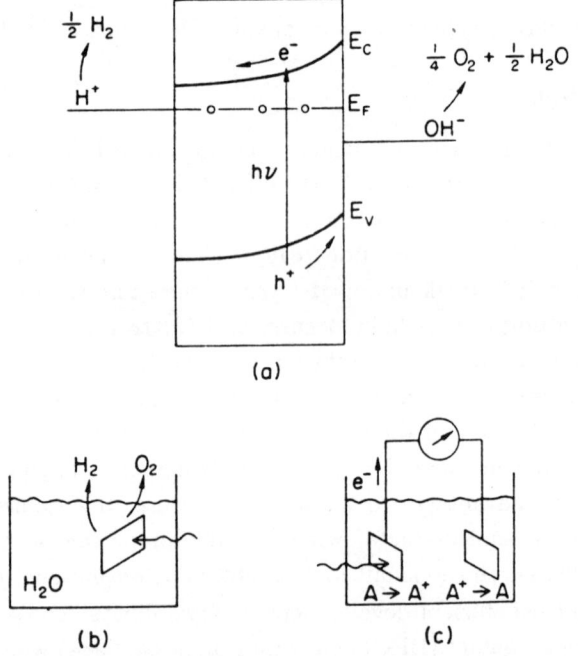

Fig. 1.11. (a) Energy-band diagram for photo-electrolysis using a semiconductor; (b) experimental arrangement for photoelectrolysis using a semiconductor; (c) experimental setup for a photoelectrochemical cell using a semiconductor. (Reprinted from R. H. Bube, "Solar Cells", in *Handbook on Semiconductors. Device Physics.* Vol. 4C. C. Hilsum, ed., 1993, p. 797, with kind permission from Elsevier Science — NL, Sara Burgerhartstraat 25, 1055 KV Amsterdam, The Netherlands.)

1.4. More Detailed Photovoltaic Models

As a first step from the idealized simple model treated in Sec. 1.1 to a more realistic model of a photovoltaic cell, we turn our attention now to several ways of introducing corrections and changes more descriptive of the actual situation.

Three major effects need to be included: (1) effects of $R_s > 0$ and $R_p < \infty$ in Fig. 1.4, (2) voltage-dependent collection effects that make the actual light-generated current density less than J_L, and (3) the possibility of a change in the major parameters J_o, A, R_s and R_p between the dark and light conditions. In the following description, some relatively simple approximations are made to include the consequences of each of these effects.

Differences between Dark and Light Conditions

The single Eq. (1.1) for the junction current in the ideal case must be replaced by two equations, one for the situation in the dark and the other for the situation under photoexcitation. In the dark, we have

$$J^d = \gamma^d \{ J_o^d \exp\left[\alpha^d (\phi - J^d R_s^d) \right] + \phi / R_p^d - J_o^d \} \tag{1.8}$$

where the superscript d denotes the dark condition, $\gamma^d = 1/(1 + R_s^d/R_p^d)$, $\alpha^d = q/A^d kT$ for transport mechanisms not involving tunneling, and $\alpha^d = \alpha'^d$ for transport mechanisms involving tunneling (as described further below). In the light, we have

$$J^l = \gamma^l \{ J_o^l \exp[\alpha^l (\phi - J^l R_s^l)] + \phi / R_p^l - J_o^l - H(\phi) J_L] \} \tag{1.9}$$

where the superscript l denotes the light condition and other definitions are similar to those in the dark. The function $H(\phi)$ is a voltage-dependent collection function that describes what fraction of the light-generated carriers are collected and contribute to the current; we may approximate $H(\phi)$ as follows (Fahrenbruch and Bube, 1974; Mitchell *et al.*, 1977a, b).

Collection Function

For the sake of simplicity, consider a p–n heterojunction with a large band-gap n-type material which transmits photons with energy less than its band gap without loss to the p-type material where they are absorbed. The collection function $H(\phi)$ can be separated into two contributions:

$$H(\phi) = g(\phi)\, h(\phi) \tag{1.10}$$

Here $g(\phi)$ describes the loss of carriers by recombination in the bulk of the p-type material before they can diffuse to the junction to be collected, and $h(\phi)$ describes the loss of carriers by recombination due to interface states at the junction interface. Qualitatively one would expect that $g(\phi)$ will depend strongly on the photon energy, decreasing for photon energies with lower optical absorption, since carriers are freed further and further from the collecting junction and have an increasing probability of recombining while diffusing to that junction; and that $h(\phi)$ will be relatively independent of photon energy but will depend strongly on the electric field at the interface.

The collection function $g(\phi)$ can be calculated as follows:

$$g(\phi) = \left\{ \int_0^w \exp(-\alpha x)dx + \int_w^\infty \exp(-\alpha x)\exp[-(x-w)/L_n]dx \right\} \Big/$$

$$\int_0^\infty \exp(-\alpha x)dx \qquad (1.11)$$

The dependence of $g(\phi)$ on the photon energy occurs through $\alpha(h\nu)$. The first term in Eq. (1.11) describes the complete collection of all carriers created in the depletion layer with width w, assuming that the drift field there assists in this collection. The second term expresses the possibility of recombination loss of such carriers with increasing distance from the depletion layer if the diffusion length of electrons in the p-type material is L_n. Integration of Eq. (1.11) gives the following expression for $g(\phi)$:

$$g(\phi) = 1 - \exp[-\alpha w(\phi)]/(1 + \alpha L_n) \qquad (1.12)$$

which depends on the variation of depletion layer width w with voltage ϕ,

$$w(\phi) = [2\varepsilon_r\varepsilon_o\,(\phi_D - \phi)/qN_{D+}]^{1/2} \qquad (1.13)$$

where N_{D+} is the density of ionized donors in the depletion layer. As $w(\phi) > 0$, $g(\phi) > 1/[1 + (1/\alpha L_n)]$.

A simple expression for the collection function $h(\phi)$ can be obtained by assuming that recombination at the interface is described by an interface recombination velocity s_I, and that the recombination probability can be considered to be a simple competition between crossing the junction without recombination and recombination at the interface:

$$h(\phi) = 1/(1 + s_I/\mu\mathbf{E}) \qquad (1.14)$$

where μ is the mobility of carriers at the interface, and $\mathbf{E}$ is the electric field at the interface, given by $\mathbf{E} = 2(\phi_D - \phi)/w(\phi)$. If there are N_I interface states per square centimeter at the interface with a capture coefficient of β_I cm^3 s^{-1}, then $s_I = N_I\beta_I$ cm s^{-1}. The carrier velocity μE must be considered to have a maximum value corresponding to the saturation of drift velocity at high fields.

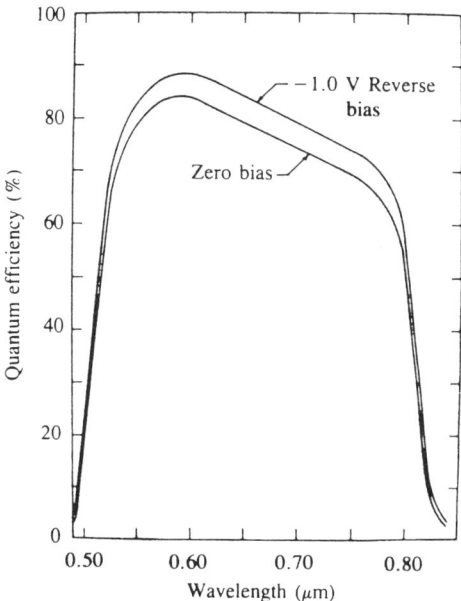

Fig. 1.12. Spectral dependence of the quantum efficiency for an n-CdS/p-CdTe heterojunction, illustrating the contributions of collection function $g(v)$ (decreasing with increasing wavelength) and $h(v)$ (increasing with applied reverse bias). (Reprinted with permission from K. W. Mitchell *et al.*, *J. Appl. Phys.* **48**, 4365 (1977). Copyright 1977, American Institute of Physics.)

An example of the qualitative effects of these collection functions is given in Fig. 1.12, showing the spectral response of quantum efficiency (density of electrons due to light in external circuit divided by density of photons absorbed) in an n-CdS/p-CdTe photovoltaic heterojunction (Mitchell *et al.*, 1977b). The short-wavelength cutoff of the spectral response is due to the absorption of shorter wavelengths by the large band gap n-type CdS, while the long-wavelength cutoff is due to failure of the small band gap p-type CdTe to absorb longer wavelengths. The effect of the collection function $g(\phi)$ is seen in the negative slope of the quantum efficiency toward longer wavelengths, with negligible dependence on voltage ϕ. The effect of the collection function $h(\phi)$ is seen in an increase in quantum efficiency with applied reverse bias (thus increasing $\mathbf{E}$ at the interface, and/or decreasing s_I through its own voltage dependence) with negligible dependence on wavelength. Experimental values of $h(\phi) = 0.84$ at $\phi = 0$ and $h(\phi) = 0.89$ at $\phi = -1$ V, indicate a relatively large value of $s_I = 2 \times 10^6$ cm s^{-1}, consistent with the fact that CdS and CdTe have a large 9% lattice mismatch.

Other Photovoltaic Parameters

When the more realistic junction model of Eqs. (1.8) and (1.9) is considered, the other photovoltaic parameters must also be recalculated. For example, the open-circuit voltage becomes

$$\phi_{oc} = (1/\alpha^l) \ln[H(\phi_{oc}) \, (J_L/J_o^l) + 1 - \phi_{oc}/J_o^l R_p^l] \qquad (1.15)$$

and the short-circuit current density becomes

$$J_{sc} = \gamma^l [J_o^l \exp(-\alpha^l J_{sc} R_s^l) - J_o^l - H(0)J_L] \qquad (1.16)$$

It can readily be shown that for the relationship between J_{sc} and ϕ_{oc} in this more realistic situation to be the same as the relationship between J^d and ϕ (as was the case in Eq. 1.4), requires seven conditions to be met: γ, R_p, α, and J_o must all be independent of light; $H(0)$ must be equal to $H(\phi_{oc})$; $J^d R_s^d$ must be much less than ϕ; and $\exp(-\alpha^l J_{sc} R_s^l)$ must be of order unity.

Junction Transport Processes

The parameters J_o and A play critical roles in determining the value of the open-circuit voltage of a solar cell. Simple models of current transport through the junction interface have been developed to indicate some of the significant factors that affect J_o and A in typical idealized cases, and to provide criteria for deciding from experimental data which transport process is active. For simplicity, we use an n^+–p heterojunction ($N_D > N_A$, $E_{Gp} < E_{Gn}$) for the following summary of these processes, and illustrate them qualitatively in Fig. 1.13.

Diffusion

The mode of junction transport corresponding to the smallest values of J_o and to $A = 1$ corresponds to a diffusion-controlled current over the junction barrier associated with the injection of electrons [Process (1) in Fig. 1.13] from the n-type material into the p-type material. After injection and diffusion, recombination finally occurs away from the junction in the p-type semiconductor bulk. The current can be expressed as

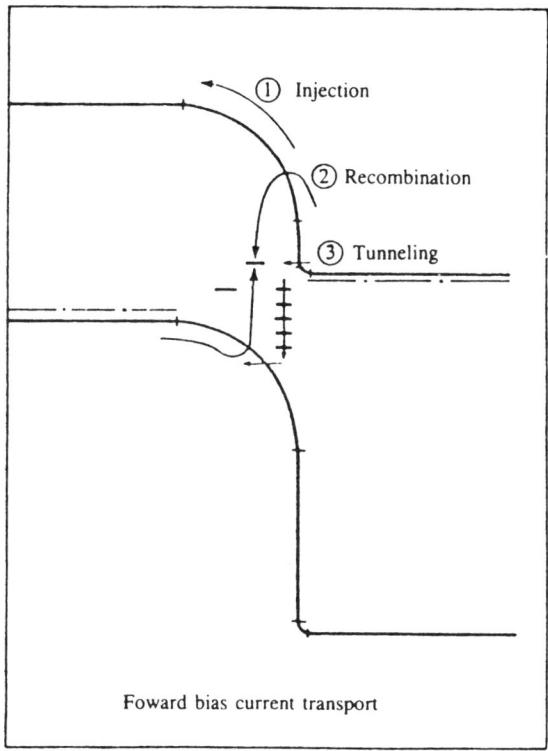

Fig. 1.13. The three major modes of forward junction current transport: (1) injection over the barrier, (2) recombination in the depletion layer, and (3) tunneling, with or without thermal assistance, through interface or imperfection states, followed by recombination. (Reprinted from R. H. Bube and A. L. Fahrenbruch, "Photovoltaic Effect", in *Advances in Electronics and Electron Physics* **56**, E. Marton, ed., p. 163. Copyright 1981, Academic Press, Orlando FL.)

$$J_{\text{diff}} = J_o[\exp(q\phi/kT) - 1] \qquad (1.17)$$

with $\alpha_{\text{diff}} = q/kT$ and $A = 1$, and

$$J_o = (n_i^2/N_A)\, q(L_n/\tau_n) \qquad (1.18)$$

where, in the p-type material, n_i is the intrinsic carrier density, N_A is the density of acceptors, L_n is the electron diffusion length, and τ_n is the electron lifetime. A plot of $\ln(J_o T^{-7/2})$ versus $1/T$ has an activation energy of E_{Gp}.

Recombination in depletion region

The second smallest mode of junction transport involves recombination in the depletion region [Process (2) in Fig. 1.13]. A recombination rate expression is used based on Shockley–Read recombination (e.g. see Ch. 4 in Bube, 1992), with integration across the depletion regions. The current can be expressed as

$$J_{\text{rec}} = J_o \left[\exp(q\phi/AkT) - 1\right] \tag{1.19}$$

with $\alpha_{\text{rec}} = q/AkT$ and $A \approx 2$, and

$$J_o = qn_i[1/(\tau_{no}\tau_{po})^{1/2}][\pi kT/4(\phi_D - \phi)]w \tag{1.20}$$

where w is the depletion layer width, τ_{no} is the minimum electron lifetime when all recombination centers are empty, and τ_{po} is the minimum hole lifetime when all recombination centers are electron-occupied. The value of A has a maximum value of 1.8 for symmetrically doped junctions if the levels at which recombination occurs lie at mid-gap; otherwise values of A between 1 and 2 may correspond to this mechanism (Sah *et al.*, 1957). A plot of $\ln(J_oT^{-5/2})$ versus $1/T$ has an activation energy of $E_{Gp}/2$. For unsymmetrically doped junctions, the value of A may be equal to or larger than 2 (Choo, 1968).

Interface recombination without tunneling

At an n^+–p interface, the density of electrons is large and current is limited by the availability of holes, which must overcome the barrier in the valence band. Such currents may be described by

$$J_{\text{int}} = J_o \left[\exp(q\phi/kT) - 1\right] \tag{1.21}$$

with $\alpha_{\text{int}} = q/kT$ and $A = 1$. Two conditions may exist: (1) the thermal velocity v_{th} of the electrons is larger than the interface recombination velocity s_I, so that the current is limited by interfacial recombination, in which case

$$J_o = qs_I N_v \exp(-q\phi_D/kT) = qN_I S_I v_{\text{th}} N_v \exp(-q\phi_D/kT) \tag{1.22}$$

where S_I is the electron capture cross-section of interface states with density N_I, and a plot of $\ln J_o$ versus $1/T$ has an activation energy of $q\phi_D$; or (2) the thermal velocity of an electron is smaller than s_I, so that the current is limited by the diffusion of holes to the interface, in which case J_o follows the expression for thermionic emission:

$$J_o = A'T^2 \exp(q\phi_D/kT) = (4\pi qm_h^*/h^3)(kT)^2 \exp(-q\phi_D/kT) \tag{1.23}$$

and a plot of $\ln(J_oT^{-2})$ versus $1/T$ has an activation energy of $q\phi_D$.

Tunneling limited recombination through interface states without thermal assistance

Electrons from the n-type material descend through interface states and then tunnel [Process (3) in Fig. 1.13] through the base of a barrier of height E_b into the valence band of the p-type material. A simple model for this process involves tunneling at the base of a parabolic barrier (Riben and Feucht, 1966). The junction current in this case is

$$J_{ti} = J_o[\exp(\alpha/kT - 1]\tag{1.24}$$

with $\alpha = (4\pi/h)\,(\varepsilon m^*/N_A)^{1/2}$ independent of temperature, and

$$J_o = qp(kT/m^*)^{1/2}\exp(-\alpha v_D)\tag{1.25}$$

so that a plot of $\ln(J_o T^{-1/2})$ versus α has an activation energy of $q\phi_D$.

Thermally assisted tunneling through interface barrier

This model is similar to the last except that holes tunnel through the barrier into electron-occupied interface states at a hole energy enhanced by thermal excitation (Padovani and Stratton, 1966). The current is given by

$$J_{tat} = J_o[\exp(\alpha\phi) - 1]\tag{1.26}$$

with

$$J_o = J_{oo}\exp[-\alpha(\phi_D + E_{fp})]\tag{1.27}$$

with $E_{fp} = E_F - E_v$,

$$\alpha = q/[E_{oo}\coth(E_{oo}/kT)]\tag{1.28}$$

$$E_{oo} = (qh/4\pi)\,(N_A/\varepsilon m_h^*)^{1/2}\tag{1.29}$$

$$J_{oo} = (\{4\pi qm^*(kT)^2\pi^{1/2}E_{oo}^{1/2}[q(\phi_D - \phi)]^{1/2}\}/\{h^3 kT\cosh(E_{oo}/kT)$$
$$\times\,[\coth(E_{oo}/kT]^{1/2}\})\exp[-E_{fp}(1/kT - 1/E_o)]\tag{1.30}$$

with $E_o = E_{oo}\coth(E_{oo}/kT)$. This model results in an α that is weakly temperature dependent. A plot of $\ln\{J_o\cosh(E_{oo}/kT)\,[\coth(E_{oo}/kT)]^{1/2}/T\}$ versus α has an activation energy of $(q\phi_D + E_{fp})$.

Tunneling processes commonly dominate junction currents in heterojunctions, particularly in experimental systems and at lower temperatures. Examples of heterojunctions in which this model appears to appropriately describe the data include n-CdS/p-Zn$_{0.3}$Cd$_{0.7}$Te (Peters *et al.*, 1988), ZnO/CdTe (Aranovich *et al.*, 1980), and ZnO/InP (Eberspacher *et al.*, 1984).

1.5. Photovoltaic Materials

The various chapter titles in the remainder of this book indicate the materials that have played a major role in photovoltaic solar energy conversion. Single crystal materials may be useful in high-technology, relatively expensive cells to be used with concentration of sunlight. In single crystal form only Si, GaAs, InP, CdTe and CuInSe$_2$ can be used in photovoltaic devices to produce efficiencies greater than 10%, and of these only Si and GaAs, and solid-solutions based on them, are considered seriously for terrestrial applications.

This situation calls attention to the importance of thin-film technology in producing thin-film photovoltaic cells for terrestrial applications. In thin-film form, a-Si:H, CdTe and CuInSe$_2$ are the leading candidates for solar-cell applications. The thin-film technology used must pay particular attention to the processing costs associated with large-area production. Not only are thin films needed for the active solar cell layers themselves, they are also needed for window materials, anti-reflection coatings, passivating coatings, and transparent-conducting contacts (e.g. CdS, ZnCdS, ZnO, SnO, SnO$_2$, In$_2$O$_3$, and indium-tin oxide (ITO)). Decisions have been needed on whether to produce these films by one of the standard methods such as vacuum evaporation, non-reactive or reactive sputtering, electron-beam evaporation, molecular beam epitaxy, and chemical vapor deposition, or by one of a set of developing techniques, such as close-spaced vapor transport (Nicoll, 1963; Saraie *et al.*, 1972; Yoshikawa and Sakai, 1974; Buch *et al.*, 1977), spray pyrolysis (Chamberlin and Skarman, 1966; Ma *et al.*, 1977; Ma and Bube, 1977), and electrochemical deposition or plating (Panicker *et al.*, 1978). The references cited here indicate some of the earlier investigative work; more recent work is described in the following chapters. An assessment of polycrystalline thin films for solar cell applications as of 1982 is summarized by Rothwarf (1982).

Because of the limited number of different materials, research has attempted to broaden the range of possible materials by focusing on solid solutions between these and related materials: e.g. a-Si:C:H, a-Si:Ge:H, Cu$_x$Ag$_{1-x}$InSe$_2$, CuGa$_x$In$_{1-x}$Se$_2$, GaInP$_2$, Zn$_x$Cd$_{1-x}$Te, and Mn$_x$Cd$_{1-x}$Te. Although in principle it is possible to design "ideal" photovoltaic systems with ideal band gap and no lattice mismatch at heterojunction interfaces, by resorting to more complicated ternary, quaternary, pentenary, and even more complex systems, the materials problems entering in these more complex systems appear to be a serious limitation.

A major attempt at increasing efficiency with the limited number of materials available has led to the development of multijunction cells, in which two (or more) different cells are used together in series to more efficiently absorb the light. Although the measured efficiency for such a multijunction cell can be expected to exceed that of either cell used separately, it is clear that efficiencies do not simply add in such a multijunction cell, since only a fraction of the incident light reaches the lower cell. The ideal situation would be to use a large number of such cells in a multijunction such that each cell could effectively absorb light only within a narrow range of its band gap. Examples of early multijunction cells with two components, and the efficiencies achieved are: GaAs/Si (31%) (Gee and Virshup, 1988); GaAs/CuInSe$_2$ (21.3%) (Stanbery *et al.*, 1977; Kim *et al.*, 1988); AlGaAs/GaAs (24–28%) (Lewis *et al.*, 1988; Virshup *et al.*, 1988; MacMillan *et al.*, 1989); a-Si:H/CuInSe$_2$ (15.6%) (Mitchell *et al.*, 1988); a-Si:H/a-Si:Ge:H (13.6%) (Guha 1989); GaInP$_2$/GaAs (25%) (Olson *et al.*, 1989). As we shall see in our later discussions, the structural complexity of even these two-component multijunction cells is often not trivial.

CHAPTER 2

CRYSTALLINE SILICON

2.1. Overview

Silicon has dominated most semiconductor applications for almost 50 years. It is the second most abundant material of the earth's crust, stable and non-toxic; its technology is highly developed; and in its crystalline form it has an almost ideal band gap for photovoltaic solar energy conversion. It is not surprising, therefore, to find that silicon has been the dominant photovoltaic material in solar cell applications over most of this time period. In 1981 the Si p–n homojunction cell was the only solar cell widely available commercially.

"Polycrystalline" silicon is the historically generic name for all non-single-crystalline silicon. In recent years, an effort has developed to distinguish between different types of material on the basis of the size of the individual single-crystal grains in the material, and to reserve the specific name "polycrystalline" for just one of them (Basore, 1994). A complete "family tree" for silicon can therefore be listed as follows: (1) single crystal — just one grain (> 10 cm) characterizes the whole material; (2) multicrystalline (mc-Si) — there are a number of different grains in the material but the grain size is relatively large, of the order of 1 mm to 10 cm; (3) polycrystalline — many grains are present in the material with grain dimensions of the order of 1 μm to 1 mm, as is also the case for polycrystalline films of CdTe (Chapter 5) or CuInSe$_2$ (Chapter 6); (4) microcrystalline (μc-Si) — single crystal regions in grains can still be identified in x-ray diffraction patterns, but the grain size is < 1 μm; (5) amorphous (a-Si) — no single-crystal regions identifiable by x-ray diffraction occur since no long-range order is present. We discuss the varieties of crystalline silicon in this chapter and amorphous silicon in Chapter 3.

The cost of single-crystal Si solar cells is greater than that of its closest competitors, thin-film multicrystalline Si cells, and the impetus is often present to trade the lower cost of the multicrystalline cells for the higher efficiency of the single crystal cells. Single-crystal Si solar cells require very high material quality; surfaces must be effectively passivated to reduce recombination there, and the bulk properties must also be of high quality because of the long optical penetration distances associated with the indirect band gap of Si. Specially designed single-crystal Si solar cells have demonstrated efficiencies greater than 28% under concentrated sunlight, as described further below, while standard single crystal Si solar cells have had an efficiency on the order of 15–17% (as described in Fig. 1.6).

Most of the solar cells exhibited in various simple demonstrations or applications are actually multicrystalline silicon cells, and are visually identifiable by the different reflection intensities from various grains in the material. Multicrystalline Si solar cells have been produced with efficiencies of about 17%, and may be less expensively produced from somewhat less pure starting materials using less expensive manufacturing methods. It has been estimated that single crystal efficiencies may ultimately reach 35%, while multicrystalline efficiencies may ultimately reach 19% (Hubbard, 1989; Hubbard and Cook, 1989).

The procedure for the growth of single-crystal Si is summarized in Fahrenbruch and Bube (1983), and the technology involved in the preparation of pure, high-quality Si has continually improved (Ciszek, 1988). Silicon single crystals have frequently been grown by one of two major growth methods (Matlock, 1979): (a) the Czochralski technique, in which a seed of known orientation is dipped into a melt of silicon, the temperature is decreased slowly, and the seed with subsequent growing crystal is pulled out of the melt, and (b) the float-zone technique in which a narrow molten region is passed slowly along a Si ingot in a vacuum or inert gas. Czochralski growth dominates the single-crystal solar cell market.

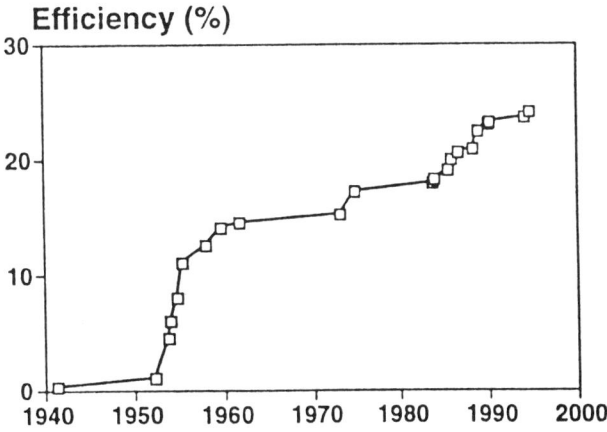

Fig. 2.1. The increase in efficiency of single-crystal silicon laboratory solar cells. (Reprinted with permission from M. A. Green, *Silicon Solar Cells: Advanced Principles & Practice*, Centre for Photovoltaic Devices and Systems, Univ. of New South Wales, Sydney, 1995.)

A developmental history, and a summary of principles and practice involved in the production of Si solar cells are given by Green (1993a, 1993b, 1995a), whose summary of the rapid improvement in single crystal silicon solar cell efficiencies is shown in Fig. 2.1. For each cell, the actual value of efficiency for an AM1.5 spectrum at 25°C and 1000 Wm^{-2} illumination intensity, the type of cell structure used, the organization where the work was done, and a relevant literature reference are given in Table 2.1.

2.2. Doping and Lifetime

Phosphorus and boron are the most widely used donor and acceptor dopants in single-crystal Si. As the density of dopants is increased in the active absorbing

Table 2.1.

Date	Eff. %	Cell Structure	Organization	Reference
3/41	< 1%	Melt-grown junction	Bell Labs	Ohl, 1941
3/52	1%	He bombardment	Bell Labs	Kingsbury et al., 1952
12/53	4.5%	Li diffused wraparound	Bell Labs	Pearson, 1985
1/54	6%	B diffused wraparound	Bell Labs	Chapin et al., 1954
11/54	8%	B diffused wraparound	Bell Labs	Bell Labs Record, 1954, 1955
5/55	11%	B diffused wraparound	Bell Labs	Bell Labs Record, 1954, 1955
12/57	12.5%	0.5 × 2 cm B diffused	Hoffman Elec.	Green et al., 1992
8/59	14%	Grid-contact B diffused	Hoffman Elec.	Green et al., 1992
8/61	15.2%	B diff. AR coat, gridded	Commercial, USASRDL	Mandelkorn et al., 1962
1/73	15.2%	Violet cell	Comsat Labs.	Lindmayer et al., 1973
9/74	17.2%	Textured non-reflecting	Comsat Labs.	Haynos et al., 1974
9/83	18.0%	MINP cell	UNSW*	Green, 1991
12/83	18.3%	PESC cell	UNSW	Green, 1991
5/85	19.0%	PESC cell	UNSW	Green, 1991
10/85	20.0%	Microgrooved PESC cell	UNSW	Green, 1991
7/86	20.6%	Microgrooved PESC cell	UNSW	Green, 1991
4/88	20.8%	Microgrooved PESC cell	UNSW	Green, 1991
9/88	22.3%	Rear point contact cell	Stanford Univ.	Verlinden et al., 1988
12/89	23.0%	PERL cell	UNSW	Green, 1991
2/90	23.1%	PERL cell	UNSW	Green, 1991
2/94	23.5%	PERL cell	UNSW	Zhao et al., 1994
9/94	24.0%	DLAR PERL cell	UNSW	Green et al., 1995d

*The abbreviation UNSW stands for University of New South Wales, Australia.

layers of solar cells, two effects occur that degrade the electronic properties of Si.

The lifetime is decreased both because of an increase in Shockley–Read type recombination (electron-hole recombination through imperfections, giving up the recombination energy as photons or phonons), and by the onset of Auger–type recombination (in which the recombination energy is given up to another free carrier — see Sec. 2.4) for impurity densities in excess of 5×10^{17} cm^{-3} (Fischer and Pschunder, 1975; Redfield, 1978, 1979, 1980).

The carrier mobility, and hence the diffusion coefficient, is decreased because of an increase in charged impurity scattering, noticeable for impurity densities greater than 10^{16} cm^{-3}. Accurate measurements of the mobility of minority-carrier holes in phosphorus-doped silicon with doping in the 10^{19} cm^{-3} range have been made using independent measurements of diffusion length and minority-carrier lifetime to obtain minority-carrier mobility values (del Alamo *et al.*, 1985; Swirhun *et al.*, 1986; del Alamo *et al.*, 1987c). Since the absorber layer in the solar cell generally has doping of less than 10^{16} cm^{-3}, the mobility effect is usually negligible.

Deep-level impurities in Si may also decrease the lifetime of carriers through recombination. Such impurities in roughly decreasing order of effect include Ti, Zr, V, Na, Au, Cu, Fe, Mg, Cr , Mn and Ni, with tolerable densities varying from 4×10^{13} cm^{-3} for Ti to 4×10^{16} cm^{-3} for Ni (Hill *et al.*, 1976; Davis *et al.*, 1980; Shimura, 1994). Particularly harmful are Na, Cu, and Fe since they also have high diffusion coefficients in Si.

Introduction of dopants into the Si is usually by the process of diffusion, which may be done in a variety of ways, often leading to a high concentration of the dopant near the surface. This highly-doped layer has a decreased lifetime and must be avoided or later removed in order not to have a "dead-layer" at the surface.

2.3. Major Sources of Recombination

Three major sources of recombination loss of photoexcited carriers in single crystal Si solars cells exist: bulk, surface, and contact recombination.

Bulk Recombination

As described in Sec. 2.2, a major source of bulk recombination is associated with imperfections in the bulk of the Si. The desired result is to minimize recombination throughout the bulk through the use of high-quality pure Si

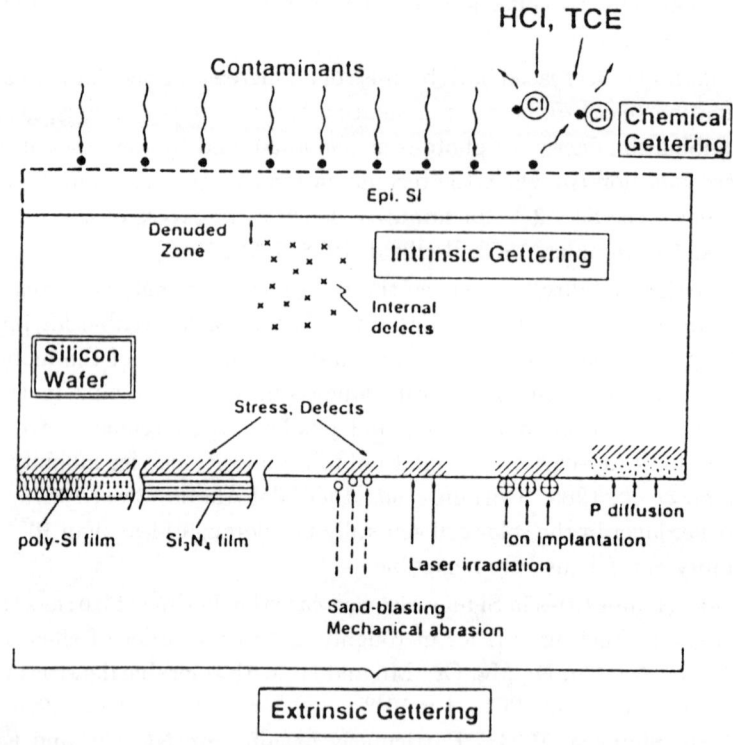

Fig. 2.2. Schematic illustration showing various gettering techniques for a silicon wafer (Reprinted with permission from F. Shimura, *Semiconductor Silicon Crystal Technology*, Copyright 1989, Academic Press, Orlando, FL.)

starting material, thinning the bulk quasi-neutral regions, and limiting the doping to the minimum levels necessary to achieve cell performance (Green, 1984).

This direct avoidance of impurities sometimes proves either too difficult or too expensive, and other methods referred to as "gettering" are used to reduce the impurity density in the bulk (Green, 1995a). These are shown schematically in Fig. 2.2 (Shimura, 1989). (a) A standard method in microelectronics is "intrinsic gettering" which involves the formation of SiO_2 precipitates in the bulk that act as gettering sites for further processing (Sardana, 1985) after depletion of oxygen from the surface by out-diffusion at high temperatures. Since this method results in very high quality material near the surface, but poor quality material in the bulk, it is not suitable for photovoltaics. (b) A second approach is "extrinsic gettering", which involves the introduction of gettering sites on one surface to which bulk defects diffuse upon high temperature treatment. (c) A third approach is "chemical gettering" in which suitable chemicals such as halogens are present in a gas stream over the crystal surface, where they react with metallic impurities in the silicon to form volatile compounds that leave the crystal.

In addition to these gettering techniques, there are also defect passivation techniques in which a mobile species, such as atomic hydrogen, is incorporated in the wafer, which can neutralize the electronic action of bulk defects (Sopori *et al.*, 1994).

Surface Recombination

Non-contacted areas of the surface of the Si provide locations where recombination is enhanced. This recombination is generally minimized by the growth of a high-quality thermal oxide. The thickness of the oxide (14 nm to 80 nm) does not appear to be an important parameter (Kopp *et al.*, 1992).

The magnitude of the effective surface recombination velocity S_{eff} for photoinduced carriers at the Si–SiO$_2$ interface as a function of injection level has been investigated (Aberle *et al.*, 1992; Glunz *et al.*, 1994a, 1994b). A strong dependence of S_{eff} at the Si–SiO$_2$ interface on injection level and doping concentration was found as shown in Fig. 2.3, which may be consistently interpreted in the following way.

Results indicate that the electron capture cross-section of interface states is much larger than their hole capture cross-section (Aberle *et al.*, 1992). Because of this difference in cross-sections, which makes the interface states behave similarly to donors with a positive charge when unoccupied, the recombination properties are quite different for n-type and p-type material. For n-type surfaces, there is a small density of holes near the surface, and hence the capture of holes is the rate-determining recombination process; the net recombination rate is small because of the small hole density and the small hole capture cross-section. For p-type surfaces, however, the situation is more complex and changes with increasing output voltage of the cell. At low voltages capture of electrons is the rate-determining step, and the recombination process is more effective than in p-type material because of the larger electron capture cross-section. With increasing voltage across the cell, however, the situation is reached where the hole capture becomes the rate-determining step, and the effective surface recombination decreases, eventually to values comparable to

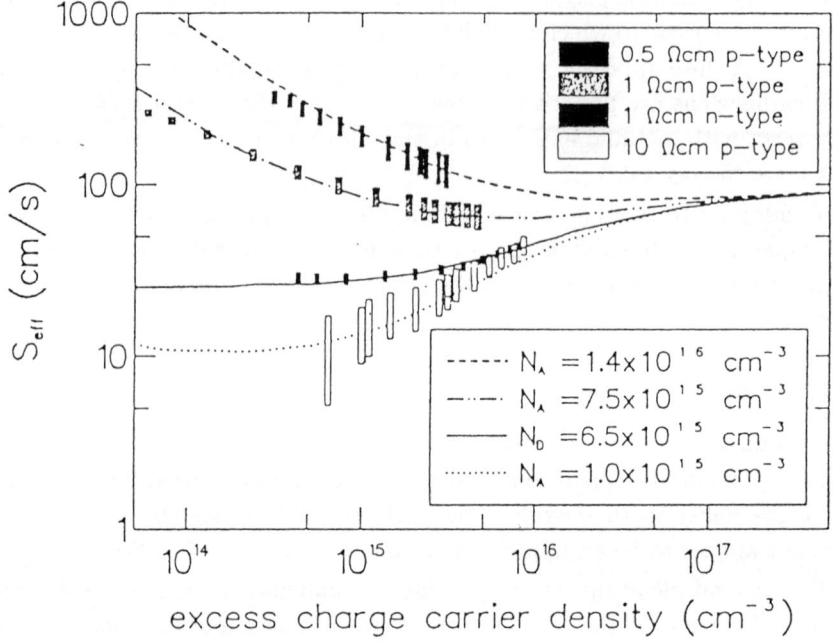

Fig. 2.3. Measured injection dependence of the effective surface recombination velocity, S_{eff}, for differently doped silicon. (Reprinted with permission from S. W. Glunz *et al.*, *12th European Photovoltaic Solar Energy Conference*, p. 492 (1994). Copyright 1994, H. S. Stephens and Associates, Bedford, United Kingdom.)

those found for *n*-type material. It must be kept in mind, however, that specific fabrication details may be significant in determining the details of surface recombination.

Recent developments in techniques for surface and bulk passivation are reported by Loelgen (1995), who describes the effects of doped layers at the surface on reducing the surface recombination, and of Al-gettering to improve the silicon bulk quality (Loelgen *et al.*, 1994a,b); and by Leguijt (1995), who investigated the effects of surface passivation by low temperature plasma enhanced chemical vapor deposition (PECVD) of silicon oxide or silicon nitride. Similarly Rosmeulen *et al.* (1995) reported a 19.5% efficient solar cell passivated by oxides deposited by low pressure chemical vapor deposition (LPCVD), and Demesmaeker *et al.* (1994) reported promising results with low-temperature surface passivation by means of a Tetra-Methyl-Cyclo-Tetra-Siloxane (TM-CTS) LPCVD oxide.

Contact Area Recombination

The regions where metal contacts are made to the Si in the construction of a solar cell to overcome the barrier to majority carrier transport, are likely to be regions of high doping and hence of high recombination. One direct solution is to keep the contact area as small as possible. Another is to strongly diffuse these contact areas in order to reduce the minority carrier densities in these areas, with effectively the same results as a "back surface field" (BSF) (Loelgen *et al.*, 1993; Leguijt *et al.*, 1995), such as is shown in Fig. 2.5. Other approaches that have been investigated include the use of an MIS contact (Green and Blakers, 1983) leading to an MINP (metal-insulator-NP junction) cell (Green *et al.*, 1984), which was the first silicon cell to demonstrate 18% efficiency; semi-insulating polysilicon (SIPOS) contacts (Yablonovitch *et al.*, 1985); or doped polycrystalline or amorphous silicon contacts (Lindholm *et al.*, 1985).

2.4. Development of Single-Crystal Silicon Solar Cells

It is the principal purpose of this book to discuss the materials properties that have been found to be significant for the design and construction of efficient solar cells. Clearly the actual cell design and construction details themselves play a major role in determining this efficiency. Although we include the significant characteristics of the various types of cells and their major properties, we do not intend to provide detailed information on cell design features themselves, and refer the reader to the bibliography and list of review papers at the end of this text for more information of this type.

Typical Single Crystal Si Cell

The form of the first modern silicon cell is shown in Fig. 2.4(a), with the form of the standard design in the '60's shown in Fig. 2.4(b). Typical single crystal Si cells consist of a slice of crystal about 200 to 500 μm thick, although cells with a thickness as small as 50–70 μm were early made with an efficiency of 12.5% using the Si available at that time period (Lindmayer and Wrigley, 1976; Chiang *et al.*, 1978). On this p-type Si crystal, for which B is a common acceptor impurity, an n-type layer about 0.1 to 0.5 μm thick is produced by diffusion of P donors, typically from PH_3. The energy band diagram for a typical single crystal Si n^+–p–p^+ cell is shown in Fig. 2.5 (for details of typical cell fabrication, see Fahrenbruch and Bube, 1983).

The n-type layer is usually fairly heavily doped in order to decrease the contribution of holes to the reverse saturation current J_o when diffusion is the

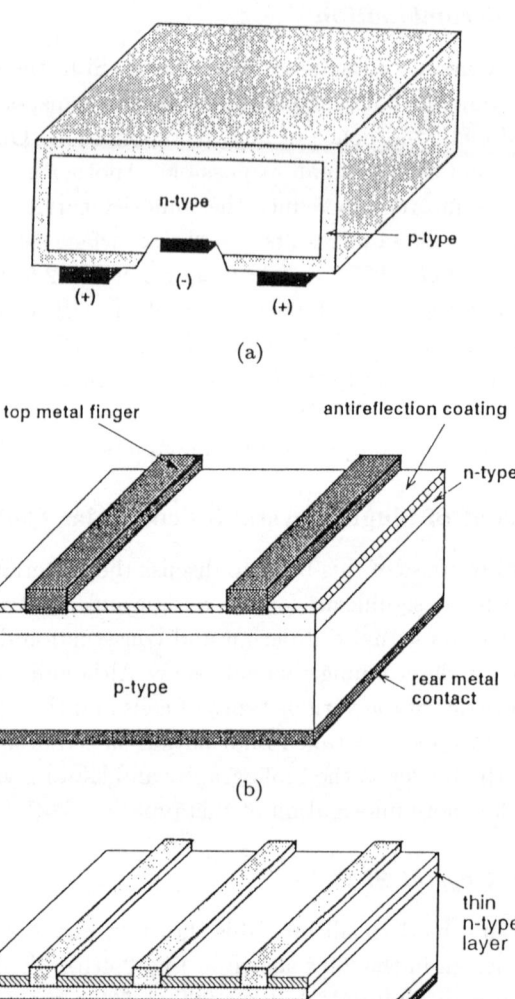

Fig. 2.4. (a) Form of the first modern silicon cell (Chapin *et al.* 1954); (b) standard design for a space silicon cell (Mandelkorn *et al.* 1962; Smith *et al.* 1963); (c) the shallow junction "violet cell" (Lindmayer and Allison, 1973). (Reprinted with permission from M. A. Green, *Silicon Solar Cells: Advanced Principles & Practice*, Centre for Photovoltaic Devices and Systems, Univ. of New South Wales, Sydney, 1995.)

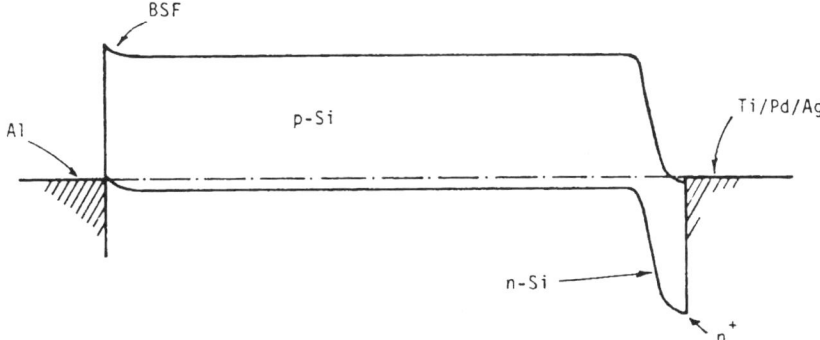

Fig. 2.5. Energy band diagram for a typical Si p^+–n–n^+ solar cell with the n-layer thickness exaggerated, and including a back surface field (BSF) created by Al diffusion. (Reprinted with permission from A. L. Fahrenbruch and R. H. Bube, *Fundamentals of Solar Cells: Photovoltaic Solar Energy Conversion*, Copyright 1983, Academic Press, Orlando, FL.)

principal junction transport mechanism (see Eqs. 1.17 and 1.18). It was this reasoning, leading to heavy doping of the n-type layer, that was responsible for the long period of fairly flat solar cell efficiencies shown in Fig. 2.1. These earlier cells had almost the same short-circuit current (≈ 40 mA/cm^2) as the best modern cells, but an appreciably smaller open-circuit voltage (0.59 V compared to 0.71 V). It is based on the more general version of Eq. (1.18) for the reverse saturation current density J_o, for uniform doping on both sides of the junction,

$$J_o = An_i^2 \left[(1/N_A)\,(D_n/\tau_n)^{1/2} + (1/N_D)\,(D_p/\tau_p)^{1/2} \right] \qquad (2.1)$$

where $n_i = (n_o p_o)^{1/2}$, N_A is the acceptor density in the p-region, N_D is the donor density in the n-region, D_n and τ_n are respectively the diffusion constant and lifetime of electrons in the p-type region, and D_p and τ_p are respectively the diffusion constant and lifetime of holes in the n-type region. Since the saturation current density is essentially the thermally generated minority-carrier flow in the dark, Eq. (2.1) indicates that J_o is reduced with increasing values of N_A and N_D. Equation (2.1), however, is no longer valid for degenerate doping (n or $p \geq 10^{19}$ cm^{-3}) since then $n_i^2 \neq n_o p_o$, the majority-carrier concentration is not equal to the impurity concentration, and Auger processes become important in determining the lifetime. Heavy doping of the n-type layer actually leads to an increase in J_o, a decrease in ϕ_{oc}, a decrease in the fill factor, and a decrease in the efficiency.

In general it is desired to have the doping in the thick p-type region to be relatively low in order to yield longer diffusion lengths for the minority carrier electrons. It is for the same reason that the thick region is chosen to be p-type rather than n-type: because of the greater diffusion length of minority carrier electrons in p-type compared to minority carrier holes in n-type.

The back contact to the p-type Si is produced by vacuum deposition of Al, followed by a heat treatment that produces a p^+-Si region by diffusion of the Al into the Si. By making a p^+ back layer, the contact resistivity is reduced, and a "back surface field" (shown in Fig. 2.5 as BSF) can be produced to drive carriers away from the back surface and hence to reduce the effects of surface recombination there. A doping technique that yields a p^+-doping profile in which it is possible to control both the doping concentration and the profile thickness independently has been described (Loelgen *et al.*, 1994) for boron doping using co-alloying with aluminum.

The front-surface contact to the n-type side is in the form of a grid with grid lines 0.03 to 0.3 cm apart, covering about 5 to 10% of the total area; the grid is formed of multiple metal layers, such as Ti/Pd/Ag, to improve bonding and inhibit undesirable electrochemical reactions. Finally, an anti-reflection coating must be used, since the reflectivity of Si over the spectral range of interest varies from 33 to 54%; a variety of materials have been used, including SiO_2, Si_3N_4, and Ta_2O_3.

The Violet Cell

An early type of Si p–n junction, called the "violet" cell (Lindmayer and Allison 1973) had a very thin n-type layer like that shown in Fig. 2.5 to increase spectral response of the cell in the blue and violet portion of the spectrum. The form of the cell is shown in Fig. 2.4(c). Such cells typically had $\phi_{oc} = 0.59$ V, $J_{sc} = 40$ mA/cm^2, $ff = 0.78$, $J_o = 10^{-12}$ A/cm^2, $A = 1$, and $\eta = 13.5\%$ for AM0 radiation.

The CNR Cell

Another early type of Si p–n junction solar cell was the "Comsat non-reflective" (CNR) cell (Haynos *et al.*, 1974; Arndt, 1975; Rittner and Arndt, 1976), sometimes also called the "black cell". It was characterized by a physically texturized front surface designed to reduce reflection losses and to cause photogeneration closer to the junction by allowing light to travel in paths that are not normal to the junction interface. Such cells typically had $\phi_{oc} = 0.59$ V, $J_{sc} = 46$ mA/cm^2, $ff = 0.77$, $J_o = 6 \times 10^{-12}$ A/cm^2, $A = 1$, $\eta = 15.5\%$

for AM0, and $\eta = 18\%$ for AM1 radiation. General issues related to coupling sunlight to solar cells in a useful way are discussed by Luque (1993).

Other Types of Early Solar Cells

Examples of other less-efficient types of solar cells made with single crystal Si include Schottky-barrier or MIS junctions, and SnO_2 or ITO junctions. Efficiencies of 8% for an Al-oxide-Si junction (Charlson and Lien, 1975) and 8.5% for a Cr-oxide-Si junction (Anderson *et al.*, 1974, 1977) were reported. Heterojunctions of n-SnO_2/n-Si have also been investigated using deposition of SnO_2 onto Si substrates followed by a short heat treatment (Mizrah and Adler, 1977; Thompson *et al.*, 1977; Nash and Anderson, 1977). Efficiencies as high as 10% for AM1 irradiation were reported, but the cells underwent degradation presumably due to the growth of the insulating oxide layer between the conducting SnO_2 and the Si.

Passivation of Top Surface

A major improvement in output voltage was achieved by improved passivation of the top surface of the cell, using thermal oxide to passivate non-contacted areas and various techniques to reduce recombination at the interface with the top metal contact. A cell known as the PESC cell (passivated emitter solar cell) had a reduced area of contact at the top surface, and ultimately an efficiency over 20% was achieved for the first time with high-quality float-zone Si wafers relatively highly doped with B (Green *et al.*, 1985).

The Back Point-Contact (BPC) Cell

A major advance in efficiency was achieved by careful attention to material control and cell design (Swanson *et al.*, 1984; Sinton *et al.*, 1985; Sinton *et al.*, 1986a; Sinton and Swanson, 1987a,b,c; del Alamo and Swanson, 1987a,b; Verlinden *et al.*, 1987, 1988) with oxide passivation of both front and rear surfaces of the cell. The cell structure is given in Fig. 2.6. Cells are formed on high-resistivity float-zone Si substrates giving a p–i–n structured solar cell. Light-trapping at the front surface is achieved by pyramidally texturing the surface, and a back-surface mirror causes reflection of most of the non-absorbed photons at angles that result in total internal reflection. This light-trapping is essential because of the need to keep the cell thin to minimize recombination and maximize collection of the photogenerated carriers. Shading or recombination associated with contacts on the front surface is avoided by the use of

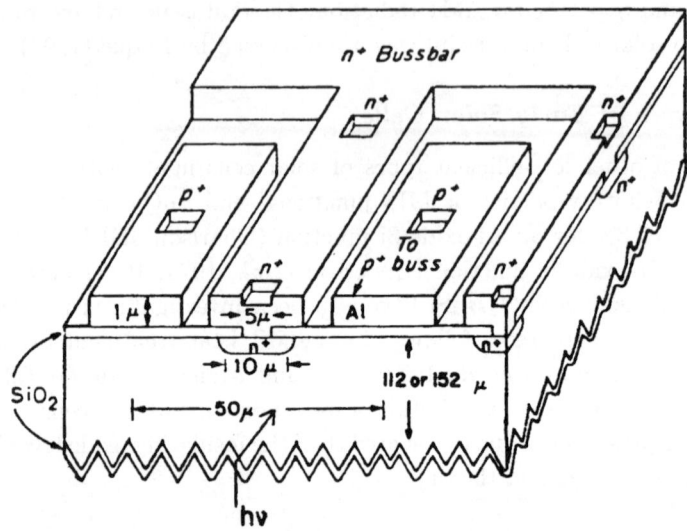

Fig. 2.6. A cross-section of a region in a Si back-point contact solar cell near one of the Al bussbars. (Reprinted with permission from R. A. Sinton *et al.*, *IEEE Electron Device Lett.* EDL **7**, 567 (1986). Copyright 1986, IEEE, NY.)

alternate *p*-type and *n*-type small point contacts only on the back surface. The small size of the contacts allows a considerable decrease in recombination, but contact resistance must be minimized. Careful attention is paid to minimizing the surface recombination velocity over the entire surface of the cell, and to maintaining maximum bulk carrier lifetimes since carriers photoexcited near the top of the cell must diffuse to the contacts at the back.

These were the first cells to exceed 22% efficiency under normal terrestrial illumination. The most efficient cell reported has a thickness of 120 μm, an area of 0.15 cm^2 with a metal-finger length of 3 mm, and an efficiency of 28% at 15 W/cm^2 of AM1.5 illumination (about 150 suns). Figure 2.7 shows how the efficiency varies with the concentration of the illumination, and Fig. 2.8 shows the calculated deconvolution of the recombination current at the maximum power point for the cell of Fig. 2.7 (Sinton *et al.*, 1986b). Over this range of concentration, the measured open-circuit voltage increases from 650 mV at 1 sun to 840 mV at 600 suns. The importance of recombination in the peripheral region, which dominates at low concentrations, decreases with increasing concentration, as does the recombination at surface and bulk defects in the illuminated regions of the cell. Recombination in the emitter (diffused) regions

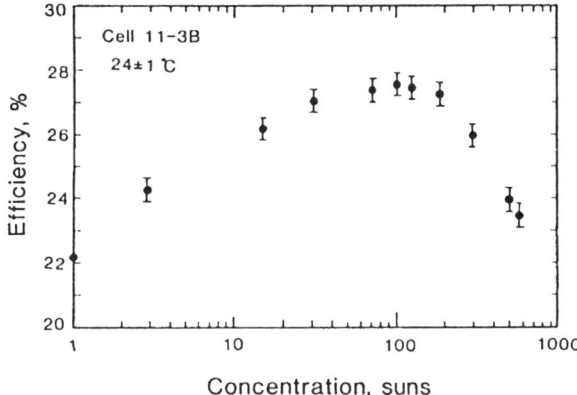

Fig. 2.7. Measured efficiency at $24°C$ versus concentration ratio for a 100 μm thick point contact solar cell. (Reprinted with permission from R. A. Sinton *et al.*, Copyright 1986, Sandia Report SAND86-0058/1, Albuquerque, New Mexico.)

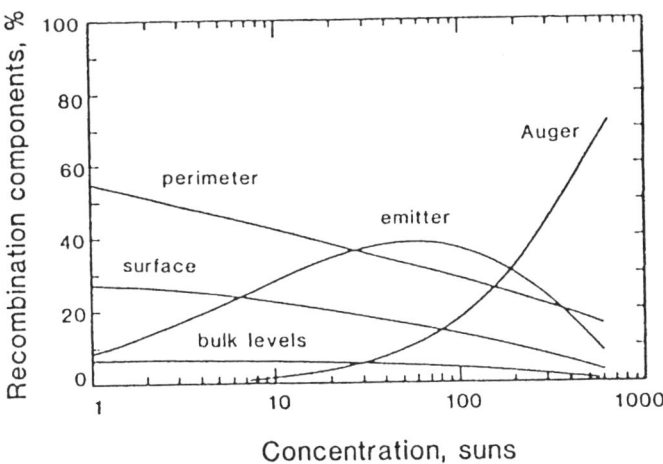

Fig. 2.8. Calculated deconvolution of the recombination current at the maximum power point of the cell of Fig. 2.7 into its components as a function of the concentration value. (Reprinted with permission from R. A. Sinton *et al.*, Copyright 1986, Sandia Report SAND86-0058/1, Albuquerque, New Mexico.)

of the cells becomes increasingly important as the concentration increases, and passes through a maximum with increasing concentration. At the highest concentration levels, bulk Auger recombination dominates.

Some caution needs to be exercised in recognizing the differences between differential (measured from a small periodic light to produce the measurement signal superimposed on a constant bias light to define the injection level) and actual surface recombination velocities (Brendel, 1995; Brendel and Wolf, 1995). The actual surface recombination velocity must be obtained by integrating measurements of the differential surface recombination velocity over all injection levels, and may be as much as several times larger than the latter.

Using a solar cell design of this type optimized for the purpose, the Auger recombination process was investigated in more detail (Sinton and Swanson, 1987) over the range of non-equilibrium carrier densities from 10^{16} to 2×10^{17} cm^{-3}. The Auger coefficient was evaluated to be 1.66×10^{-30} cm^6/s, a value some four times larger than that most commonly assumed. These results suggested that the lifetimes in the very best quality doped material may already be approaching this fundamental Auger recombination limit and that a common mechanism exists for recombination in both doped and highly injected materials.

The initial rear point contact cells proved to be unstable due to optical degradation of the silicon/oxide interface in an effect apparently similar to the optical instability of amorphous silicon solar cells due to the photogeneration of defects, described in Chapter 4. In the case of the BPC cell, diffusion of the exposed surface with P and deposition of an ultraviolet-absorbing material such as TiO$_2$ on the top surface appears to have removed the problems of instability (Verlinden *et al.*, 1993).

A simplified BPC cell for use without concentration has been described, which can be fabricated by a less complex process yielding a cell with an efficiency of 21.9% at one sun (Sinton and Swanson, 1990). Developments with the BPC cells are summarized in several more recent publications (Verlinden *et al.*, 1994a,b; Verlinden *et al.*, 1995; Sinton *et al.*, 1995).

The Passivated Emitter, Rear Locally-Diffused (PERL) Cell

A record high efficiency for terrestrial illumination of greater than 23% has been achieved in a cell that combines the best features of the PESC and BPC cells, shown in Fig. 2.9 (Green, 1991). Inverted pyramids covered with an anti-reflection coating along the top surface increase the fraction of the incident light effectively used in carrier generation in the cell. Weakly absorbed light that passes through the cell is reflected by the rear oxide layer covered by an aluminum layer. These two optical features of the cell increase the path length of weakly absorbed light within the cell, giving an effective absorption

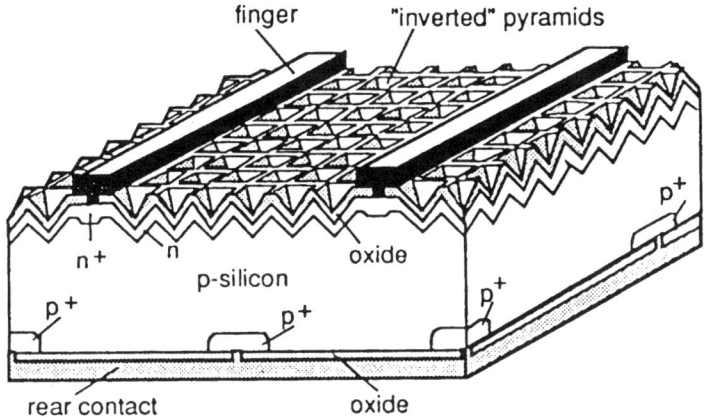

Fig. 2.9. Schematic of a passivated emitter, rear locally-diffused cell (PERL cell) which has shown an efficiency for terrestrial illumination of 23% (Reprinted with permission from M. A. Green, *Silicon Solar Cells: Advanced Principles & Practice*, Centre for Photovoltaic Devices and Systems, Univ. of New South Wales, Sydney, 1995.)

thickness about 30 times that of the physical thickness of the cell. Typical output values for PERL cells are $J_{sc} = 41$ mA/cm^2, $\phi_{oc} = 710$ mV, $ff = 0.83$ and $\eta = 24\%$. An efficiency greater than 45% has been measured with this structure for 1.02 μm monochromatic light (Green *et al.*, 1992) and of 46.3% for 1.04 μm wavelength light (Zhao *et al.*, 1995).

As far as recombination losses in these cells are concerned, it is believed that bulk recombination can be made negligible using present designs with appropriate processing conditions (Green, 1993a). Rear surface recombination in non-contacted areas is not important at open circuit (the effective recombination velocity decreases from large values of 10^4 cm/s to low values below 30 cm/s as the voltage increases from short-circuit to open-circuit values), but is important at the maximum power point voltage, resulting in a reduction in fill factor. At the open-circuit voltage, recombination in the top diffused layer and its contact region dominates the behavior. A detailed investigation of the limiting loss mechanisms in PERL cells has been made by Aberle *et al.* (1995).

High Efficiency Solar Cells from FZ, CZ and mc-Silicon: A Processing Summary

Developments in the production of high-efficiency solar cells using silicon prepared by different methods (FZ — float zone (21.1%), CZ — Czochralski

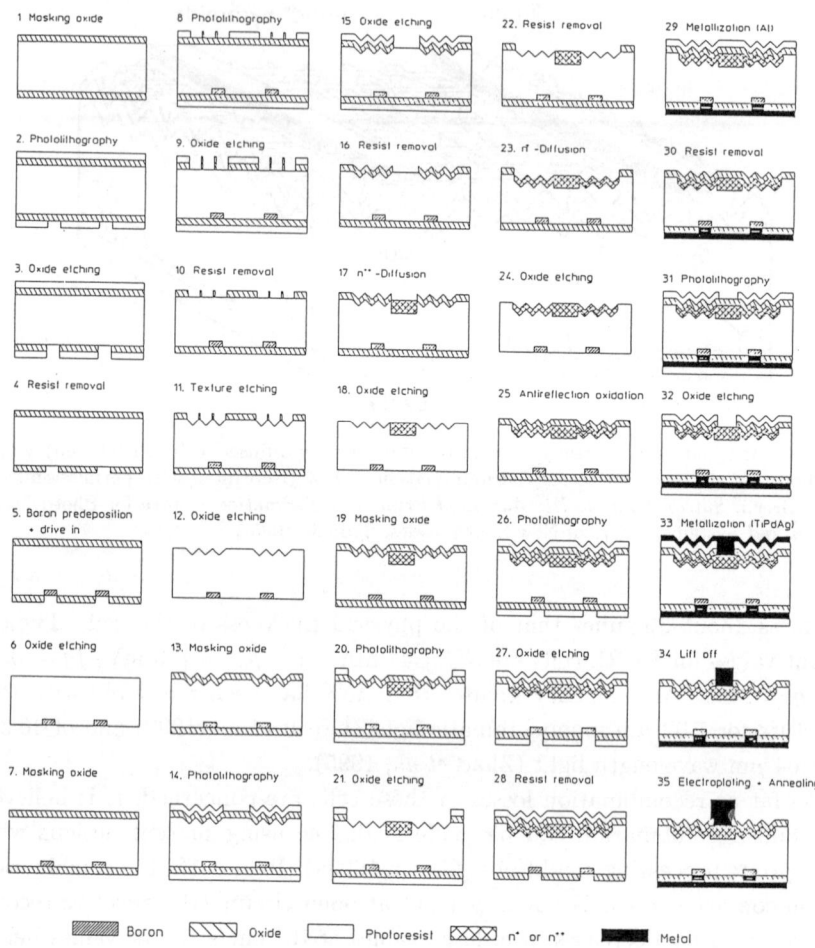

Fig. 2.10. Illustration of a processing scheme for a high-efficiency silicon solar cell with a boron back surface field. (Reprinted with permission from J. Knobloch *et al.*, *23rd IEEE Photovoltaic Specialists Conference*, p. 271 (1993). Copyright 1993, IEEE, NY.)

(19.3%), and mc — multicrystalline (16.2%)) provide specific examples of the typical processing steps in modern silicon solar cell development (Knobloch *et al.*, 1993). These LBSF (local back surface field) cells (Knobloch *et al.*, 1989) are similar to the PERL cell described above. Subsequent work with an optimized emitter, based on an understanding of the heavy-doping effects described above, raised the efficiency of CZ-silicon cells to 21.3% (Knobloch *et al.*, 1995), and of FZ-silicon cells to 22.3% (Sterk *et al.*, 1994; Glunz *et al.*, 1995). A large-area PERL silicon cell module with area of 743 cm^2 has also been reported with an efficiency of over 20% (Zhao *et al.*, 1993).

Figure 2.10 summarizes the processing procedure in cleanroom conditions with a boron-doped BSF on FZ silicon. (1)–(6). Local boron diffusion. (7)–(12). Surface texturing. (13)–(24). n^{++} diffusion with phosphorus from POCl$_3$, followed by n^+ diffusion with phosphorus. (25) SiO$_2$ passivation of front and back surfaces, the front passivation layer also acting as an anti-reflection coating. (26)–(30). Open the oxide at rear side at the points with the local boron BSF, evaporate Al on the whole rear side. (30)–(35). Open the windows on the front side for the metal grid, evaporate Ti, Pd, and Ag, lift off the excess metal in acetone, and electroplate silver on the front surface. Anneal the cells at 400–450°C for 20–30 minutes in forming gas consisting of 5% hydrogen in nitrogen. The current-voltage characteristics for the best cells of each type are shown in Fig. 2.11. Boron-diffused BSFs proved superior to aluminum diffused BSFs for both FZ- and CZ-crystals. The specific efficiency of the mc-silicon cells depended, not surprisingly, on the defect structure of the silicon, being highest for the low defect structures, which behaved more like the single crystal cells.

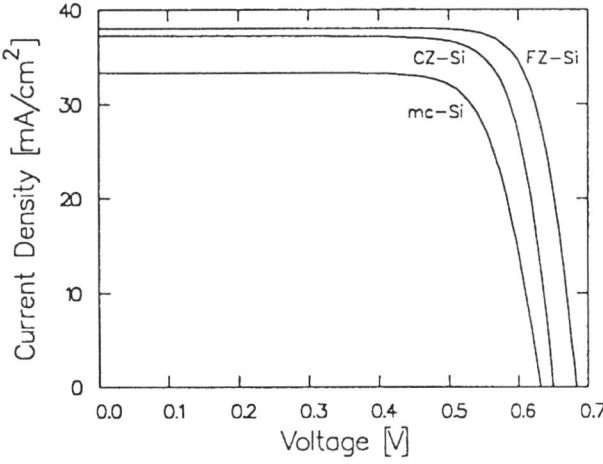

Fig. 2.11. Current-voltage characteristics for the best solar cells from FZ-, CZ- and multicrystalline silicon prepared by the processing schedule of Fig. 2.10. (Reprinted with permission from J. Knobloch *et al.*, *23rd IEEE Photovoltaic Specialists Conference*, p. 271 (1993). Copyright 1993, IEEE, NY.)

A simplified processing method for high-efficiency silicon cells involving an emitter diffusion process with a near-ideal doping profile and a passivating oxide in a single furnace treatment is described by Basore *et al.* (1994). Limiting the processing to a single high-temperature treatment preserves the bulk lifetime, important particularly in lower-cost material with a high concentration of oxygen or carbon. FZ efficiencies over 19%, and CZ efficiencies over 18% are routinely obtained.

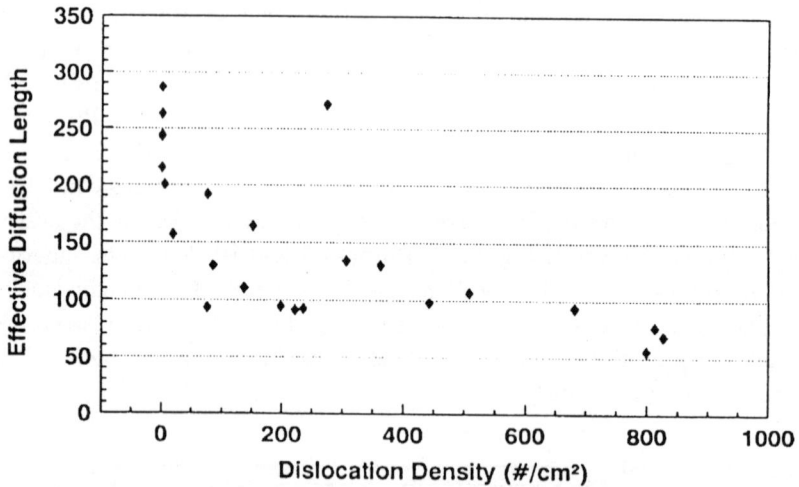

Fig. 2.12. Dependence of effective diffusion length on the dislocation density in CZ silicon solar cells. (Reprinted with permission from K. L. Pauls *et al.*, *23rd IEEE Photovoltaic Specialists Conference*, p. 209 (1993). Copyright 1993, IEEE, NY.)

Recent developments in all areas of CZ technology have been described with the goal of achieving module costs under \$2/Wp (Mitchell, 1994; Mitchell *et al.*, 1994).

Rapid thermal processing (RTP), using incoherent light sources with a wide spectrum, has been used in the fabrication of large area n^+pp^+ silicon solar cells with an efficiency of 15% on 10 cm^2 CZ silicon wafers (Sivoththaman *et al.*, 1995). Two short (50 s) RTP steps were used for emitter formation, BSF formation, and surface oxidation. Initial work was also reported on the use of spectral selection of the light in rapid thermal processing (Noel *et al.*, 1995). By selecting parts of the spectrum of incoherent RTP light, it was possible to "tailor" the profiles of phosphorus doped emitters, thus allowing selectively doped areas without masking or etching. Several other discussions have been given of the potential application of RTP to multicrystalline silicon solar cells (Hartiti *et al.*, 1993, 1994; Schindler *et al.*, 1993; Ventura *et al.*, 1994; Hahne *et al.*, 1994).

Slip dislocations are known to occur in CZ material, and the importance of these dislocations for solar cell applications has been investigated by Pauls *et al.* (1993). Dislocations are in some ways similar to grain boundaries, discussed in Sec. 2.7, but the difference is that dislocations consist of a line of charged defects whereas grain boundaries consist of a sheet of charged defects. Effective minority carrier diffusion lengths were calculated from the measured quantum efficiency of solar cells using standard spectral response equations for specimens with different dislocation densities, with the results shown in Fig. 2.12. For low dislocation densities, the diffusion length is about 250 to 300 μm, but it decreases to less than 100 μm if the dislocation density increases to 800 cm^{-2}. In spite of this the cell efficiency decreases only from 15–16% for low dislocation densities to about 14% for a dislocation density of 800 cm^{-2}, but then decreases rapidly for higher dislocation densities due to the reduced carrier collection from the bulk.

2.5. Increasing the Efficiency of Silicon Cells

General considerations leading to possibilities of increasing the efficiency of silicon solar cells have been summarized by Green (1993a).

Increasing the Current

An increase in the solar cell current can be obtained by having effective absorption of a greater portion of the solar spectrum. One way to achieve this is by having regions of the cell consist of material with a lower band gap than silicon. Such effective band-gap lowering can in principle be achieved by selective doping, alloying, and use of electric fields to reduce the band gap (Franz–Keldysh effect). Of these conceivable approaches, only alloying appears to be really promising; alloying with germanium is the most likely approach, and initial research is being undertaken (Ruiz *et al.*, 1994; Wollweber *et al.*, 1994; Borne *et al.*, 1994; Losada *et al.*, 1995).

A mechanism for an increased quantum efficiency is the utilization of carrier multiplication by hot carriers (Auger generation) in the wavelength regime of visible light (Werner *et al.*, 1994a,b, 1995a). $Si_{1-x}Ge_x$ alloys with $x = 0.6$–0.7 have been suggested as possible solar cell materials in which carrier multiplication by Auger generation would lead to an increased current (Kolodinski *et al.*, 1995), and experimental verification of this possibility has been achieved

(Werner *et al.*, 1995b). For a terrestrial cell, the predicted theoretical maximum efficiency including this process is 44.2%.

Another way to increase the solar cell current is to make use of multiple-step absorption through defect levels in the forbidden gap in silicon. The danger here, of course, is that an increase in defect density corresponds to an increase in recombination loss. It has been concluded, however, that impurity levels lying closer to the band edge than to midgap can in principle improve the current output without major loss in cell voltage.

These various methods of increasing the current in a Si cell by the effective use of subgap absorption emphasize the need for an accurate knowledge of the absorption of intrinsic silicon over the whole solar spectral range. Keevers and Green (1995) used sensitive subgap photocurrent measurements on high efficiency PERL silicon solar cells to determine the absorption coefficient between 1.19 and 1.45 μm. These measurements have been augmented by previous data for shorter wavelength light and the absorption over the whole range from 0.25 to 1.45 μm is listed by Green (1995a). A plot of the spectral dependence of the absorption coefficient corresponding to these data is given in Fig. 2.13. They cover 14 orders of magnitude of the absorption coefficient, and the long wavelength fine structure giving rise to inflection points at 1.18 μm, 1.25 μm, and 1.34 μm corresponds to the 1-phonon edge, the 2-phonon edge, and the 3-phonon edge respectively.

Increasing the Voltage

The magnitude of the open-circuit voltage in high-efficiency PERL cells is limited by recombination along the top surfaces of the cell. Improvements in design and processing should allow the voltage to reach limits determined by Auger recombination in the bulk of the cell (Green, 1984). It is predicted that the highest voltages will be obtained in lightly doped material with the maximum open-circuit voltage being given by:

$$\phi_{oc} = (2/3)\,(kT/q)\ln[J_L/(qn_i^3(C_n + C_p)\,W)] \tag{2.2}$$

where J_L is the light-generated current density at ϕ_{oc}, n_i is the intrinsic carrier concentration, W is the cell thickness, and C_n and C_p are the electron and hole Auger recombination coefficients in silicon (Green, 1987). Using current values of $n_i = 1 \times 10^{10}$ cm^{-3} (Sproul and Green, 1991) and $C_n = C_p = 1.66 \times 10^{-30}$ cm^6 s^{-1} at 300 K, gives a value for the maximum ϕ_{oc} of 748 mV for a 280 μm thick cell. Experimental values on actual cells are approaching this value.

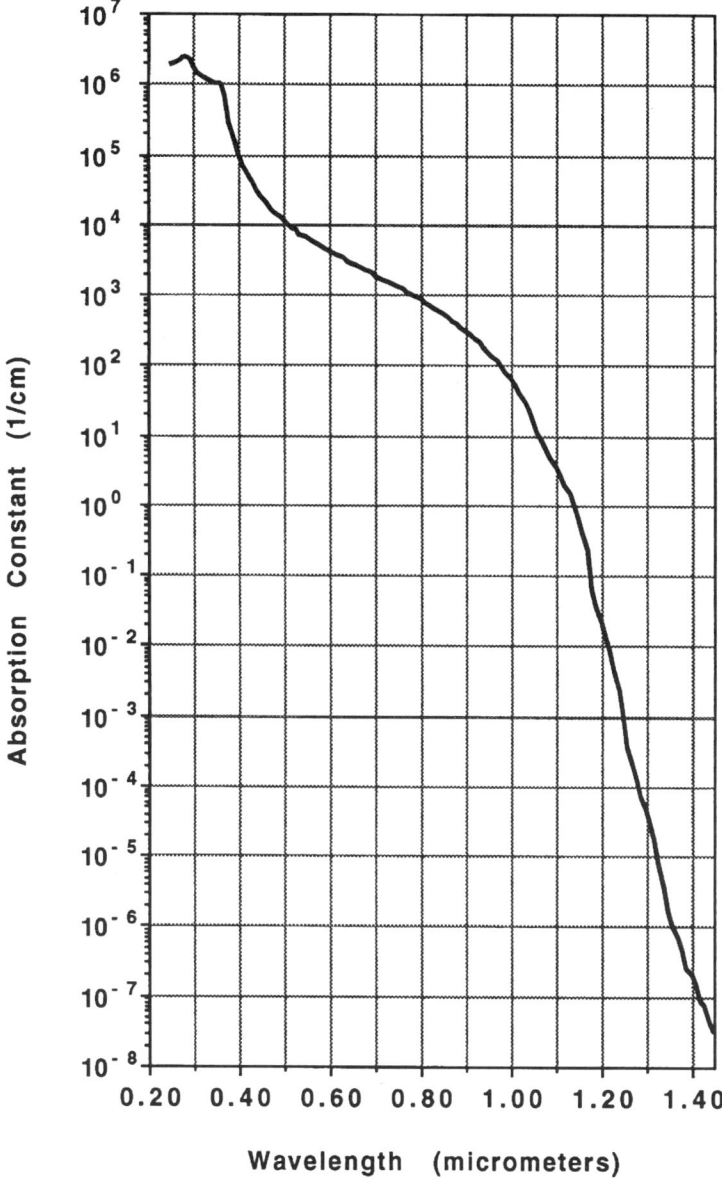

Fig. 2.13. Optical absorption coefficient for intrinsic silicon determined by photocurrent measurements using a PERL solar cell structure. Long wavelength data are listed by Keevers and Green (1995), and all of the data are listed by Green (1995a).

2.6. Novel Structures

Several different novel structures have been proposed to increase the efficiency and utility of single-crystal silicon solar cells.

Vertical Multijunction Cells (*VMJ*)

This term has been used to describe two different types of cells: (a) *a stacked VMJ*: an edge-illuminated stack of discrete *p–n* junction cells so that the open-circuit voltages add, used to increase the open-circuit voltage for concentrating systems (Goradia *et al.*, 1976), and (b) *a convoluted VMJ*: a single *p–n* junction cell with a convoluted junction plane, designed to increase radiation tolerance and for use in concentrating systems (Frank and Goodrich, 1980).

Tandem Junction Cells

Tandem cell structures in general consist of two of more cells each involving material with a different band gap, so that the light which passes through the upper cell of the tandem structure with a larger band-gap material, can be more efficiently absorbed in the lower cell with a smaller band-gap material. Silicon has an appropriate band gap to be used as the lower cell in such a tandem structure. The problem, however, is that there are not many materials with a suitable lattice constant and band gap to serve as an epitaxially grown top cell on the silicon (Corkish, 1991), so that a buffer layer of higher band gap than silicon is needed to take up the lattice mismatch and also possibly to provide electrical connection between the top cell and the lower-lying silicon cell (Yamaguchi, 1992).

Silicon Spheres

Solar cells have been fabricated using inexpensive material as feed stock, and then growing single crystal 1-mm-diameter spheres of silicon by melting silicon particles (Levine *et al.* 1991; Levine, 1992; Ahrenkiel *et al.*, 1996). The process in which small-diameter spheres are grown provides a purification step by directional solidification, allowing for the use of highly impure starting material. The starting material is melted and then single crystal spheres are formed by the surface tension from small droplets of melted silicon. Impurities segregate to the surface of the spheres, which is removed in a repeated process to produce sufficient purification for solar cell use. The spheres are commonly B-doped *p*-type, and a *p–n* junction is formed on the entire outer surface by P diffusion. The spheres are pressed into small orifices in an Al foil that makes contact with the n-type surface, the top of the spheres are etched back to reveal the p-type interior doping, and a contact is made to the inner p-type material with a second foil, the two foils being electrically insulated from each other by deposition of a polymer. Because of their geometry the cells are particularly efficient at collecting diffuse light. A power module consisting of forty 100 cm^2 cells has demonstrated an efficiency of 10.3%.

Recombination lifetime studies on these silicon spheres has been carried out using radio-frequency photoconductivity decay (Ahrenkiel *et al.*, 1996), comparing cells made from metallurgical grade silicon with those made from electronic grade silicon. After two melting cycles, the lifetimes of cells made from electronic grade material are an order of magnitude larger than those made from metallurgical grade material.

Superlattices

The possible application of superlattice design to silicon solar cell developments should be included in this discussion. To date most attention has been given to silicon/germanium superlattices, in which the presence of strain introduces new effects of possible ultimate interest (Barnham *et al.*, 1990, 1991).

2.7. Thin Multicrystalline Silicon Solar Cells

Detailed discussions of multicrystalline silicon thin films for solar cells are given in (Fahrenbruch and Bube, 1983; Van Overstraeten and Mertens, 1986; Werner *et al.*, 1994c; Green, 1993b, 1995a; Bergmann *et al.*, 1997). Werner *et al.* (1994c) provide a comprehensive review of the literature.

General

The use of thin multicrystalline silicon cells offers five major advantages: (1) less material is needed without loss of current provided that light-trapping is adequate; (2) somewhat lower quality material can be used since smaller diffusion lengths can be tolerated in thin cells; (3) theoretically thin cells can produce higher open circuit voltages and hence efficiencies than thick cells with the same bulk diffusion length; (4) thin cells are more radiation-resistant; and (5) a major advantage in economy of growth.

The use of multicrystalline materials in solar cells, however, introduces several new challenges. Major considerations concern the control of nucleation and growth of crystalline Si on foreign substrates at low temperatures, the need for a high-quality light-trapping scheme since the thickness of the material available for a one-pass absorption of the light is reduced (Redfield, 1974), more effective surface passivation as the film thickness decreases and the effective ratio of surface to bulk increases, and passivation of grain boundaries to eliminate their deleterious effects.

Werner *et al.* (1994c) summarize the following physical characteristics required for an effective thin film cell: (1) cell thickness W between 10 μm and 100 μm, (2) effective light-trapping with transparent material in direct contact with both sides of the Si absorber, (3) bulk diffusion length L greater than the cell thickness, (4) electrically inactive grain boundaries, or grain sizes much greater than the diffusion length, and (5) back surface recombination velocity less than or equal to the diffusion velocity $v_d = D/L$, where D is the diffusion constant.

Grain Boundary Effects

In addition to an increased density of defects in general (dislocations, stacking faults, and twinning planes) compared to single-crystal silicon because of reduced control of the crystallization process, a multicrystalline material is made up of grains with different crystalline orientations. The grain boundaries between these grains are characterized by interface states leading to a potential barrier between the grains, which have some similarity to the type of structure expected near a surface, i.e. in some ways a grain boundary can be thought of as a kind of "internal surface".

Figure 2.14 shows a simplified energy band diagram of an intergrain barrier in an n-type polycrystalline material. It is likely that grain boundary states are distributed across the gap with a high density near the band edges and probably a local maximum about one-third of the way across the gap (Jackson *et al.*, 1983), similar to the density of states often associated with amorphous silicon (see Chapter 3). The grain boundary interface states below the Fermi level are negatively charged, and the grain boundary interface is surrounded on both sides by semiconductor depletion regions with a positive charge. It is evident from the figure that the minority carrier holes in this case experience a local field at the grain boundary which enhances their probability of being captured at a grain boundary interface state, thus decreasing the minority carrier lifetime. It is also evident that majority carrier electrons experience a potential barrier at the grain boundary which decreases their effective mobility and increases the resistivity of the film for majority carriers. The effect of photoexcitation is generally to increase the density of majority carriers and to

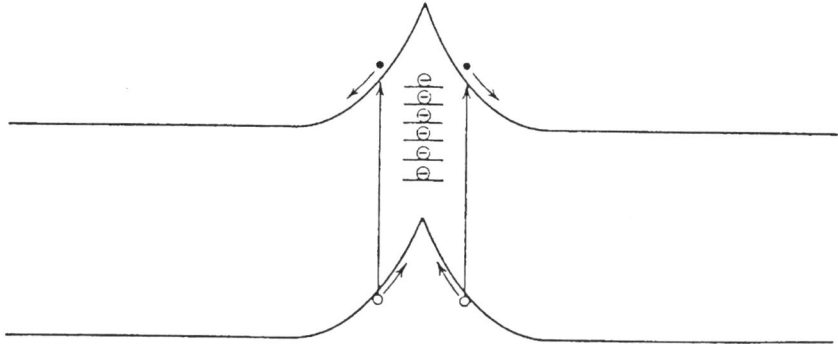

Fig. 2.14. Energy band diagram for a grain boundary in an *n*-type polycrystalline material.

decrease the height of the grain boundary potential barriers, thus increasing the majority carrier mobility, and the film conductivity (Bube, 1992).

A general schematic diagram for a thin film polycrystalline solar cell is given in Fig. 2.15, illustrating the difference between horizontal grain boundaries lying within *n*- or *p*-type material, and vertical grain boundaries crossing the interfaces between *n*- and *p*-type regions. Horizontal grain boundaries may reduce the short-circuit current because of a decrease in minority carrier lifetime through an increase in the density of sites for recombination, and vertical grain boundaries may reduce both the short-circuit current and the open-circuit voltage if they intersect the junction boundary and provide effective shorting paths that increase the reverse saturation current. In addition, grain boundaries may also increase the diffusion of impurities by rapid diffusion along grain boundaries, and provide favorable sites for impurity segregation due to a kind of gettering effect.

It is believed that grain boundaries in Si are symmetrically depleted with about the same activity in *n*-type and *p*-type material , and that the electronic properties of these grain boundaries are not dominated by such intrinsic properties as dangling bonds, but by extrinsic (impurity-related) properties (Werner *et al.*, 1994c). The presence of oxygen and carbon, as well as many transition metals within the boundary plane, strongly alters the properties of Si grain boundaries.

The formation of photoinduced metastable defects associated with dangling bonds has been reported for hydrogenated polycrystalline silicon (Nickel *et al.*, 1993; Redfield and Bube, 1996) with many similarities to the effects observed in amorphous silicon as described in Chapter 3. Electron spin resonance

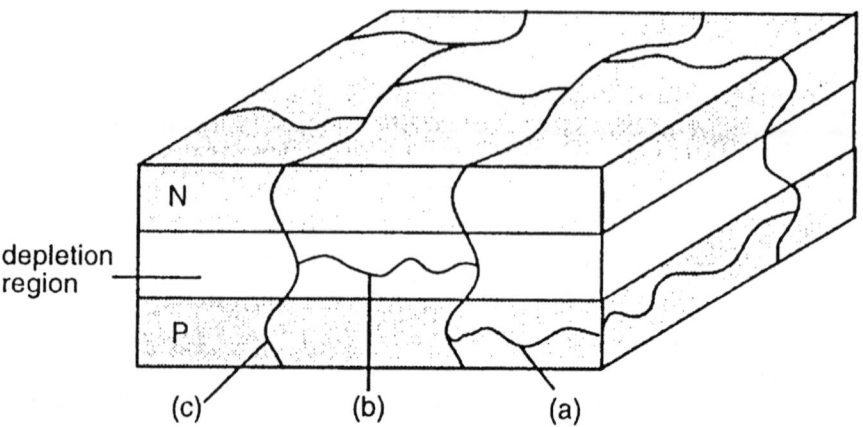

Fig. 2.15. Thin film polycrystalline silicon solar cell showing grain boundaries: (a) horizontal boundary in bulk region; (b) horizontal boundary in depletion region; (c) vertical grain boundary. (Reprinted with permission from M. A. Green, *Silicon Solar Cells: Advanced Principles & Practice*, Centre for Photovoltaic Devices and Systems, Univ. of New South Wales, Sydney, 1995.)

identifies the defects as dangling bonds (Johnson *et al.*, 1982), hydrogenation decreases the density of dangling bonds (Kamina and Marcoux, 1980), and hydrogen decreases the concentration of weak Si-Si bonds at or near the grain boundaries (Jackson *et al.*, 1983). There are, however, definite differences between polycrystalline Si and amorphous silicon (a-Si:H): (a) In polycrystalline silicon, the defects are essentially confined to the two-dimensional grain boundaries, (b) grain boundaries in polycrystalline Si show long-range order, whereas a-Si:H displays only short-range order, (c) the total density of metastable defects in polycrystalline silicon is generally smaller than in a-Si:H since the defects are confined to the grain boundary regions, and (d) unlike the case in a-Si:H, the density of photoinduced defects decreases with repeated illumination and anneal cycles, and is restored upon re-exposure to monatomic hydrogen. Although the production of photoinduced defects in a-Si:H solar cells is a principal mechanism for cell degradation (see Chapter 3), a similar effect has not been reported for multicrystalline Si solar cells.

Cast Multicrystalline Wafers

One approach to making thin-sheet silicon solar cells is a conventional casting process (CC) that produces multicrystalline ingots that may then be sawed into wafers (Hubbard and Cook, 1989; Watanabe *et al.*, 1990). In a process known as the "Heat Exchanger Method" (HEM), a heat exchanger is positioned at one end of the reservoir of molten Si to establish an almost one-dimensional temperature gradient; the castings produced by HEM are of excellent quality with grain sizes of several millimeters or more, and all grain boundaries are perpendicular to the plane of slices made from the casting (Khattak and Schmid, 1980). An example of a production process for low-cost multicrystalline silicon solar modules with 16% encapsulated efficiency is given by Verhoef *et al.* (1994). Improvements in the homogeneity of multicrystalline silicon ingots with columnar grain structure are described by Schaetzle *et al.* (1993). A continuous casting method (CCM) designed to produce low-cost silicon wafers, known as the "Drip-controlled method", has been developed (Goda *et al.*, 1994). A record high efficiency of 18.6% has been reported for mc-Si solar cells fabricated by a process that involves impurity gettering and back surface passivation of 0.65 ohm-cm mc-Si grown by HEM (Rohatgi *et al.*, 1996).

A process called electro-magnetic casting (EMC) has been developed (Kaneko *et al.*, 1994; Kawamura *et al.*, 1994) to produce a potentially inexpensive material characterized by a small grain size associated with a short diffusion length, but containing less oxygen and other impurities because of the absence of contact between the walls of the crucible and the molten silicon during growth. Hydrogen passivation is reported to be effective in improving the quality of EMC material, and an efficiency of 16% is reported on 4 cm^2 cells by combining surface oxide passivation, hydrogen passivation by remote plasma from the front side and by rf plasma from the back side (Elgamel *et al.*, 1994a,b,c, 1995; Rosmeulen, 1994). It was found that more hydrogen could be incorporated in the EMC multicrystalline silicon than in the CC material, probably because of the larger oxygen content in the CC material (Ghannam *et al.*, 1994).

"Ribbon" Technology

Various procedures belonging to variations on a "ribbon" technology in which sheets/ribbons/foils of Si for solar cells are produced, have been developed to allow lower cost of multicrystalline Si solar cells, while giving up some of the efficiency (Goetzberger and Rauber, 1988). They include (1) dendritic web (WEB) with shaping by two dendrites (Hopkins *et al.*, 1987), (2) edge-defined film-fed growth (EFG) with shaping by a graphite die (Wald, 1987), (3) melt-spinning (MS) with casting in a graphite die (Madea *et al.*, 1987), (4) ramp-assisted foil technique (RAFT) with foil growth on a reusable ramp (Beck *et al.*, 1987), (5) horizontal supported web (HSW) with wedge-shaped growth initiated by a graphite net (Falckenberg *et al.*, 1987), (6) silicon sheets from powder (SSP) with zone melting of powder layers using optical heating (Eyer *et al.*, 1987), and (7) embedding a matrix of small, spherical solar cells into perforated aluminum foil (Levine *et al.*, 1991).

Only the WEB (efficiencies of about 17%) and EFG (efficiencies about 15%) techniques have reached the stage of pilot plant activity. In the WEB technique, two naturally arising parallel dendrites form the boundaries of a web or ribbon pulled from a supercooled Si melt. The quality of the ribbon is equal to good Czochralski single-crystal material. In the EFG technique a carbon die with a slot-shaped aperture is immersed part way in a vessel of molten Si. The liquid Si wets the die and flows through the slot to feed the solid zone above. The shape of the ribbon is controlled by the shape of the top surface of the die, by surface tension, temperature gradients, and the pull rate. Ribbons are flexible enough to be wound on large drums and ribbons 20 m long have been grown in a single run. No polishing is required. EFG ribbon Si has rather high dislocation and twinning densities, which limit the efficiencies achievable in solar cell applications.

Thin Film Cells on Silicon Substrates

One approach to thin film Si cells is to use Si itself as a substrate. There have been two main ways that this has been attempted: by thinning a thicker Si wafer (Uematsu *et al.*, 1990; Nunoi *et al.*, 1990; Somberg, 1990; Hezel and Ziegler, 1993), or by epitaxial deposition of Si on a Si substrate using techniques like liquid phase epitaxy (LPE) and chemical vapor deposition (CVD) (Wagner *et al.*, 1993; Rodot *et al.*, 1987; Ciszek *et al.*, 1993; Zheng *et al.*, 1994; Fujimoto *et al.*, 1993; Oelting *et al.*, 1993; Bergmann *et al.*, 1994; Werner *et al.*, 1994d). Efficiencies of cells produced in this way range about 11% to 17%, with the highest value of 17.3% achieved for a diffused p–n junction formed by CVD deposition on a (100) p^+–Si substrate (Werner *et al.*, 1993, 1994d).

Thin film solar cells have also been fabricated with efficiencies as high as 9.3% by recrystallizing (using a large area heater, or by zone melting) 30–50 μm thick Si layers on 2 mm thick SiO_2 intermediate layers, deposited on Si substrates (Hebling *et al.*, 1996).

Thin Film Cells on Foreign Substrates

The fabrication of both low-cost and efficient thin film cells has several fundamental requirements (Werner *et al.*, 1994c): (1) deposition of Si on foreign substrates, (2) efficient light trapping, (3) passivation of grain boundaries and surfaces, (4) gettering techniques, and (5) development of low-temperature processing techniques suitable for the substrate used.

The substrates themselves should have a low manufacturing cost, chemical and mechanical stability during Si deposition and processing, sufficient purity to avoid contamination of the active Si film, a thermal expansion coefficient close to that of Si, and should be optically transparent and electrically insulating. Materials suitable for high-temperature processing (> 1000°C) include graphite, quartz glass, most ceramics and oxidized Si (which could be considered as a "foreign" substrate if deposition takes place on the oxide).

Glass is a good candidate for a low-temperature substrate (< 600°C). A procedure for low-temperature surface passivation suitable for use with silicon on glass substrates has been reported (Langguth *et al.*, 1995), involving growing a thermal pre-oxide by a short anneal at $T < 680°C$ to obtain a low defect density at the Si/SiO_2 interface, and depositing a top-oxide by reactive evaporation of SiO_2 at room temperature to supply fixed oxide charge. The solid phase crystallization of amorphous Si on glass substrates has been investigated to grow large-grained polycrystalline Si (Bergmann *et al.*, 1995).

A high-temperature-resistant glass substrate (stable up to 1000°C) was coated with a thin film of amorphous Si, which was then solid phase crystallized during atmospheric pressure chemical vapor deposition of a several micron thick light absorbing film (Bergmann *et al.*, 1996, 1997). Thin crystalline silicon solar cells have been deposited by direct high-temperature (1000°C) CVD from $SiHCl_3$ on temperature-resistant glass substrates (Brendel *et al.*, 1997). The measured diffusion length is only 0.6 μm in these cells, but theoretical modeling indicates that a sufficiently thin (0.4 μm) film of this diffusion length could yield 10% efficient cells, and a cell thickness of 1 μm could yield efficiencies of 15% for material with a diffusion length of 5 μm.

Examples of processes showing promise in the development of thin-film Si solar cells involve the deposition of thin solar grade Si layers on a rigid ceramic substrate, which has shown large area (78 cm^2) efficiency of 8.5%, and small area (1 cm^2) efficiency of about 15% (Barnett *et al.*, 1989, 1991, 1993), and the recrystallization of Si thin films deposited on oxide layers on single-crystal Si substrates by a special surface melting technique (Reis *et al.*, 1988; Arimoto *et al.*, 1993; Ishihara *et al.*, 1993). There are also several interesting attempts to deposit Si on low temperature substrates, but further research is needed for an adequate evaluation. In the effort to find a lightweight, low-cost substrate for thin polycrystalline Si solar cells, a 13.4% efficient solar cell has been produced by a thin layer of polycrystalline Si grown directly on a filament-based graphite fabric substrate (Rand *et al.*, 1994). Multicrystalline Si layers have also been grown on graphite substrates in an atmospheric pressure by rapid thermal chemical vapor deposition (RTCVD) method using trichlorosilane and the dopant trichloroborine diluted in hydrogen (Monna *et al.*, 1996).

Passivation procedures

The grain boundary potential barriers in both n- and p-type silicon are reduced by the diffusion of monatomic hydrogen into the grain boundaries (Aucouturier, 1991; Sopori *et al.*, 1991; Seager *et al.*, 1982; Rohatgi *et al.*, 1993). There is a decrease in the density of trapping sites for majority carriers, and a decrease in the recombination velocity by a couple of orders of magnitude for minority carriers. Still the fundamental microscopic mechanism is not completely understood. The extent of the passivation depends on the specific origin of the boundaries, and results not only from removing dangling bonds, but also electronic defects related to impurity segregation. Both grain boundaries and dislocations in LPE-grown Si appear to have a much lower recombination velocity than those in the substrate, suggesting intrinsic gettering and bond reconstruction during the near-equilibrium LPE growth of the film.

Another possibility for effective passivation of thin-film solar cells is the plasma enhanced chemical vapor deposition (PECVD) of silicon nitride (Michiels *et al.*, 1990; Szlufcik *et al.*, 1994; Schetter *et al.*, 1995). It can serve as an antireflection coating and passivating layer, and can be deposited in existing commercial systems. Strong enhancement of the bulk and surface passivation from a plasma nitride layer has been reported during a fast firing for screen printing metallization.

Gettering techniques

Effective gettering for a thin-film solar cell where the entire volume is electronically active, requires the movement of impurities to the surface. Techniques with P and Al gettering have been reviewed (Verhoef *et al.*, 1990; Gee, 1991; Rohatgi *et al.*, 1993; Porre *et al.*, 1993; Perichaud *et al.*, 1993), as has gettering with P and Cl (Jastrzebski 1995). Verhoef *et al.* (1990) reported diffusion experiments indicating that impurities such as copper, gold and nickel are effectively gettered at the aluminum-doped layer. Gettering in single- and multicrystalline silicon has been compared, using P, Al, and Cl gettering, as well as back-side hydrogen gettering, and ultrasound treatments (Sopori *et al.*, 1996).

Multilayer Thin Film Silicon Solar Cells

Another approach to the development of high efficiency silicon solar cells on lower quality material is that associated with multilayer thin film cells (Wenham *et al.*, 1994; Sproul *et al.*, 1994; Green and Wenham, 1994; Green 1995c) made with multicrystalline silicon. The process consists of depositing a dielectric layer such as silicon oxynitride/nitride on a cleaned glass superstrate, followed by the deposition of five to ten layers of silicon. Alternate layers are doped with a different dopant impurity to give *p*- or *n*-type properties. A process like chemical vapor deposition can be used for the deposition of this multilayer stack. Finally, a silicon oxynitride/nitride layer is deposited on the top surface as a capping layer, producing a total thickness between 10 and 20 μm. Then, using the technique for making buried contact cells, a laser is used to form one set of grooves of one type of polarity connecting all the *n*-type layers within the stack in parallel, and a second set of grooves which join all the *p*-type layers, followed by metallization of the grooved areas.

The structure can be thought of as a number of very thin cells all connected in parallel, with the thickness of each layer chosen to be smaller than the carrier diffusion length within that layer, thus insuring that most of the carriers photogenerated in the stack are collected. The theoretically expected efficiency increases from less than 8% for a single junction to over 15% for six or more junctions, assuming that the primary loss mechanism can be attributed to uniformly distributed defects. The stacked cells should also be much less sensitive to the effects of grain boundaries.

Contactless Techniques for Characterization

In evaluating the properties of materials for multicrystalline silicon solar cells, it is advantageous to be able to make a spatially resolved measurement of the lifetime or diffusion length of carriers in the material without contacts to the material.

Modulated free-carrier absorption (MFCA)

A technique known as "modulated free-carrier absorption (MFCA)" has been developed (Glunz *et al.*, 1994c, 1995) which is able to probe the excess minority carrier distribution over a wide area. In this method, photogenerated free carriers are optically monitored by the absorption of infrared light. A phase shift between the generation light and the IR transmission occurs due to the lifetime of the carriers. An effective lifetime can be determined from this phase shift, and in order to reach a high spatial resolution, the infrared laser beam is focused on the sample (light-spot diameter < 100 μm).

Light-beam induced luminescence (LBIL)

A standard technique for the measurement of the diffusion length L of carriers in a solar cell is electron-beam or light-beam induced conductivity (EBIC or LBIC) (e.g. see Fahrenbruch and Bube, 1983). The current is measured as a small region of the material is excited by an electron beam or a light beam at a varying distance x from the junction. For moderate to large values of x/L, the short-circuit current of the cell is given by $I_{sc} = A \exp(-x/L)$. It has been shown that a contactless method for obtaining similar information can be obtained from light-beam or electron-beam induced luminescence in silicon (Daub *et al.*, 1994), which is applicable both to unprocessed wafers and to completed solar cells.

2.8. Microcrystalline Silicon

Applications of p-type μc-Si:H and μc-Si:C:H in a-Si:H solar cells are described further in Secs. 3.9 and 3.10. A number of results have been reported for solar cells involving μc-Si:H deposited by the Very High Frequency Glow Discharge (VHF) method (Finger *et al.*, 1993) which is also described in connection with a-Si:H deposition in Sec. 3.6. Structural investigations of μc-Si:H indicate that it is a composite material made up of crystalline grains, grain boundaries and an amorphous phase. The VHF technique involves deposition in a glow discharge at an excitation frequency of 70 MHz from silane (1.5%) diluted in hydrogen, with phosphine and diborane as dopant gases. Initial investigation indicated that electrical transport in μc-Si:H cannot be described by means of a single activation energy, but instead evidence exists for a distribution of activation energies probably correlated with a distribution of grain sizes and grain boundary defect densities.

A number of reports exist on the use of these VHF-deposited μc-Si:H layers in p–i–n solar cells (Flueckiger *et al.*, 1993, 1995a, 1995b; Meier *et al.*, 1994a, 1994b; Meier *et al.*, 1996). The as-deposited μc-Si:H was slightly n-type, and compensation was obtained by boron-doping. Entirely μc-Si:H p–i–n type solar cells were fabricated incorporating the compensated layer, with first results yielding an efficiency of 3.8% and apparently better stability than standard a-Si:H cells. The μc-Si:H cells showed larger infrared response beyond 750 nm compared to a-Si:H cells because of a smaller effective band gap of about 1 eV.

Subsequent work raised the efficiency to 4.6% and indicated that exposure to solar illumination did not degrade the totally μc-Si:H cells. An a-Si:H/μc-Si:H stacked cell showed an efficiency of 9.1%, which did not degrade after 280 hours of intensive illumination with a sodium lamp with brightness of about six suns. It is suggested that the electronic behavior of the μc-Si:H is controlled by the crystalline phase in these materials. A beginning has also been made on understanding the transport properties of compensated μc-Si:H (Wyrsch *et al.*, 1996) and the enhanced optical absorption of μc-Si:H (Beck *et al.*, 1996). The potential utility of μc-Si$_{(1-x)}$C$_x$ interface layers in a-Si:H and mc-Si solar cells has also been investigated (Ma *et al.*, 1994b).

Microcrystalline-crystalline Si heterojunction solar cells were made using a thin p^+ μc-Si:H window layer on top of 1 ohm-cm n-type single-crystal Si (van Cleef *et al.*, 1996). These cells had the following structure: Ag/ITO/p^+ μc-Si:H/a-Si:H/n-Si/Al. A buffer layer consisting of intrinsic a-Si:H deposited at low temperature was found to be essential in achieving high open-circuit voltage and efficiency (12.2%).

CHAPTER 3

AMORPHOUS SILICON

3.1. Overview

Amorphous silicon is a thin-film material without long range atomic order that can be deposited by a variety of techniques suitable for large area production. In this form, however, amorphous silicon has a high density (of the order of 10^{19} cm^{-3}) of coordination defects ("dangling bonds") corresponding to departures from the local tetrahedral coordination between four silicon atoms (see Sec. 3.2). They act as recombination centers to greatly reduce the carrier lifetime and the carrier diffusion and drift lengths, and to pin the Fermi energy in such a way that the material cannot be effectively doped n- or p-type and is not electronically useful. It was discovered, however, that the incorporation of about 10% hydrogen (from glow-discharge decomposition of silane gas) during the deposition process greatly reduces the density of these defects to about 10^{16} cm^{-3} (Chittick *et al.*, 1969), and this hydrogenated material can then be doped n- or p-type to make useful semiconductor devices (Spear and LeComber, 1976). A promising new thin-film silicon alloy, a-Si:H, had been discovered.

Perhaps of greatest significance for photovoltaic applications, the hydrogenated amorphous silicon alloy (a-Si:H) also has a higher-energy absorption edge and a larger optical absorption constant for solar radiation than crystalline silicon. a-Si:H has a well-defined optical threshold at about 1.75 eV, considerably larger and sharper than the indirect band gap of 1.1 eV of crystalline Si, as shown in Fig. 3.1, and only 1 to 2 μm thickness of a-Si:H is required to absorb virtually all of the light above the absorption edge.

In the 20 years since this initial work, a voluminous literature has grown up related to amorphous silicon. As examples we cite a general discussion of amorphous silicon in a four-volume series edited by Pankove (1984), a two-volume set edited by Fritzsche (1989), a book by Street (1991), and a two-volume series edited by Neber-Aeschbacher (1995) containing thirty-six review papers and sixteen original papers on many aspects of hydrogenated amorphous silicon and its alloys. A portion of a book by Redfield and Bube (1996) is

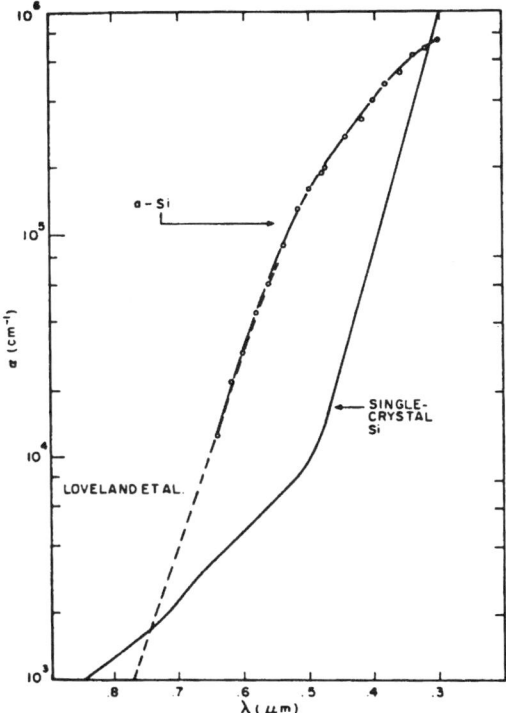

Fig. 3.1. A comparison between the optical absorption coefficient for single crystal Si and a-Si:H as a function of wavelength. Long-wavelength data are from Loveland *et al.* (1973/74). (Reprinted with permission from D. E. Carlson and C. R. Wronski, *Appl. Phys. Lett.* **28**, 671 (1976). Copyright 1976, American Institute of Physics.)

concerned particularly with the properties of the defects and their induction by photoexcitation.

It appears, therefore, that a-Si:H should be an excellent choice for thin-film solar cells: It is related to the general silicon technology and it can be relatively inexpensively deposited in thin film form with high uniformity over large areas on inexpensive substrates (e.g. glass) at low temperatures. Problems are associated with the high density of defects still present even in the hydrogenated material, and with the observation that the density of these defects increases upon a variety of treatments of the material (see Sec. 3.3), particularly to exposure to high-intensity light (optical degradation). In spite of these instabilities, a-Si:H has often been considered the front-runner in competition with polycrystalline materials (see Chapters 5 and 6) for thin-film solar cells.

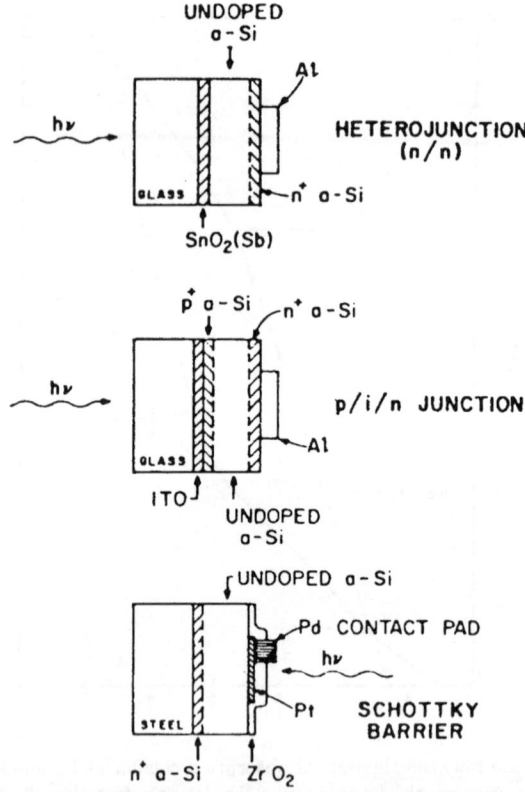

Fig. 3.2.   Three types of a-Si:H junctions: (top) a heterojunction between SnO_2 and a-Si:H, (center) a p–i–n junction with p^+- and n^+-a-Si:H layers with undoped a-Si:H between them, and (bottom) a Schottky barrier between Pt and n^+-a-Si:H doped with P. (Reprinted with permission from D. E. Carlson, *IEEE Trans. Electron Devices* **ED-24**, 449 (1977). Copyright 1977, IEEE, NY.)

A variety of junction structures has been tested for solar cells using a-Si:H, as indicated in Fig. 3.2. The most effective is generally agreed to be a p–i–n structure, consisting of an undoped insulating region between doped p- and n-type end regions, which makes it possible for the main absorber portion of the cell to have the longer lifetimes of undoped a-Si:H, and for the carrier collection process to be dominated by drift rather than diffusion. Initial efficiencies as high as 12% had been obtained for single-junction cells by the late '80's (Wronski, 1986; Hubbard, 1989; Hubbard and Cook, 1989). The general improvements in a-Si:H solar cell efficiency are summarized in

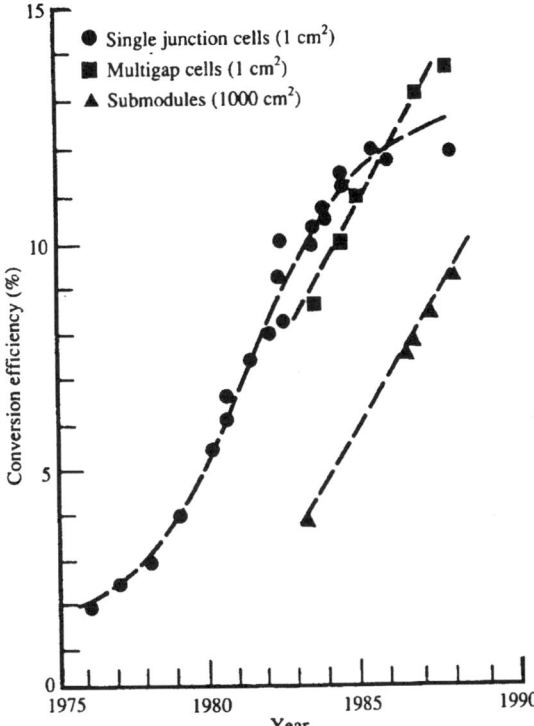

Fig. 3.3. Improvement of a-Si:H based solar cells from 1975 to 1989. (Reprinted with permission from E. S. Sabisky and J. L. Stone, *20th IEEE Photovoltaic Specialists Conference*, p. 39 (1988). Copyright 1988, IEEE, NY.)

Fig. 3.3 with data shown for small area single and stacked cells and for large area modules (Sabisky and Stone, 1988). It is projected that the efficiency of a multijunction cell may reach 20% by the year 2000 (Maruyama *et al.*, 1995).

It is the effects of optical degradation that have limited the practical efficiency of a-Si:H solar cells, and many attempts have been made to determine the exact mechanisms involved in this process, to overcome defect formation itself, or to redesign the solar cells so that the optical degradation effects are minimized. These effects are discussed further in Secs. 3.3 and 3.8.

3.2. Electronic Structure and Dangling Bond Defects in a-Si:H

As an introduction to the use of a-Si:H in solar cells, we give in this section a brief summary of some of the basic properties of the defects in a-Si:H. The formation and properties of defect states in a-Si:H are discussed in fifteen review papers in the volumes edited by Neber-Aeschbacher (1995).

Comparison with Crystalline Silicon

In spite of the absence of long range order in a-Si:H, many of the electronic properties are similar to those of crystalline Si. As shown in Fig. 3.1, a sharp optical absorption edge exists in a-Si:H, although it is displaced some 0.65 eV toward higher energies compared to crystalline Si. Potential fluctuations in the amorphous material, however, create electronic states which form a nearly exponential distribution of *band-tail* states, which prevent a true density-of-states band gap from existing in the amorphous material. In these tails carrier mobility decreases to low values at low energies and a threshold known as the

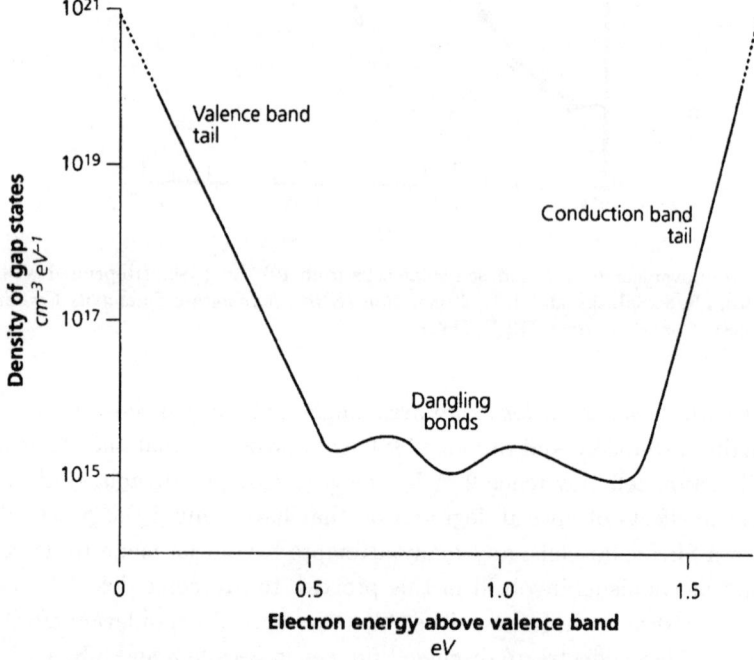

Fig. 3.4. Density of electron states in the band gap of a-Si:H, characterized by exponentially varying valence and conduction band tails, and states associated with dangling bond defects near the middle of the gap. (Reprinted with permission from D. E. Carlson and S. Wagner, in *Renewable Energy: Sources for Fuels and Electricity*, eds. T. B. Johansson *et al.*, p. 403, Copyright 1993, Island Press, Washington, D.C.).

mobility edge exists in the energy spectrum of mobility. The gap that does exist, between the mobility edge for electrons in the conduction band and the edge for holes in the valence band, is a *mobility gap*, which is quantitatively similar to the optical absorption threshold. Figure 3.4 gives a schematic plot of the density of defect states in a-Si:H as a function of electron energy above the valence band (Carlson and Wagner, 1993).

Diffraction patterns of quality a-Si:H show that nearly every Si atom is bonded to four others in the same tetrahedral coordination as in crystalline Si, with first and second nearest-neighbor distances similar to the values in crystalline Si. The major difference is that in the amorphous material there is an approximately 10° range of bond angles centered at the tetrahedral value of approximately 109°. The densities of amorphous and crystalline Si are also approximately the same. The ideal structure for an amorphous material has been called a *continuous random network* in which a complete amorphous network of tetrahedrally bonded atoms can be formed, retaining the short-range order except for small deviations in bond angles. The existence and properties of hydrogen complexes in a-Si:H are discussed by Jackson and Zhang (1990).

Native Defects

In this context a native defect in amorphous silicon can be seen as a departure from the tetrahedral coordination of some particular Si atom to four other Si atoms: a broken Si bond, or a *dangling bond*. These defects are common to all samples of amorphous silicon prepared around the world, and are identified by a characteristic electron spin resonance (ESR) $g = 2.0055$ (Brodsky and Title, 1969; Taylor, 1984).

Actually ESR can detect only the paramagnetic species of dangling bond with one unpaired electron; other possible states with zero or two unpaired electrons are diamagnetic. The charge state of the defect in a particular material depends on the location of the Fermi energy in that material; if the Fermi energy is high in the gap then most defects will be in the diamagnetic D^- state occupied by two electrons, whereas if the Fermi energy lies close to the valence edge, most defects will be in the diamagnetic D^+ state, unoccupied by electrons. When the Fermi energy lies close to the middle of the gap, as it does in undoped, high quality a-Si:H, most defects will be in the paramagnetic D^0 state with one unpaired electron. Both those defects that are present in the dark after long times of thermal annealing, the so-called *built-in* defects, and those that are induced by photoexcitation have the same ESR behavior.

Another issue of considerable importance (see Sec. 3.4) is whether the defects are intrinsic to the nature of the amorphous material, or whether they are directly related to some extrinsic property such as a physical or chemical abnormality (Stafford and Sabisky, 1987; Redfield and Bube, 1990; Redfield, 1991; Stafford, 1991).

Metastable defects in doped a-Si:H have been investigated by photomodulation spectroscopy (Chen *et al.*, 1992; Kocka *et al.*, 1993). In addition to information on the dangling bond defect (with state D_i^0 in the dark, becoming D_i^- or D_i^+ after capturing a photoexcited electron or hole, respectively) in undoped material as summarized above, deep defects have been investigated associated with phosphorus doping in a-Si:H:P (with state D_P^- in the dark, becoming D_P^0 after trapping a photogenerated hole), and with boron doping in a-Si:H:B (with state D_B^+ in the dark, becoming D_B^0 after trapping a photogenerated electron). Exposure of a-Si:H:P to light produces both D_P defects and D_i defects.

3.3. Modes of Defect Generation

In this section we summarize the different treatments of a-Si:H that lead to an increase in the density of metastable dangling bond defects. Several reviews have been given (e.g. see Wagner and Smith, 1988; Neber-Aeschbacher, ed., 1995; Redfield and Bube, 1996).

Doping with Standard Dopants (Beyer *et al.*, 1977; Austin *et al.*, 1979)

Doping a-Si:H with standard Si dopants such as phosphorus (n-type) or boron (p-type) leads to an increase in the density of dangling bond defects if the dopants are uncompensated, but actually to a decrease in defect density if they are compensated (Amer and Jackson, 1984).

Quenching from Elevated Temperatures

If a sample of a-Si:H is heated to an elevated temperature ($\approx 220°C$) and then quenched, it is found that the density of dangling bonds is larger than the initial room temperature value (Street *et al.*, 1986). The density of dangling bonds has been thermally increased, and the excess can be annealed away by subsequent heating. A detailed investigation of thermal equilibrium processes induced by thermal quenching effects in undoped a-Si:H has been reported by Meaudre *et al.* (1991), who interpreted the results as indicating that the equilibration temperature of the films is about 180–200°C.

Exposure to High-Intensity Light (Staebler and Wronski, 1977)

When a-Si:H is exposed to high-intensity light such as sunlight, the density of dangling bond defects increases until reaching a saturation value N_{sat} after long times of exposure. Both the dark conductivity and the photoconductivity of the a-Si:H decrease; these effects can be annealed away at about 150°C. This effect occurs for photon energies down to about 1 eV, even though the band gap of a-Si:H is ≈ 1.7 eV. Also, the density of light-induced defects in a material increases with the density of defects induced in the material in the dark by previous doping, as shown in Fig. 3.5 (Skumanich *et al.*, 1985). Metastable defects are formed with comparable efficiency at all temperatures below 300 K down to 4.2 K, showing that the creation mechanism is not thermally activated; about 40% of the defects induced at 4.2 K anneal between 150 K and 300 K, indicating that low-T exposure creates defects that would not be stable for high-T exposure (Stradins and Fritzsche, 1993).

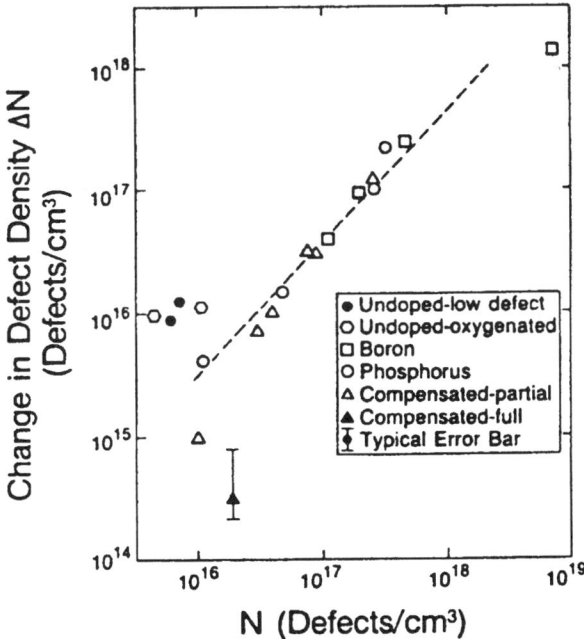

Fig. 3.5. Correlation of the densities of light-induced defects ΔN with the density of defects N present in the starting a-Si:H material for various doping conditions. (Reprinted with permission from A. Skumanich *et al.*, *Phys. Rev. B* **31**, 2263 (1985). Copyright 1985, American Institute of Physics, College Park, MD.)

In a number of different cases the description of the kinetics of defect formation under illumination appears to fit very well with a so-called stretched-exponential description, usually associated with dispersive phenomena having a distribution of time constants (Redfield and Bube, 1996). It was shown that the time-dependence of defect density during thermal annealing of these light-induced defects can be described by a stretched exponential (Kakalios *et al.*, 1987), and subsequently similar results were demonstrated for the time-dependence of defect formation under illumination based on a symmetrical rate equation including both optical and thermal, formation and annealing terms (Redfield, 1988, 1989). Without dispersion this rate equation was proposed to be

$$dN/dt = C_1 R(N_T - N) - C_2 RN + \nu_1 (N_T - N) - \nu_2 N \qquad (3.1)$$

where R is the total carrier capture (or recombination) rate, C_1 and C_2 are effectiveness coefficients for photoinduced generation and recovery of the metastable condition, N_T is the maximum density of possible defect centers and ν_1 and ν_2 are analogous thermally-activated coefficients for thermally induced generation and recovery, respectively. The quantity $(N_T - N)$ is the density of centers that are in their ground state and are thus available for defect formation. Including dispersion, Eq. (3.1) becomes

$$dN/dt = (t/P)^{(\beta-1)} [C_1 R(N_T - N) - C_2 RN + \nu_1 (N_T - N) - \nu_2 N] \qquad (3.2)$$

with solution

$$N(t) = N_{\text{sat}} - (N_{\text{sat}} - N_0) \exp[-(t/\tau)^\beta] \qquad (3.3)$$

Here β is the "stretch parameter" related to the breadth of the distribution of contributing processes $(\beta < 1)$ and P is a scaling factor that preserves the dimensions. N_0 is the initial density of metastable defects, and τ is an effective time constant for the transient (Redfield, 1992). N_{sat} and τ contain all four of the coefficients in Eq. (3.1).

An example of the application of this formulation is provided by the measurements of Park *et al.* (1989) using two different intensities of photoexcitation on two different samples, as analyzed by Bube *et al.* (1990), and shown in Fig. 3.6(a). A convenient way to test the appropriateness of a stretched-exponential description, and to obtain the value of the fitting parameter K (writing the exponential in Eq. (3.3) as $\exp[-Kt^\beta]$) is to rewrite Eq. (3.3) as

$$\log f^*(N) = \log K + \beta \log t \qquad (3.4)$$

where $f^*(N) = \ln[(N_{\text{sat}} - N_0)/(N_{\text{sat}} - N)]$ and is plotted in Fig. 3.6(b) for the data of Fig. 3.6(a). For an actual stretched-exponential behavior, a plot of

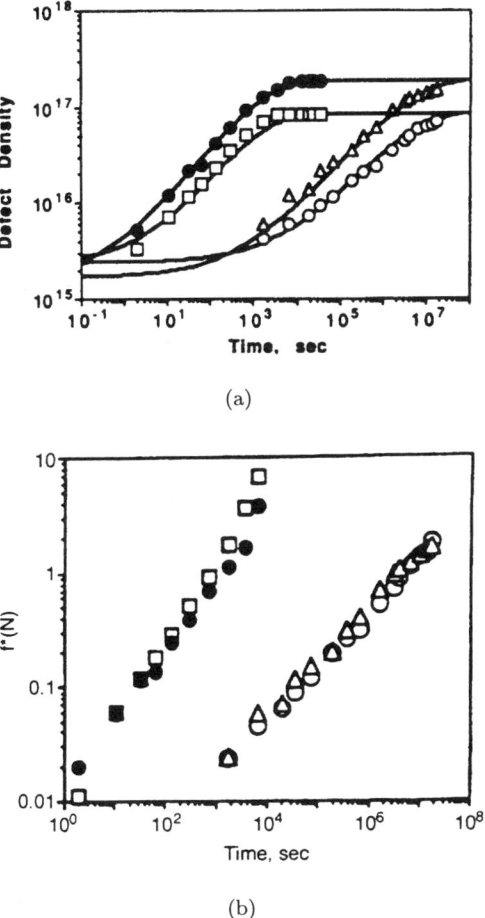

(a)

(b)

Fig. 3.6. (a) Comparison of experimental data points from Park *et al.* (1989) for defect density (cm^{-3}) in a-Si:H with curves calculated from a stretched-exponential description of optical degradation with a single fitting constant for all four curves (•) Sample A44, HI (high-intensity Kr$^+$ laser light), 343 K; (□) sample A62, HI, 343 K; (△) sample A44, LI (lower-intensity tungsten-halogen light) 308 K; (○) sample A62, LI, 308 K. (b) The data of Park *et al.* (1989) given in (a), plotted using the form of Eq. (3.4) to test the closeness of the data to stretched exponentials, which produce straight lines in this kind of plot. (Reprinted with permission from R. H. Bube *et al., Appl. Phys. Lett.,* **57**, 79 (1990). Copyright 1990, American Institute of Physics.)

$\log f^*(N)$ versus $\log t$ yields a straight line with slope β and intercept $\log K$. Such stretched-exponential descriptions appear to be appropriate for a variety of defect kinetics data obtained in recent years (Bube and Redfield, 1989).

An irreversible optical degradation of a-Si:H solar cells exposed to intense illumination (50 suns) at elevated temperatures ($> 130°C$) has recently been reported (Carlson and Rajan, 1995, 1996). Unlike the optical degradation induced at lower temperatures, this optical degradation is not reversed by annealing and is not suppressed by a strong bias. Measurements indicate that the effect is due to the diffusion of hydrogen at elevated temperatures both in the dark and under intense illumination.

Exposure to Electron Beams (Street *et al.*, 1979; Schade, 1984; Schneider *et al.*, 1987; Scholz and Schroeder, 1990, 1991; Grimbergen *et al.*, 1993)

Metastable defects are formed in a-Si:H by high-energy (typically 20 keV) electron beam irradiation similar to those caused by strong optical illumination. The major differences between the two defect production processes are (a) the electron beam produces a higher density of defects, (b) the kinetics for electron-beam defect generation follow a simple exponential rather than a stretched exponential, although the kinetics for annealing of electron-beam-induced defects does follow a stretched exponential, (c) the higher density of defects induced by electron beam anneal more rapidly than defects induced by light, (d) both electron-beam and light excitation cause a considerable decrease in the photoconductivity, but the dark conductivity, which decreases for photoexcitation, actually increases for electron-beam excitation in undoped material (Voget-Grote *et al.*, 1980).

Passage of Forward Currents in Device Structures (Staebler *et al.*, 1981; Crandall, 1987, 1991)

The observation that dangling-bond defects, the same as those induced by light, are formed in the dark by passage of forward current in a p–i–n device led to the proposal that defects are not produced directly by the light, but rather by the energy released when excess carriers recombine or are captured.

An investigation of light and current formation of defects in a-Si:H p–i–n, n–i–n, and p–i–p structures indicated that defects are formed both by recombination and by injection of holes, but not by injection of electrons (Ostendorf *et al.*, 1993).

Formation of a Space-Charge Region (Hepburn *et al.*, 1986; van Berkel and Powell, 1987; Deane and Powell, 1993)

When a voltage is applied to the gate electrode in a field-effect transistor (FET), charges are trapped near the a-Si:H/dielectric interface. Analysis of the somewhat complicated phenomena involved indicate that metastable defects are formed in the a-Si:H itself by this field-induced effect without requiring recombination to occur.

3.4. Models for Dangling Bond Defects

Because of their importance in determining the electronic properties and stability of a-Si:H, here we consider briefly a few representative models used to describe the dangling-bond defects (Carlson, 1987; Redfield and Bube, 1996) before proceeding to more general descriptions of solar cell development.

Configurational Coordinate Diagram

An appropriate description of the dependence of defect energies on local atomic order or disorder calls for what has been named a "configurational coordinate" representation (Baraff *et al.*, 1980; Redfield and Bube, 1996). A simplified version is given in Fig. 3.7, showing a plot of energy as a function of some appropriately chosen configuration coordinate, e.g. the distance between two

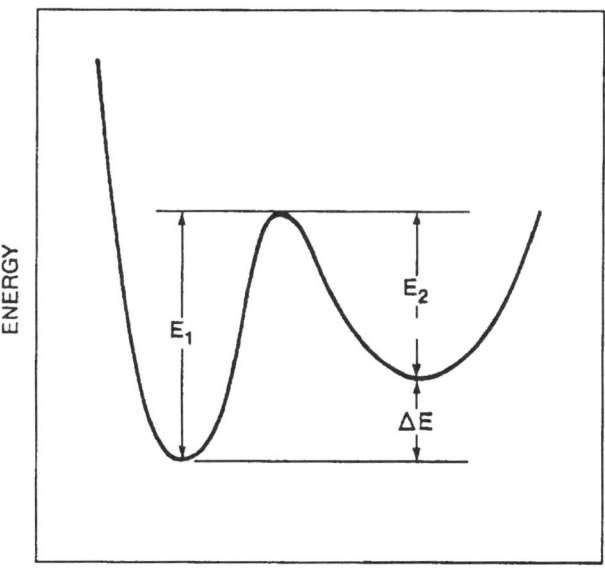

CONFIGURATION COORDINATE

Fig. 3.7. Simplified configuration-coordinate diagram that is generally used for the dangling-bond defect in undoped a-Si:H. The minimum on the left represents the ground state, and that on the right represents the dangling-bond metastable state.

neighboring Si atoms. The left energy minimum in the figure corresponds to the inactive ground state (often referred to as a "weak bond") which sees an energy barrier E_1 for passing over into the metastable ("dangling bond") state, with energy barrier E_2 for return to the ground state. The difference $\Delta E = (E_1 - E_2)$ can be inferred from high-temperature measurements of the equilibrium density of defects and appears to have a value of about 0.2 eV.

Weak-Bond Breaking

Some variation of the concept of weak-bond breaking underlies most of the models describing the origin of metastable dangling bond defects in a-Si:H. It appears to be a natural inference from the fact that the normal Si–Si bond strength is about 2.2 eV, but it is known that photoexcitation can produce dangling bonds with photons with energy as low as 1 eV. The first, and perhaps the simplest model, for the formation of dangling bonds is one in which weak bonds are broken by the energy released by the recombination between free electrons and holes formed by photoexcitation (Adler, 1984; Stutzmann *et al.*, 1985). This model predicted a variation of defect density as $t^{1/3}$ during optical degradation, but later examination showed that the apparent agreement of the data, more completely described by a stretched exponential as described in Sec. 3.3, with such a $t^{1/3}$ behavior was coincidental (Redfield, 1989; Bube *et al.*, 1990; Bube and Redfield, 1992).

Distribution of Energies: Defect Pool Model

Because of the nature of the amorphous material, it is appropriate to consider a distribution of energy values, and this has been done in a number of models. One of the most thoroughly investigated is the "defect pool model", which assumes a broad distribution of possible energies for the dangling bonds within the energy gap (Smith and Wagner, 1987, 1989; Winer, 1990, 1991; Powell *et al.*, 1991, Branz *et al.*, 1991, Hata and Wagner, 1992; Deane and Powell, 1993; Schumm, 1994; see also a review in Redfield and Bube, 1996). A weak-bond state is associated with a state of the valence-band tail, and the valence-band-tail width is taken to be a measure of the structural disorder in the material. An equilibrium description is used for defect reactions in terms of a chemical-equilibrium equation relating the densities of weak bonds and dangling bonds. Because of the broad distribution of possible defects, charged as well as neutral defects can occur for any value of the Fermi energy. Hydrogen atoms play a key role in bond switching. A representative equation is:

$$\text{(Si–Si)}_{\text{wb}} + \text{Si–H} \rightarrow [D^0 + \text{(Si–H)}] + D^0 \tag{3.5}$$

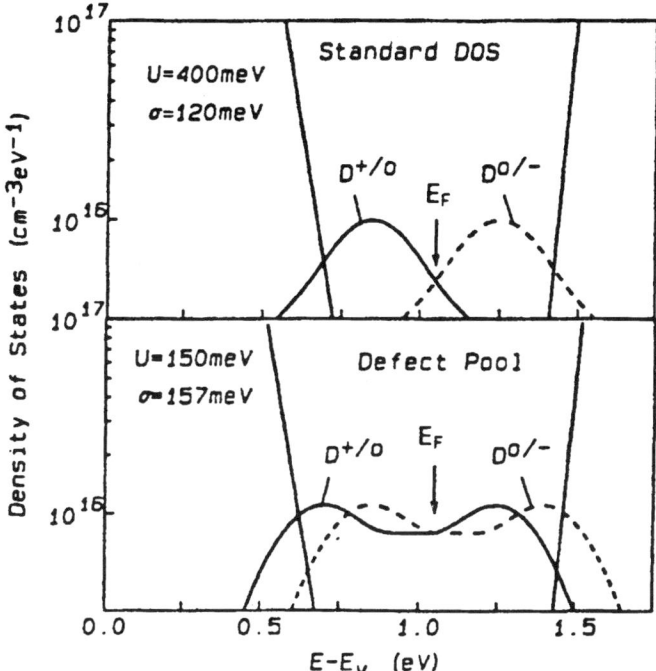

Fig. 3.8. Calculated densities of states in the gap of a-Si:H using the defect-pool model approach, showing the difference between the simple two-peaks (above), similar to that obtained for a Gaussian distribution of energies for a single type of defect, obtained when the correlation energy U is large, and the "defect-pool" result (below) when U is smaller. (Reprinted with permission from G. Schumm, *Phys. Rev. B* **49**, 2427 (1994). Copyright 1994, American Institute of Physics, College Park, MD.)

where $(Si–Si)_{wb}$ is the initial, weak bond state; Si–H is a separate complex of a Si atom that is bonded to one H atom; and D^0 is a neutral dangling bond state formed when the weak bond breaks.

A comparison is given in Fig. 3.8 between two energy distributions of the density of dangling bonds calculated according to the same defect-pool formulae, on top when $\sigma \ll U$, and at the bottom when $\sigma \approx U$, where σ is the width of a Gaussian distribution of energies, and U is the defect correlation energy (the energy difference between when the defect is negatively charged and when neutral, assumed the same for all defects to be of the order of 0.2 to 0.3 eV). The results shown on top in Fig. 3.8 are quite similar to those expected when the dangling bond is described by a simple Gaussian shape in a discrete energy-level model. The results shown on the bottom in Fig. 3.8 indicate the consequences when the dangling bond is described by a typical defect-pool model (Schumm, 1994). The occurrence of various charge states at different places in the material is the consequence of a range of local environments.

Some apparent problems with the defect-pool model (DP) include (a) DP predicts a greater density of charged defects than neutral defects even when the Fermi level is at midgap, but measurements of ESR and capacitance are essentially equal, indicating that most defects are neutral (Unold *et al.*, 1994); (b) DP calculations start with equilibrium relationships, and then apply these procedures to steady-state, nonequilibrium conditions (Winer, 1990; Deane and Powell, 1993); (c) Ganguly and Matsuda (1993) report that the bulk defect density is determined by surface-controlled processes during material deposition, and not by bulk equilibration; (d) relaxation of charged defect states have been interpreted as ruling out a DP model (Cohen *et al.*, 1993).

Impurity-Induced Defects

It is known that uncompensated doping of a-Si:H increases the density of metastable defects, as mentioned earlier. One of the early models for meta-stable defects, proposed by Street (1982b), involved the formation of defects by the process of self-compensation analogous to the behavior in partially ionic, crystalline materials, in which, for example, acceptor-type cation vacancies might be formed to lower the total energy when donor impurities are incorporated. Experimental tests of this model of doping have been reported by Kroetz *et al.* (1991). It is uncertain, however, how similar such processes in a-Si:H should be considered as compared to those in crystalline materials, and how important such self-compensation processes are even in crystalline materials in view of developments in recent years (Neumark, 1992; Chadi, 1994).

Rehybridized Two-Site (RTS) Model

This model suggests that extrinsic sources are the cause of the existence of weak bonds in a-Si:H, and that therefore reduction of impurities might result in material with fewer dangling bonds produced by photoexcitation (Redfield and Bube, 1990; Redfield, 1991). The RTS model was suggested by the success-ful description of the DX and related centers in III–V and II–VI compounds by a two-site model for a foreign atom (Chadi and Chang, 1988, 1989a,b), and is consistent with a number of properties of the dangling bond defects in a-Si:H. Figure 3.9 illustrates the RTS model; in this case the ground state of a foreign atom at a substitutional site has fourfold coordination, and the center is electronically inert; in the metastable state one bond to a neighboring Si atom is broken, and the bonds of the foreign atom rehybridize to form three-fold coordination. A critique of the RTS model includes the fact that the DX-type center that was used as its basis does not occur in crystalline Si, and particularly that dangling bond defects in a-Si:H occur in good-quality undoped material in which the density of impurities (O: 2×10^{15} cm^{-3}, C: 7–10×10^{15} cm^{-3}, and N: 5×10^{14} cm^{-3}) is much less than the density (5×10^{17} cm^{-3}) of dangling bond defects (Nakata *et al.*, 1993; Kamei *et al.*, 1996).

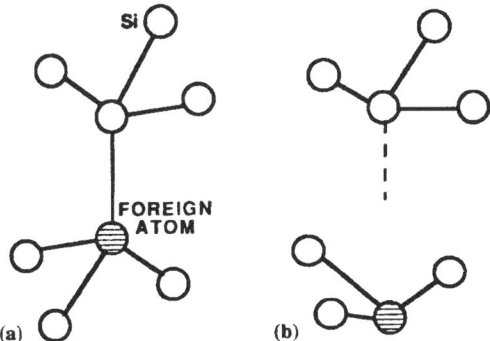

Fig. 3.9. A proposed RTS model for the metastable center based on a foreign atom in analogy with the DX center in crystalline compounds; (a) ground state, (b) metastable state. (Reprinted with permission from D. Redfield and R. H. Bube, *Phys. Rev. Lett.* **65**, 464 (1990). Copyright 1990, American Institute of Physics, College Park, MD.)

Carrier-Induced Model

Some of the characteristics of dangling-bond defect formation in a-Si:H do not appear to fit with the general theme of the above models, i.e. that recombination energy is the driving force. Typical examples are (a) the formation of defects in accumulation layers where carrier recombination is negligible, (b) photoinducing of defects as effectively at temperatures as low as 4 K as at 300 K, and (c) enhancing the annealing rate of excess defects by the addition of photoexcitation (Street, 1991; Meaudre and Meaudre, 1991a,b; Isomura and Wagner, 1992; Gleskova *et al.*, 1993a,b). An open-ended proposal has been advanced (Redfield and Bube, 1996) in terms of the configuration coordinate diagram of Fig. 3.10, which resembles that for the EL2 defect in GaAs (Dabrowski and Scheffler, 1989).

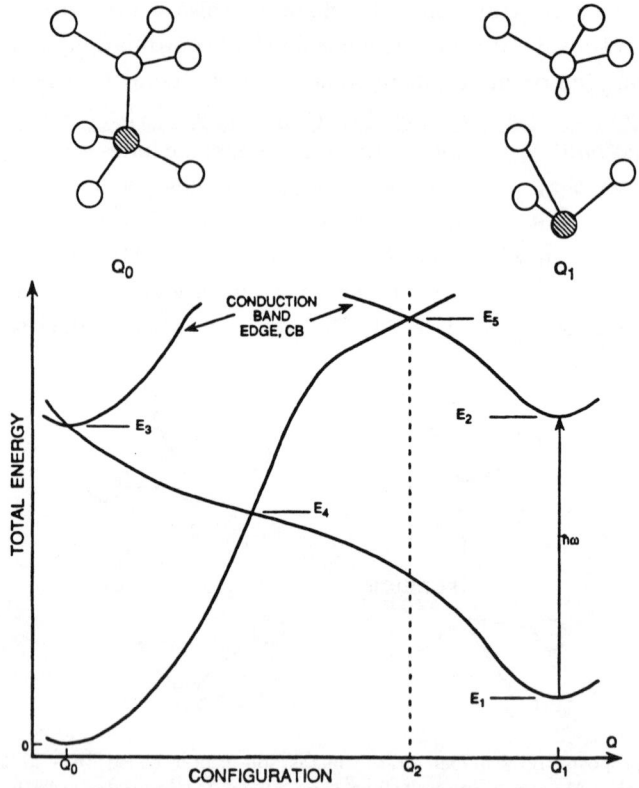

Fig. 3.10. (Lower) A proposed configuration-coordinate diagram for a carrier-induced defect-formation mechanism . It is assumed that the conduction band is higher at a defect than elsewhere, and that at E_3 the latent center on the left can convert to the defect on the right with no energy barrier, simply by capture of a carrier from the conduction band. The photoinduced process starts with the excitation of an electron at an existing defect; the excited electron can move to a latent center to form another defect. (Upper) Atomic arrangements at configurations Q_0, the latent-defect configuration, and Q_1, the dangling-bond configuration. (Reprinted with permission from D. Redfield and R. H. Bube, *Phil. Mag. B* **74**, 309 (1996). Copyright 1996, Taylor & Francis Ltd., London.)

Discrete Energy Level Models

In spite of the possible complications of the amorphous structure, striking examples of the success of a discrete energy level model for dangling bond defects in describing a variety of measurable effects can be cited. It is at least helpful to think in terms of such a discrete-level model for describing the electronic behavior of such defects. Such a discrete-level model can readily be generalized to more complex distributions — for example, discrete levels corresponding to several different types of multivalent defects, or a Gaussian distribution of level density with energy, similar to the top of Fig. 3.8.

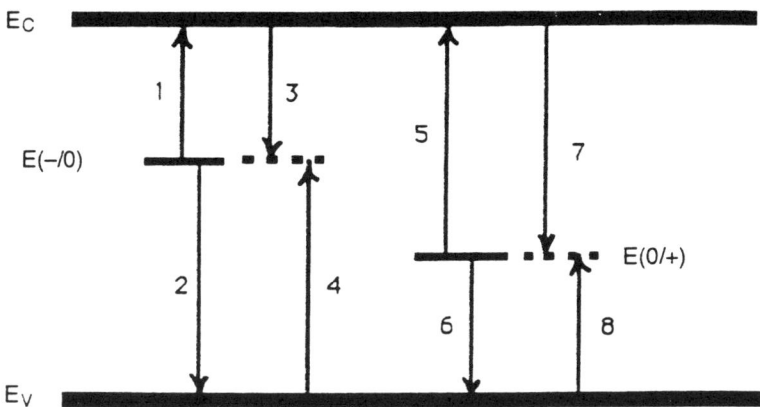

Fig. 3.11. Simple discrete-energy-level model for a trivalent defect, such as a dangling bond defect in a-Si:H, with possible charge states of −, 0, and +. The dashed portion of each defect level indicates the condition in which the level is in its state of lower electronic occupancy. The arrows represent possible optical or thermal excitation energies, and recombination energies as described in the text.

Model for a trivalent defect

A representation of such a model for a trivalent defect such as a dangling bond in a-Si:H is given in Fig. 3.11. The upper level in the gap corresponds to the negatively charged defect when occupied by two electrons (solid line) and the neutral defect when occupied by only one electron (dashed line). For the sake of simplicity in discussion take $E_v = 0$ as the reference energy. To indicate the dual-charge nature of the energy level, it is designated as lying at $E(-/0)$ above the valence edge. It is a simplification to put both of these levels at the same energy, but one that appears to be applicable to covalent semiconductors such as a-Si:H, as further described below. The lower level corresponds to the neutral defect when occupied by one electron (solid line) and the positively charged defect when electron unoccupied (dashed line). Again, to indicate the dual-charge nature of the energy level, it is designated as lying at $E(0/+)$ above the valence edge. The model is completed by the inclusion of exponential conduction-band-tail states and valence-band-tail states, but generally considers only capture transitions involving these recombination centers and electrons or holes in extended states. The density of trapped electrons or holes in tail states is calculated simply from the location of the dark or quasi-Fermi levels; the assigned role of the tail states in this model is therefore to behave as electron or hole traps.

This model using a single set of discrete energy levels with $E(-/0) = 0.90$ eV and $E(0/+) = 0.45$ eV, was reasonably successful in describing the published variation of photoconductivity with doping (LeComber and Spear, 1986) and changes in photoconductivity versus photoexcitation intensity under

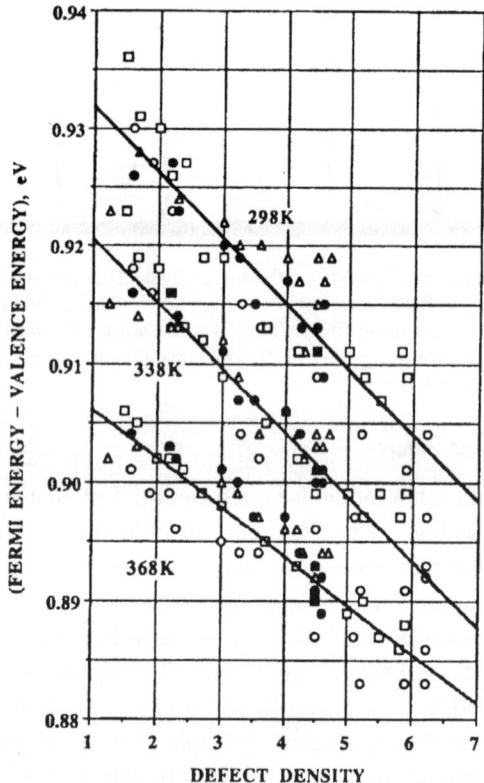

Fig. 3.12. The Fermi energy as a function of the defect density (in units of 10^{16} cm^{-3} and measured at room temperature) at three different measurement temperatures: 298, 338 and 368 K, for a sample of undoped high-quality a-Si:H. At each of these temperatures data are plotted from optical degradation versus time runs at 45°C (○), 80°C (□) 95°C (●), and 110°C (△). The lines are empirical least-square linear fits to all the data at a fixed temperature. (Reprinted with permission from R. H. Bube *et al.*, *J. Appl. Phys.* **75**, 1571 (1994). Copyright 1994, American Institute of Physics.)

optical degradation (Staebler and Wronski, 1980), as well as in providing the framework for making corrections to measured defect densities by the constant photoconductivity method (CPM) because of the change in Fermi energy with optical degradation (Bube and Redfield, 1989; Bube *et al.*, 1992, 1994; Mettler *et al.*, 1994; Sauvain *et al.*, 1994; Hubin *et al.*, 1995).

Model with two trivalent defects

A particularly interesting development from attempts to use simple discrete-level models arises from the consideration of a large amount of data relating dark conductivity, photoconductivity and defect density measured with a single sample of undoped, high-quality a-Si:H (Benatar *et al.*, 1992; Benatar, 1993) as a function of time of optical degradation at four different temperatures. The order of events was as follows:

(1) Examples of the empirical relationships shown by the data are given in Fig. 3.12 (Bube *et al.*, 1994). The results of the analysis show that given any two of the quantities, Fermi energy, temperature, and metastable defect density, the third is very closely specified, regardless of the history of the sample. It may be concluded that the defects measured by CPM in this undoped, high-quality a-Si:H are identical to the defects that control the Fermi energy.

(2) Fairly stringent conditions are imposed on a simple model that can describe the above phenomena. No simple model consisting of a single trivalent defect is able to do this, but a model containing two trivalent defects reproduces the data very well, if their CPM density, as a function of the total measured defect density, varies as shown in Fig. 3.13. Higher-lying defects have a density N_1 that is relatively constant as the total defect density increases, whereas lower-lying defects have a density N_2 that increases monotonically with total defect density. At 298 K in this model $E_1(-/0) = 1.057$ eV, $E_1(0/+) = 0.857$ eV, $E_2(-/0) = 0.965$ eV, and $E_2(0/+) = 0.765$ eV, where $U = 0.2$ eV has been arbitrarily chosen for both defects. Because of the high sensitivity of the results, a small temperature dependence of these energies must be included in the model, which can be related to the dependence of the Fermi energy on temperature.

A number of other literature references exist in which evidence for at least two-different kinds of defects are indicated (Bennett *et al.*, 1987; McMahon, 1992; Sakata *et al.*, 1992; Saleh *et al.*, 1992, 1993; Grimbergen *et al.*, 1993; Tran *et al.*, 1993; Zhang *et al.*, 1994).

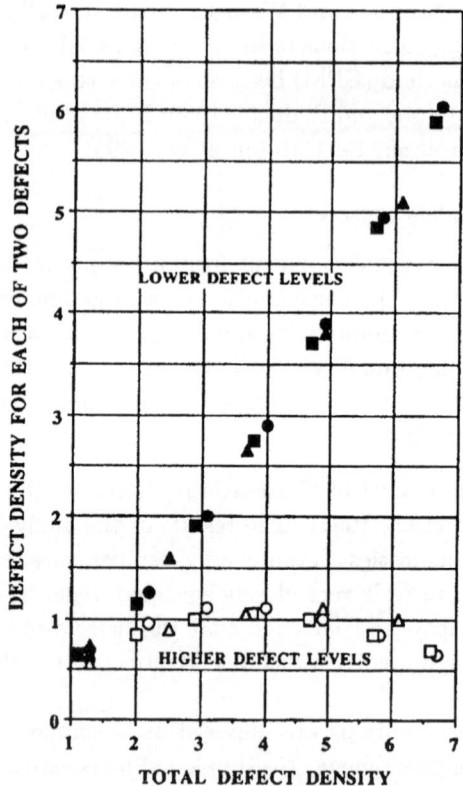

Fig. 3.13. Calculated variation of the densities (all densities in units of 10^{16} cm^{-3}) of two kinds of trivalent defects as the defect densities are increased with optical degradation, as required to fit the data of Fig. 3.8 using the model of Fig. 3.7 for two kinds of defects. Open points are calculated for the higher-lying defects, and solid points are calculated for the lower-lying defects. Measurements of Fermi energy were made at 298 K ($\circ$), 338 K ($\square$), and 368 K ($\triangle$). (Reprinted with permission from R. H. Bube *et al., J. Appl. Phys.* **75**, 1571 (1994). Copyright 1994, American Institute of Physics.)

(3) Using the two-defect model, the variation with optical degradation of defect densities, photoconductivity, and Fermi energy was used to obtain information about the values of the electron capture cross sections of charged and neutral dangling bond defects at room temperature. The following values were obtained: $S_1^0 \approx (1 \pm 0.09) \times 10^{-16}$ cm^2, $S_2^0 \approx (0.6 \pm 0.05) \times 10^{-16}$ cm^2, $S_1^+ \approx (1.8 \pm 0.5) \times 10^{-16}$ cm^2, and $S_2^+ \approx (18 \pm 6) \times 10^{-16}$ cm^2.

These values for capture cross sections are reasonably consistent with earlier attempts to determine them. Measurements of the neutral cross sections had indicated values of $S^0 \approx 4 \times 10^{-15}$ cm^2 (Street, 1982), 10^{-15} cm^2 (McElheny *et al.*, 1991), or at least one order of magnitude smaller (Pandya and Schiff, 1985); measurements of S^+/S^0 indicated a ratio of approximately unity (Gunes *et al.*, 1991; Street, 1991), but other reports indicate $S^+/S^0 \approx$ 50–100 (Beck *et al.*, 1993, 1996).

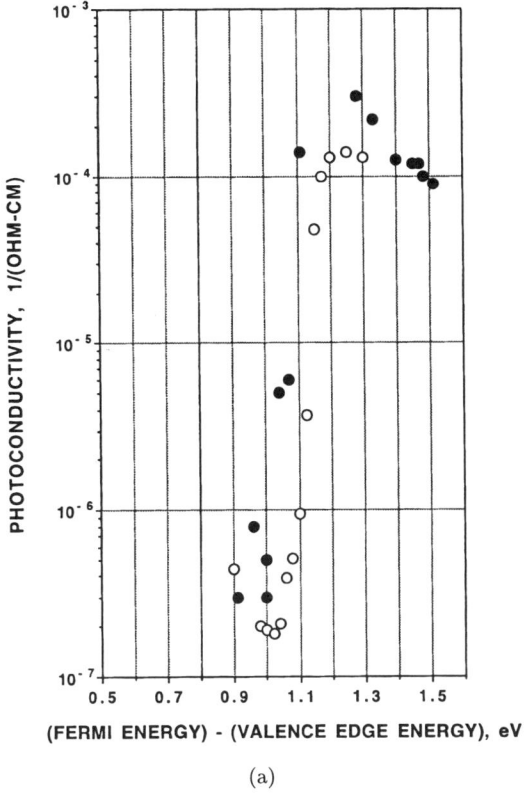

(a)

Fig. 3.14. (a) Variation of the photoconductivity in a-Si:H as a function of the location of the Fermi energy as varied by doping. The solid circles are experimental data from LeComber and Spear (1986), and the open circles are the results of applying the two-defect model of Fig. 3.13, using a typical variation of total defect density with doping from Street (1985). (b) Comparison of the measured values of electron lifetime ($\times 10^{-8}$ s) corresponding to optical degradation at 353 K (solid circles) with the values of electron lifetime calculated from the two-defect photoconductivity model of Fig. 3.13 using measured total defect densities and the capture cross sections determined as above (open circles). (Reprinted with permission from R. H. Bube *et al.*, *J. Appl. Phys.* **79**, 1926 (1996). Copyright 1996, American Institute of Physics.)

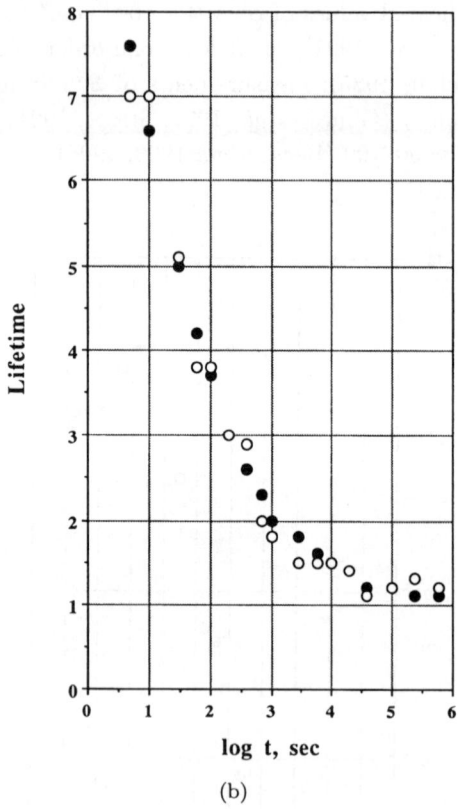

(b)

Fig. 3.14. (*Continued*)

(4) Using the two-defect model and the capture cross sections just described, comparison with different kinds of experimental data showed excellent correlation. Figure 3.14(a) shows the dependence of photoconductivity on Fermi energy according to the data of LeComber and Spear (1986), which fits with the two-defect model using typical total defect densities as a function of doping level from Street (1985), considerably better than for the single trivalent model mentioned above (Bube and Redfield, 1989). And Fig. 3.14(b) shows a close fit between the experimental optical degradation of lifetime at 353 K and the values of electron lifetime calculated from the photoconductivity model with the two trivalent defects and inferred capture cross sections.

(5) Finally, it can be shown that sublinear photoconductivity, i.e. $\Delta\sigma \propto G^{\gamma}$ where G is the photoexcitation rate and $\gamma < 1$, commonly observed at lower light intensities at room temperature, results directly from an increase in the density of positively charged defects corresponding to both higher-energy and lower-energy defects in the two trivalent defect model upon photoexcitation, the effect of increased defect density being amplified by the ratio $R \gg 1$ of charged to neutral capture coefficients for the lower-lying defects.

3.5. Overview of a-Si:H Solar Cells

Summaries of solar cells fabricated from a-Si:H before 1985 are given by Fahrenbruch and Bube (1983), Chopra and Das (1983), Carlson (1984), Hamakawa (1985), and Van Overstraeten and Mertens (1986). The photovoltaic effect was first observed in a-Si:H in 1974 using films deposited by the DC glow-discharge decomposition of silane (SiH_4) and both Schottky-barrier and p–i–n device structures (Carlson, 1977a). Initial efficiencies were limited to less than 1% because of poor contacts, but by 1976 the efficiency of the p–i–n structure had been improved to 2.4% (Carlson and Wronski, 1976), and in 1977

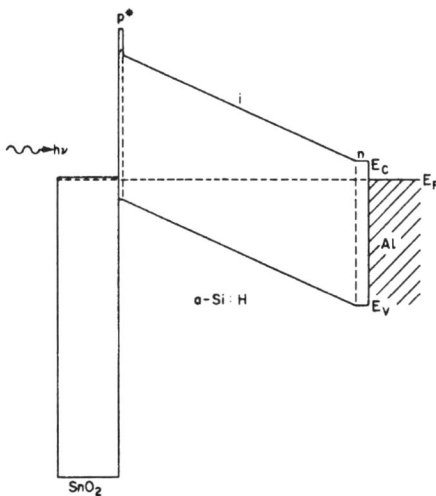

Fig. 3.15. An energy band diagram for a p–i–n cell in the short-circuit mode. (Reprinted from D. E. Carlson, "Solar Cells" in *Semiconductors and Semimetals*. Vol. 21. *Hydrogenated Amorphous Silicon, Part D. Device Applications*, J. I. Pankove, ed., p. 7 (1984). Copyright 1984, Academic Press, Orlando, FL.) '

an efficiency of 5.5% was obtained in a small Schottky-barrier device (Carlson, 1977b). It was during this time that the importance of H in the structure, forming effectively a Si-H alloy, became fully appreciated (Triska *et al.*, 1975; Connell and Pawlik, 1976). An energy-band diagram for a typical p–i–n cell is given in Fig. 3.15, and schematic diagrams of typical solar cell structures are given in Fig. 3.16.

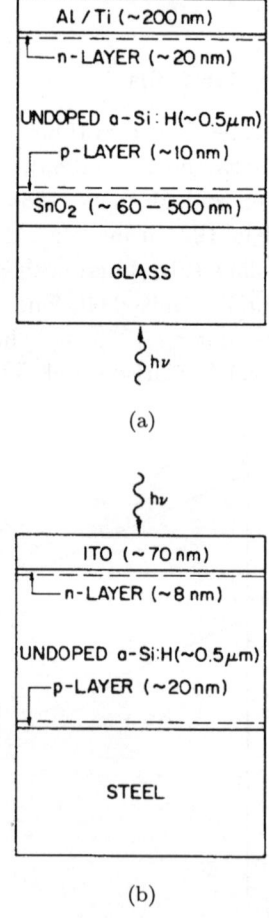

(a)

(b)

Fig. 3.16. (a) A p–i–n solar cell structure on a glass substrate. (b) A p–i–n solar-cell structure on a steel substrate. (c) A stacked or multijunction structure where the i^* layer is a narrow-band-gap alloy such as a-Si:Ge:H. (Reprinted from D. E. Carlson, "Solar Cells" in *Semiconductors and Semimetals.* Vol. 21. *Hydrogenated Amorphous Silicon*, Part D. *Device Applications*, J. I. Pankove, ed., p. 7 (1984). Copyright 1984, Academic Press, Orlando, FL.)

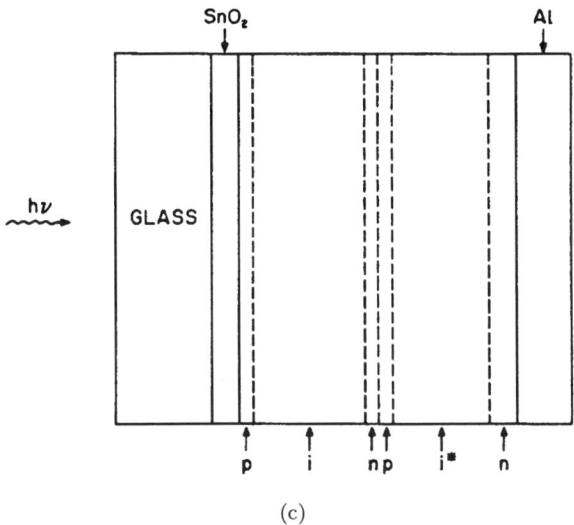

(c)

Fig. 3.16. (*Continued*)

Continued improvements in deposition conditions led to an initial efficiency of 6.1% in p–i–n cells by 1980 (Carlson, 1980), to 7.5% in 1981 using a p-type layer window-layer that was a larger-bandgap, boron-doped a-Si:C:H alloy (Tawada *et al.*, 1981) and then to 10.1% in 1982 with a similar structure (Catalano *et al.*, 1982). The output parameters for this 1.09 cm^2 cell were $\phi_{oc} = 0.84$ V, $J_{sc} = 17.8$ mA/cm^{-2}, and $ff = 0.68$. Boron doping of a-Si:H results in increased absorption and lower carrier lifetime, leading to a loss in photoinduced current; alloying the a-Si:H with C to produce an a-Si:C:H p-type layer decreases this effect, as shown in Fig. 3.17(a). A stacked-junction structure using a smaller-bandgap a-Si:Ge:H alloy in the back junction of three stacked p–i–n junctions gave an efficiency of 8.5% (Nakamura *et al.*, 1982). The output parameters for this 0.09 cm^2 cell were $\phi_{oc} = 2.2$ V, $J_{sc} = 6.74$ mA/cm^{-2}, and $ff = 0.57$. Optical absorption spectra comparing a-Si:H, a-Si:Ge:H, and a-Si:C:H are shown in Fig. 3.17(b).

A general summary of solar cells utilizing a-Si:H is given in Fig. 3.18 (Maruyama *et al.*, 1995). They include (a) the standard all a-Si:H p–i–n structure, (b) replacement of the p-type a-Si:H by p-type a-Si:C:H, (c) textured transparent conducting coating for better optical absorption, (d) use of a graded p-type a-Si:C:H layer, (e) two-cell tandem p–i–n all a-Si:H structure, (f) three-cell tandem p–i–n cell structure using larger band gap a-Si:N or a-Si:C and smaller bandgap a-Si:Ge or a-Si:Sn, (g) two-cell tandem using an

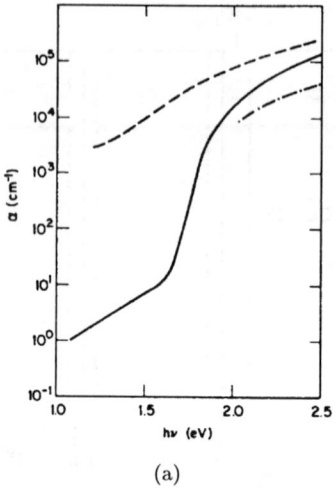

(a)

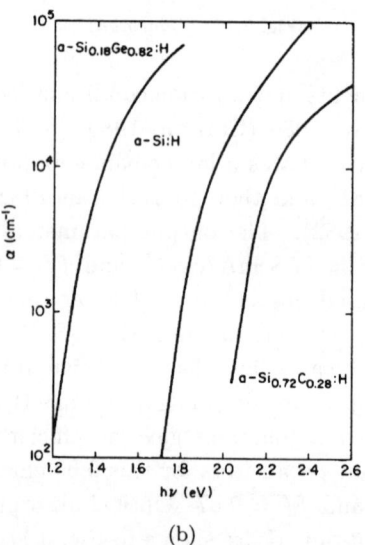

(b)

Fig. 3.17. (a) Optical absorption as a function of photon energy for undoped a-Si:H (solid curve), *p*-type a-Si:H (dashed curve), and *p*-type a-Si:C:H (dashed-dotted curve.) Both *p*-type films contain a few atomic percent of boron, and the a-Si:C:H film also contains 20 at. % of carbon. (b) Optical absorption coefficient as a function of photon energy for a-Si:H, a-Si$_{0.18}$:Ge$_{0.82}$:H (von Roedern *et al.* 1982), and a-Si$_{0.72}$:C$_{0.28}$:H (Morimoto *et al.*, 1982). (Reprinted from D. E. Carlson, "Solar Cells" in *Semiconductors and Semimetals.* Vol. 21. *Hydrogenated Amorphous Silicon*, Part D. *Device Applications*, J. I. Pankove, ed., p. 7 (1984). Copyright 1984, Academic Press, Orlando, FL.)

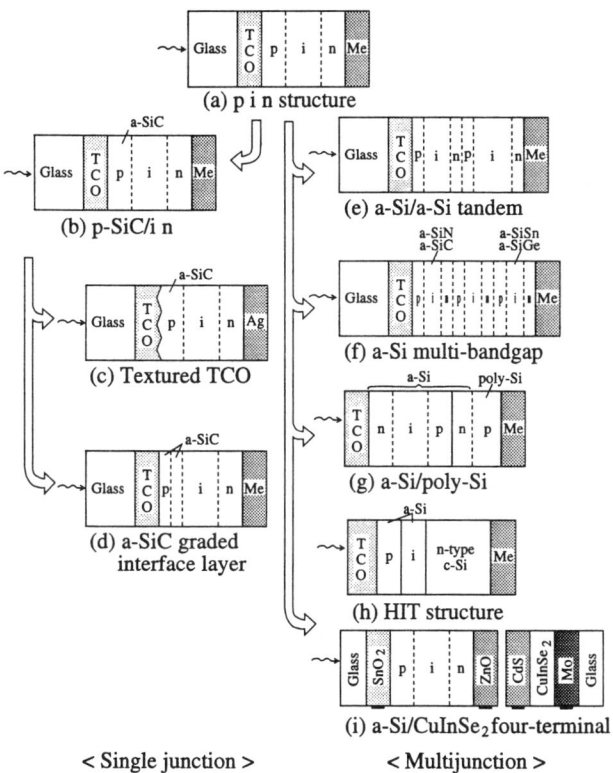

(a) p i n structure

(b) p-SiC/i n

a-SiC

(c) Textured TCO

a-SiC

(d) a-SiC graded
interface layer

(e) a-Si/a-Si tandem

(f) a-Si multi-bandgap

(g) a-Si/poly-Si

(h) HIT structure

(i) a-Si/CuInSe₂ four-terminal

< Single junction >   < Multijunction >

Fig. 3.18. Schematic structure of a variety of cells involving a-Si:H. (Reprinted with permission from E. Maruyama *et al.*, *Solid State Phenomena* **44-46**, 863 (1995). Copyright 1995, Scitec Pub., Switzerland.)

a-Si:H p–i–n cell in tandem with a microcrystalline Si p–n cell, (h) so-called "Heterojunction with Intrinsic Thin Layer" (HIT) between p- and i-type a-Si:H and n-type crystalline Si, and (i) two-cell stacked tandem using a p–i–n a-Si:H cell in tandem with a CuInSe₂ cell. Variations of the transparent conducting oxide (TCO) layers that are used in all of these cells led to the proposal of a composite TCO consisting of glass/SnO₂:F/TiO₂:Nb as a good electrical contact for a-Si:H solar cells (Hegedus *et al.*, 1995).

For single-junction solar cells, recent progress has increased the initial efficiency to 13.2% (Miyachi *et al.*, 1992) and the stabilized efficiency after long-time optical exposure to 8.5% (Xi *et al.*, 1994) or 8.8% (Hishikawa *et al.*, 1994). These more advanced cells include such efficiency-enabling features as composition-graded layers, multilayers, and optical coatings to increase light absorption (Schropp, 1995). With the use of thinner undoped absorbing layers of a-Si:H, multijunction a-Si:H-based solar cells using a-Si:H alloys have achieved a stabilized efficiency of over 10% (Nomoto *et al.*, 1993; Arya *et al.*, 1993; Guha *et al.*, 1993).

A significant development has been the fabrication of a-Si:H solar cells on a thin, flexible, stainless steel substrate, making possible a continuous "roll-to-roll" manufacturing technology (Izu *et al.*, 1983, 1993, 1994; Ovshinsky 1988; Nath *et al.*, 1988). This process benefits from the mechanical strength of the stainless steel, transport of modules during manufacturing is simple and less expensive, and heating and cooling of the substrate during deposition can be achieved quickly. An a-Si:H solar cell with a plastic film substrate has also been described (Ichikawa *et al.*, 1994; Kuo *et al.*, 1994, Shinohara *et al.*, 1994).

A detailed summary of some sixteen optimization techniques is given by Schropp (1995) to increase the initial efficiency, increase stability, achieve optical enhancement, and optimize tandem structures. A review of the physics and technology of a-Si:H solar cells and modules is given by Kusian (1995). Consideration of seven research areas for the improvement of solar cell efficiencies using amorphous semiconductors is given by Paul *et al.* (1993).

3.6. Variations in Deposition Techniques

General information on the deposition and structure of a-Si:H films can be found in several chapters in Vol. 21A of the book edited by Pankove (1984), and in six review papers in the volumes edited by Neber-Aeschbacher (1995). Although a variety of techniques were used to grow a-Si:H films for solar cells (Carlson, 1984), the highest-efficiency cells were made from films grown in radio-frequency (rf) glow discharges in SiH_4 using a p-type a-Si:C:H layer (Catalano *et al.*, 1982). Comparable performance was reported for cells made from glow discharges in silicon tetrafluoride (SiF_4) and hydrogen (Madan and Ovshinsky, 1979; Hack and Shur, 1982).

Other deposition techniques that were early investigated include DC glow discharge in SiH_4 (Uchida, 1984), chemical vapor deposition (CVD) using SiH_4 and also higher-order silanes such as Si_2H_6 (Dalal, 1982), and homogeneous CVD (HOMOCVD) (Scott *et al.*, 1981). Sputtered a-Si:H films have been investigated but the density of defects is sufficiently high that cells made from sputtered films have only limited efficiency (Anderson *et al.*, 1977; Moustakas *et al.*, 1977). Heterojunction solar cells were fabricated by DC magnetron sputtering of a-Si:H from an n-type Si target, onto p-type single-crystal Si, with an efficiency of 10.7% (Jagannathan and Anderson, 1996).

The glow-discharge decomposition of silane remains the standard deposition procedure for a-Si:H. A variety of other methods have also been tested in recent years. (a) Shah *et al.* (1988), Finger *et al.* (1992) and Kroll *et al.* (1992) explored a very high frequency (VHF) glow discharge as a high rate deposition method for a-Si:H using plasma excitation frequencies in the range of 70–100 MHz compared to the standard frequency of 13.6 MHz. The higher frequency increases the deposition rate by a factor of three to five to 1.2 to 1.5 nm/sec, without reducing film quality. The use of this method for the production of μc-Si:H films was discussed in Sec. 2.8. (b) Fabrication of a-Si:H p–i–n solar cells by CVD using disilane has been reported (Hegedus *et al.*, 1987), but they exhibit a rather low efficiency of 4%, limited by the quality of the intrinsic CVD material. (c) Materials more nearly equivalent to standard a-Si:H have been deposited by photo-assisted CVD (Hegedus *et al.*, 1988; W. Y. Kim *et al.*, 1988a,b). (d) Device quality a-Si:H was also reported as deposited by DC magnetron reactive sputtering (Pinarbasi *et al.*, 1988) with deposition rates of 0.055 to 0.33 nm/sec. (e) High quality i-layers of a-Si:H were deposited with the hot-wire deposition technique (Nelson *et al.*, 1994; Mahan *et al.*, 1996), and also were incorporated into p–i–n solar cells.

An attempt to modify reactions at the growing a-Si:H surface through the use of an inert gas plasma treatment, using rf plasma CVD, has been investigated (Maruyama *et al.*, 1996). It appears that the inert gas plasma treatment reduces the hydrogen content, promotes the rearrangement of the Si-network, and reduces the ratio SiH_2/SiH. High-quality films have been produced with a high, stabilized photoconductivity.

3.7. Material Transport Properties Important for Solar Cells

A discussion of electronic structure and transport properties in a-Si:H is given in eight review papers in the volumes edited by Neber-Aeschbacher (1995). In a-Si:H p–i–n solar cells, collection of photoexcited carriers near short-circuit conditions is primarily by drift in a built-in electric field $\mathbf{E}$, and high efficiency requires a drift length ($\mu\tau\mathbf{E}$) several times larger than the film thickness, where μ is the carrier mobility, τ the carrier lifetime, and $\mathbf{E}$ the electric field in the i-region. Crandall (1982) proposed that transport in p–i–n solar cells can be characterized by a collection length that is the sum of the electron and hole drift lengths. In short-circuit conditions, the collection length is typically ≥ 5 μm in the short-circuit mode and decreases as the cell goes into forward bias, while transport by diffusion dominates as the cell approaches the open-circuit condition. High efficiency requires a diffusion length several times larger than the thickness of the active region, which in turn requires a mobility lifetime product $\mu\tau > 10^{-7}$ cm^{2}V^{-1} for a thickness of 0.5 μm, a condition that is met in good-quality a-Si:H.

Large diffusion lengths in a-Si:H require a low density of defect states in the gap. Such defects are most commonly associated with dangling-bond defects as described in Sec. 3.4. They may arise during film growth by hydrogen out-diffusion (Fritzsche *et al.*, 1978), or may be associated with microstructural imperfections such as polymer chains (Knights *et al.*, 1979), or with impurities such as oxygen (Pontuschka *et al.*, 1982) and carbon (Morimoto *et al.*, 1982). It has been suggested that transport in a-Si:H may be strongly affected by inhomogeneities that give rise to potential fluctuations (Overhof and Beyer, 1981). Metastable dangling-bond defects formed by photoexcitation (see Sec. 3.4) also decrease the diffusion length.

High doping levels in the n- and p-type layers of a p–i–n a-Si:H solar cell are desired to order to obtain a large built-in potential, approaching the band gap in magnitude. Such high doping levels, however, are limited both by the existence of the band-tail states and by low doping efficiency in a-Si:H, presumably because most dopant atoms do not occupy electronically active sites. Estimates of the doping efficiency are in the range of 0.1 to 1.0% (Faughnan and Hanak, 1983; Dresner, 1983). The Fermi level is about 0.2 eV below the conduction band in a-Si:H:P (Spear, 1977) and about 0.5 eV above the valence band in a-Si:H:B (Jan *et al.*, 1980), limiting the built-in potential of a-Si:H p–i–n solar cells to about 1.0 eV (Williams *et al.*, 1979). This built-in potential is increased by using p-type a-Si:C:H layers (Tawada *et al.*, 1982), but the resistivity of the layer increases with the band gap, i.e. with the C/Si ratio. The conductivity of the p-type layer in a boron-doped a-Si:H layer can be significantly increased by forming microcrystalline-doped Si:H films (Matsuda *et al.*, 1980) but without significant increase in the built-in potential (Carlson and Smith, 1982). The n-type layer can also be improved by making it microcrystalline (see Sec. 2.8), which is formed if the silane gas is heavily diluted with hydrogen (Smith, 1991). The microcrystalline material has a higher conductivity than a-Si:H, thus reducing resistive loss and improving the fill factor.

Under the assumption that the electric field in the i-layer is almost uniform, if both the trapped charge and the free-carrier space charge are assumed to be negligible, a simple model was developed that related the fill factor of p–i–n cells to the collection length (Crandall, 1982; Faughnan and Crandall, 1984) with the results shown in Fig. 3.19. They found that the collection length at

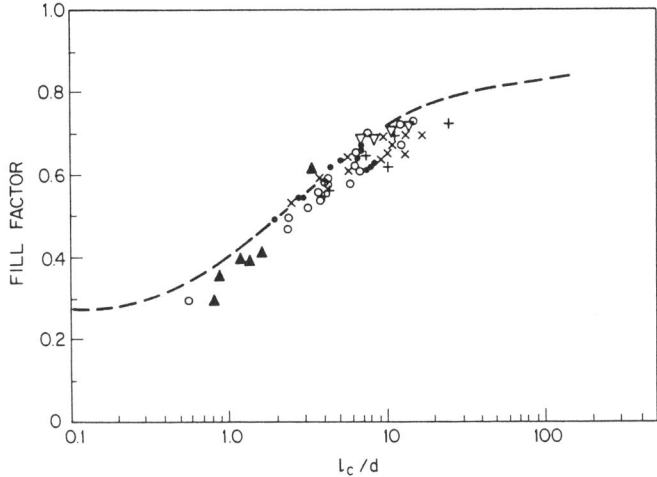

Fig. 3.19. Dependence of the fill factor as a function of collection length for a-Si:H *p–i–n* cells. Comparison of theoretical variation (dashed line) and experimental points on a number of *p–i–n* solar cells. (Reprinted with permission from B. W. Faughnan and R. S. Crandall, *Appl. Phys. Lett.* **44**, 537 (1984). Copyright 1984, American Institute of Physics.)

short-circuit must be 25 times the *i*-layer thickness to obtain a fill factor of 0.72. In many low-performance a-Si:H cells, the net space charge in the *i*-layer is not negligible, and a quasi-neutral region is present in the central portion of the *i*-layer, resulting in depletion regions or space-charge fields near the *p–i* and *i–n* interfaces and loss of photoexcited carriers by recombination.

One of the transport properties of a-Si:H of interest for solar cells is the interpretation of the mobility-lifetime, $\mu\tau$, product. A problem was raised by the observation that the $\mu\tau$ product deduced from steady-state photoconductivity is about 100 times larger than the $\mu\tau$ product deduced from time-of-flight charge collection . This conflict was apparently resolved (Kocka *et al.*, 1991) by the proposal that the $\mu\tau$ product deduced from time-of-flight measurements is the result of single-carrier, trap-limited transport, whereas the $\mu\tau$ product deduced from steady-state secondary photoconductivity is a two-carrier, recombination limited quantity.

The investigation of the interpretation of the $\mu\tau$ product in a-Si:H has been continued with a combination of steady-state photoconductivity, time-of-flight, and steady-state photocarrier grating measurements (Wyrsch *et al.*, 1995; Beck *et al.*, 1996). They define a correlation quantity $\mu^0\tau^0$, deducible from combining steady-state photoconductivity and steady-state photocarrier grating measurements, which corresponds to the value of the $\mu\tau$ product when all the defects in the material are neutral and therefore does not depend on the actual dangling bond occupation in the measured sample. The values of the $\mu^0\tau^0$ products evaluated on films are reported to correlate with the efficiencies of corresponding solar cells.

3.8. Solar Cell Stability

In spite of the promise and initial high efficiency of a-Si:H solar cells, their widespread use has been slowed by the decrease in their efficiency with time of photoexcitation. Most solar cells of a-Si:H lose about 30% of their generating capacity and must be derated accordingly. Typical dependence of efficiency on time for cells of the types shown in Fig. 3.16(a) and 3.16(b) are given in Fig. 3.20. The efficiency can be recovered by thermally annealing the

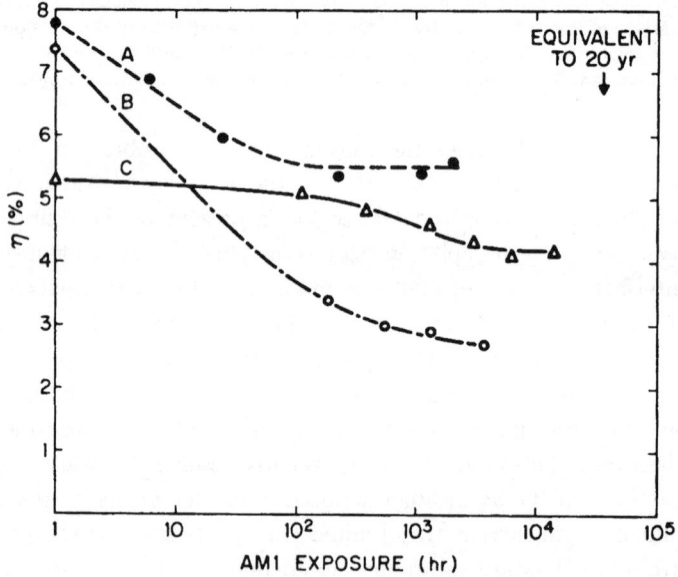

Fig. 3.20. Efficiency as a function of exposure time to AM1 radiation for a-Si:H solar cells. Curves A and B are data for cells of the type shown in Fig. 3.16(a), and curve C is for a cell of the type shown in Fig. 3.16(b). For curves A and B, the p-type region was a-Si:C:H, with the carbon content for curve B (1.7×10^{22} cm^{-3}) being about 100 times larger than that associated with curve A. (Reprinted from D. E. Carlson, "Solar Cells" in *Semiconductors and Semimetals.* Vol. 21. *Hydrogenated Amorphous Silicon*, Part D. *Device Applications*, J. I. Pankove, ed., p. 7 (1984). Copyright 1984, Academic Press, Orlando, FL.)

a-Si:H cells at about 200°C for several minutes. It was reported early that the decrease in a-Si:H solar cell efficiency with time of photoexcitation was similar to the formation of metastable defects in thin film a-Si:H under photoexcitation (Staebler *et al.*, 1981). The formation of new defects by exposure to light decreases the lifetime, diffusion length, and drift length. It was also found that the net space charge often increases with photoexcitation (Carlson *et al.*, 1983), contributing by itself to a decrease in efficiency.

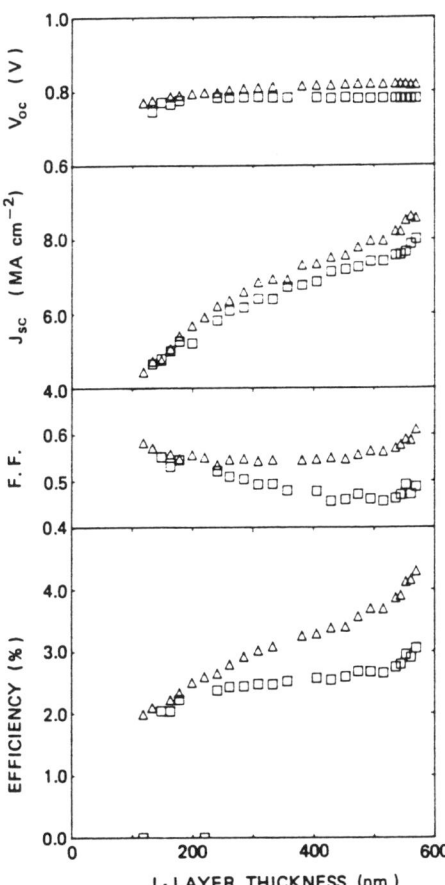

Fig. 3.21. Photovoltaic parameters as a function of the thickness of the *i*-layer in *p–i–n* cells before (△) and after (□) 67 hours of exposure to AM1 illumination. (Reprinted with permission from J. J. Hanak and V. Korsun, *16th IEEE Photovoltaic Specialists Conference*, p. 1381 (1982). Copyright 1982, IEEE, NY.)

Three significant observations may be noted (Staebler *et al.*, 1981): (a) photoinduced degradation of a *p–i–n* a-Si:H solar cell can be reduced by operating the cell short-circuit rather than open-circuit, and can be prevented by applying a reverse bias to the cell during photoexcitation; (b) consistent with this is the observation that photodegradation is reduced in cells with thinner *i*-layers since the electric field in the *i*-layer is increased; and (c) cells degrade similarly in the dark under forward current flow.

Point (b), the dependence of cell stability on *i*-layer thickness, is illustrated by Fig. 3.21 (Hanak and Korsun, 1982). No degradation is observed in *p–i–n* cells with thin *i*-layers, but when the *i*-layer thickness exceeds 0.2 μm, all the photovoltaic parameters exhibit some degradation, particularly the fill factor.

A variety of at least partly effective methods have been used to reduce the optical degradation effects in a-Si:H solar cells (Maruyama, 1995). These include the reduction of impurities such as oxygen and nitrogen (Tsuda *et al.*, 1995), the reduction of Si–H_2 bonds in the a-Si:H film (Kuwano *et al.*, 1987), a multijunction structure with thin *i*-type a-Si:H layers (Nakamura *et al.*, 1985), a tandem structure using blocking layers (Ichikawa, 1990), the use of "hot-wire" deposition techniques resulting in a lower hydrogen concentration and an altered microstructure (Vanecek *et al.*, 1992), the use of hydrogen-diluted silane ($[H_2]/[SiH_4] = 10$) in the rf-PECVD deposition of a-Si:H layers (Lee *et al.*, 1996), and the use of hydrogen electron-cyclotron-resonance (ECR) plasma-CVD growth techniques (Dalal *et al.*, 1994, 1996).

An investigation of the light-saturated defect density in a-Si:H as a function of oxygen content, microstructure and hydrogen bonding, indicated no increase in stability with decrease in the concentration of oxygen down to 5×10^{17} cm^{-3}, but an increase in stability with a decreased hydrogen content due to a change in material microstructure (Haage *et al.*, 1994; Vanecek *et al.*, 1995).

A comprehensive study of solar-cell degradation was carried out by Chen and Yang (1991). Their data on the efficiency of a-Si:H solar cells as a function of time of photoexcitation are shown in Fig. 3.22 with fits to their data using stretched exponentials with essentially the same parameters as those which apply to defect densities in undoped, homogeneous films. This is one of several indications that the optical degradation of a-Si:H solar cells is dominated by an increase in the density of metastable defects in the *i*-layer of the cells, thus decreasing the carrier lifetime and possibly causing other effects directly related to the density of defects, such as a decrease in carrier mobility or a change in the field distribution in the cell. Recent data comparing the effects of optical degradation in films of a-Si:H and solar cells based on a-Si:H have been discussed

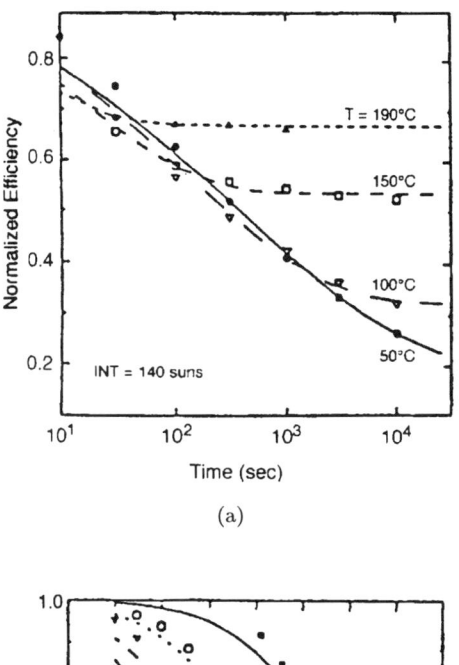

(a)

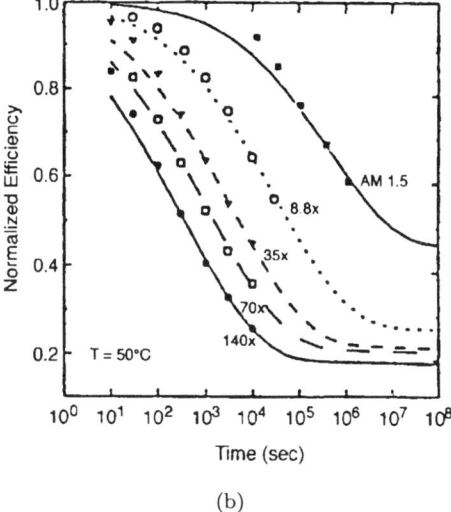

(b)

Fig. 3.22. Efficiency of a-Si:H solar cells as a function of time of exposure to light. Points are the measured efficiencies, normalized to their initial values, and curves are fits to the data using stretched exponentials. In (a) a fixed intensity of 140 suns was used with various temperatures, whereas in (b) and (c) fixed temperatures of 50°C and 100°C, respectively were used with various intensities. (Reprinted with permission from L. Chen and L. Yang, *J. Non-Cryst. Solids* **137/138**, 1185 (1991). Copyright 1991, North-Holland, Amsterdam.)

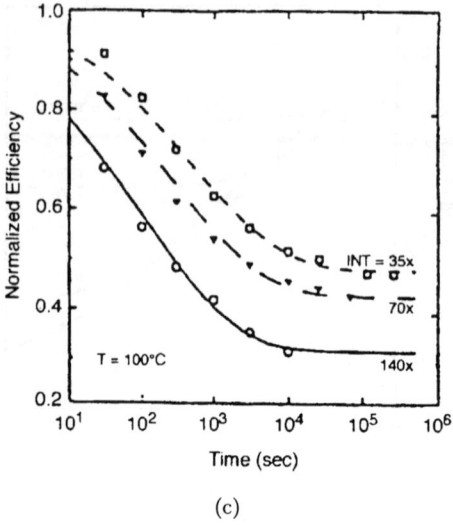

Fig. 3.22. (*Continued*)

(Redfield and Bube, 1991), based on annealing kinetics, existence and properties of saturation, degradation kinetics, thickness dependence of degradation,
and performance modeling of films and cells, with the conclusion that degradation in a-Si:H solar cells is an *i*-layer effect directly related to the formation
of metastable defects.

A number of other investigations have concluded that the degradation of
solar cell performance is directly related to the increase in the density of
metastable defects under illumination. (1) The relationship between fill factor
and carrier collection length shown in Fig. 3.19 has been tested by relating the
collection length to the defect density, with the result that changes in solar cell
performance due to changes in fill factor are directly correlated with defect
density (Frammelsberger *et al.*, 1991). (2) A computer model of cell degradation (Pawlikiewicz *et al.*, 1991) showed good correlation between measured fill
factor and bulk defect density. (3) A comprehensive computer model of similar Schottky barrier MIS cells, for different *i*-layer thicknesses and durations of
light exposure, showed good agreement between observed degradation and calculated quantum efficiency spectra based on the density of defects only (Malone
et al., 1991). (4) Measurements of efficiency versus time of optical degradation
for large monolithic solar-cell modules involving a-Si:H show behavior quite
similar to that for films and single cells (Luft *et al.*, 1991). (5) Optical degradation at 60°C and modeling of a-Si:H *p–i–n* cells well into saturation of their

conversion efficiency were interpreted to mean that the light-induced defects in the i-layer control the optical degradation of $p-i-n$ cells (Wagner *et al.*, 1991). (6) An investigation of $p-i-n$ and $n-i-p$ solar cells after optical degradation, current injection and keV-electron irradiation led to the conclusion that the degradation of the i-layer bulk is the dominant effect in the optical degradation of such cells (Herbst *et al.*, 1993). (7) At least as far as a-Si:H deposited by the VHF method (see Sec. 3.6) is concerned, the electron and hole drift mobilities remain unchanged near room temperature by light-induced degradation (Wyrsch and Shah, 1991). This result contradicts the suggestion that it is long range potential fluctuations that affect the electron mobility and limit solar cell performance, rather than the increase in the density of deep defects (von Roedern and Madan, 1991). (8) Experiments focusing on the p/i interface and designed to separate interfacial from bulk defect-related degradation indicated that optical degradation effects can be modeled without including any interfacial losses (Vasanth *et al.*, 1994). (9) By means of an extensive characterization of bifacial a-Si:H solar cells, using a well-defined sequence of degradation steps, and using alternatively both strongly absorbed and uniformly absorbed light, it was concluded that optical degradation is controlled mainly by an increase of the deep defect density in the bulk of the i-layer (Wyrsch and Shah, 1994).

Still disagreements about basic degradation mechanisms exist. As indicated above, some of these have been addressed in recent research and shown to be probably inapplicable. (1) Evidence has often been advanced that other phenomena, e.g. interfacial recombination effects at the $p-i$ or $i-n$ boundaries, are responsible for at least some of the degradation (Mitchell *et al.*, 1985). (2) Some research indicates contributions to degradation both from defects in the i-layer and at the p/i interface (Li *et al.*, 1992). (3) A comparison of current-induced and light-induced defects in a-Si:H indicates that the same dangling bond defects are formed in $p-i-n$ diodes by the two techniques, but optical excitation forms a high density of defects close to the $p-i$ interface, while current flow yields a more uniform distribution (Street *et al.*, 1993). (4) Analysis of thin film stability by itself does not adequately correlate with the stability of solar cells involving such films (Schropp *et al.*, 1993). (5) Improvement in the stabilized efficiency of a-Si:H solar cells has been reported by profiling the i-layer of the solar cell to prevent a modification of the spatial electric field distribution upon photoinduction of defects (Schropp *et al.*, 1993a,b, 1994). (6) Reports have been made of cases where materials with lower spin-determined defect densities do not automatically produce better solar cells (von Roedern *et al.*, 1992), and it is proposed that solar cell performance is determined and limited by the density of charged dangling bonds that are not measured by ESR (von Roedern, 1993). (7) Significant differences in the stabilized performance are reported depending on the conditions used to illuminate a-Si:H multijunction modules (von Roedern *et al.*, 1995). (8) Empirically a relationship has been found between a greater electric field dependence of the drift mobility (μ_d increases with electric field) of an a-Si:H film in the annealed state and a poorer stability of the photoconductivity upon photoexcitation (Tang *et al.*, 1996). (9) Along with the decrease of photoconductivity and lifetime during exposure to light, a continuous decay of the drift mobility was also found, suggesting enhanced scattering associated with charged optically induced defects (Tang and Braunstein, 1996). It is reported that different generation kinetics are measured using stretched-exponential kinetics for defects leading to increased recombination (decrease of lifetime) and those leading to decrease of drift mobility.

3.9. Amorphous Silicon Alloys

As pointed out elsewhere in this chapter, alloys of a-Si:H, particularly with C or Ge, play an important role in the overall solar-cell development picture (Bauer, 1995; Demichelis and Pirri, 1995). A review of deposition processes for amorphous silicon alloys is given by Luft and Tsuo (1993).

The performance of a-Si:Ge:H solar cells were investigated as a function of the band gap between 1.2 and 1.7 eV, with the finding that the solar cell performance drops sharply if the band gap is smaller than 1.5 eV (Smith *et al.*, 1987). Such a decrease in performance with increasing Ge/Si ratio may be due to simply the decrease in band gap, or it may be intensified by an increase in defect density or a widening of the valence band tail with increasing Ge (Turner *et al.*, 1987).

Variations have been made in the doping of the *p*-type a-Si:C:H window layer of the a-Si:H *p–i–n* solar cell. Mohring (1987) replaced the standard use of diborane in the doping of the *p*-type layer, and substituted plasma assisted B diffusion from thin SiB-layers on the substrate into the growing intrinsic film with promising results reported. Ouwens *et al.* (1994) deposited a high band-gap p^+-a-Si:C:H layer without deteriorated electronic properties using trimethylboron as a dopant gas.

Investigation indicates that improvements are achieved by introducing doped microcrystalline Si (μc-Si) and its carbon alloys (μc-Si:C) at the TCO/*p*-a-Si:C interface, resulting in an increase in the built-in potential (Ma *et al.*, 1994; Sannomiya *et al.*, 1994; Rath *et al.*, 1996).

Shibata *et al.* (1988) used the technique of "δ-doping" to increase the peak boron concentration to 2×10^{21} cm^{-3} in the p-type layer, with apparently beneficial results, producing a cell with initial efficiency of 11.5% The δ-doped p-layer consists of one or several very thin boron layers (0.1 to 0.5 atomic layer) and undoped a-Si:H layers.

Measurements of electron transport and recombination (Nebel *et al.*, 1988a; Bauer *et al.*, 1988), and diffusion lengths from optical grating techniques (Nebel *et al.*, 1988b) on a-Si$_{1-x}$:Ge$_x$:H alloys have been reported for values of x between 0 and 0.3. It is reported that the electron mobility in a-Si:Ge:H decreases with increasing hydrogen concentration and germanium concentration (Fortmann *et al.*, 1991). Capacitance techniques have been applied to investigate the midgap density of states in a-Si:H and a-Si:Ge:H p–i–n solar cells and Schottky junctions (Hegedus and Fagen, 1992).

Optical degradation has been observed for a-Si:Ge:H films (Aljishi *et al.*, 1987) with effects depending on the band gap of the alloy. If the band gap is larger than 1.4 eV, an increase in subgap absorption, a decrease in photoconductivity, and a decrease in dark conductivity with an increase in activation energy was observed. For band gaps less than 1.4 eV, an increase in subgap absorption, constant photoconductivity, and an increase in dark conductivity with a decrease in activation energy was observed. Chu *et al.* (1988) carried out a similar investigation concentrating on band gaps between 1.2 and 1.4 eV of interest for multijunction cells.

Variations in deposition techniques have also been tested with a-Si:Ge:H alloys. Weller *et al.* (1987) used dc-glow discharge, and Hegedus *et al.* (1988) used photo-assisted CVD.

Of particular interest is the series stack of two or three solar cell junctions with alloys a-Si:C:H with larger band gap, a-Si:H itself, and a-Si:Ge:H with smaller band gap, shown in Fig. 3.18(f) and in more detail in Fig. 3.23. Since the i-layer thickness in these cells is narrower than in a simple a-Si:H cell, it is expected that optical degradation effects would be reduced. Offsetting this gain is the apparent increased instability of a-Si:C:H compared to a-Si:H, and uncertainty about the apparent decreased instability of a-Si:Ge:H alloys because of cited evidence that there is a lack of correlation between defect density in the bulk and stability of these devices (Xu *et al.*, 1993). Still, research on the effects of H, Si, and Ge in alloy films has been used to produce what is claimed to be the most stable a:Si:Ge:H solar cell (Terakawa *et al.*, 1994). A variety of approaches are being considered to achieve more stable multijunction cells (Haku, 1991; Maruyama *et al.*, 1993a,b; Hishikawa *et al.*, 1994; Sayama, 1994).

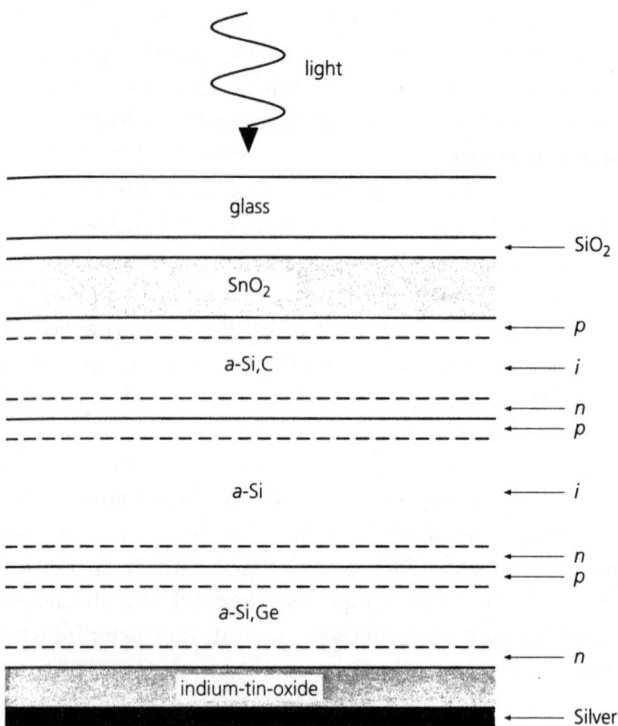

light

glass

SiO$_2$

SnO$_2$

a-Si,C

p

i

n

p

a-Si

i

n

p

a-Si,Ge

n

indium-tin-oxide

Silver

Fig. 3.23. More detail on the device structure of an a-Si:C:H/a-Si:H/a-Si:Ge:H multijunction solar cell shown in Fig. 3.18(f). (Reprinted with permission from D. E. Carlson and S. Wagner, in *Renewable Energy: Sources for Fuels and Electricity*, T. B. Johansson *et al.*, eds., p. 403 (1993). Copyright 1993, Island Press, Washington, D.C.).

Properties of a-Si:H/a-Si:Ge:H quantum-well multilayers have also been investigated, with the demonstration that the optical and transport properties of such multilayers can be tuned over a wide range by varying the geometry of the multilayer, without changing the chemical composition of the individual layers (Conde *et al.*, 1994).

Although not strictly an alloy, mention may well be made at this point of attempts to make a red-sensitive a-Ge:H solar cell (Kusian *et al.*, 1991). Improvements were achieved by increasing the substrate temperature, decreasing the pressure during deposition, reducing the i-layer thickness and using a p-Si:C/p^+-S front layer, but the final result was still only a 2% cell ($\phi_{oc} = 0.26$ V, $J_{sc} = 16$ mA/cm^2, $ff = 0.47$). A limiting factor seemed to be the relatively high density of defects, of the order of 9×10^{16} cm^{-3}.

3.10. Hybrid Solar Cells Involving a-Si:H

Types of heterojunction solar cells involving both a-Si:H and other forms of Si promise increased efficiency.

Tandem Cell with Multicrystalline Si

In the tandem cell configuration, an a-Si:H/mc-Si heterojunction cell has been investigated as a bottom cell structure (Hamakawa *et al.*, 1983). Its advantage is that it does not require high temperature processing for junction formation, and the top a-Si:H cell can be fabricated continuously. A (*p*-type a-Si:H/ *n*-type mc-Si) heterojunction, with a 10 μm thick mc-Si film fabricated by solid phase crystallization yielded an efficiency of 8.5% (Matsuyama, 1994).

A high-efficiency a-Si:H/mc-Si four-terminal tandem solar cell has been developed using a p-μc-Si:C/n-mc-Si/n-μc-Si heterojunction bottom solar cell with a conversion efficiency of 20.3% and good stability. (Yoshimi *et al.*, 1992)

Tandem Cell with Microcrystalline Si

It has been reported that microcrystalline (μc) Si:H p–i–n junctions have an extended infrared response, and are entirely stable under photoexcitation with an efficiency of 7.7% (Fischer *et al.*, 1996). The tandem arrangement of an a-Si:H solar cell with a μc-Si:H solar cell also appears promising A remaining problem is the insufficient deposition rates for the μc-Si:H layers.

Heterojunction with Intrinsic Thin Layer (HIT)

Figure 3.18(h) shows a typical HIT structure. A thin undoped a-Si:H layer (5 nm thick) is inserted into the p–n junction in order to reduce the density of states near the interface. A conversion efficiency of 20% has been achieved with a single crystalline silicon (c-Si) wafer using a low-temperature process.

Multijunction Cells with CuInSe$_2$

There have been several investigations of a multijunction involving an a-Si:H solar cell and a CuInSe$_2$ solar cell (Mitchell *et al.*, 1988; McCandless, 1988) as in Fig. 3.18(i). Figure 3.24 shows the solar cell structure for a 14.2% efficient a-Si:H/CuInSe$_2$ multijunction four-terminal cell (Morel, 1988). Thin film tandem modules with 0.4 m^2 area have been reported with 41.5 watt and an efficiency of 10.5% (Mitchell *et al.*, 1990).

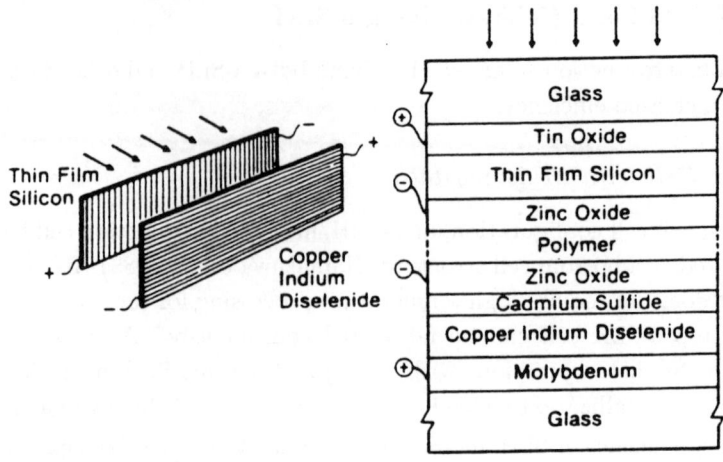

Fig. 3.24. Solar cell structure for a 14.2% efficient multijunction cell consisting of an a-Si:H *p–i–n* cell and a ZnO/CdS/CuInSe₂ cell. (Reprinted with permission from D. L. Morel, Solar Cells **24**, 157 (1988). Copyright 1988, Elsevier Sequoia S. A., Lausanne, Switzerland.)

CHAPTER 4

GALLIUM ARSENIDE AND OTHER
III–V MATERIALS

4.1. Overview

It is useful to place gallium arsenide and other III–V materials relevant to solar cell applications in the context of the Periodic Table of the Elements and their relationship to the element silicon that formed the basis for the previous two chapters. A pertinent subsection of the Periodic Table is shown in Fig. 4.1. Noting the usual trend that elements and compounds have increasing ionic binding and band gaps as the rows lie higher in the periodic table, and as the difference between the columns increases, we see the effect described in the previous chapter that the band gap of Si, in the same column as Ge but lying

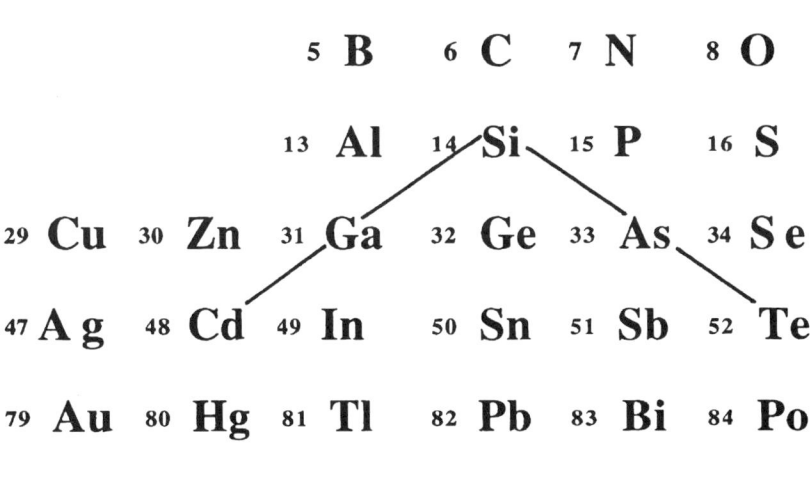

Fig. 4.1. A subsection of the Periodic Table of the elements showing the relationship between Group IV elements like Si and Ge, Group III–V compounds like GaAs and InP, and Group II–VI compounds like CdTe.

one row higher, is greater than the band gap of Ge. GaAs consists of Ga in the column to the left of Ge, and As in the column to the right of Ge, giving rise to a band gap in GaAs, from these considerations only, which is greater than that of Ge, and would be expected to be more like the band gap of Si. Actually, since GaAs is a direct-band gap material, its band gap (1.43 eV) is somewhat larger than that of crystalline Si (1.1 eV) with an indirect band gap. It may be concluded, therefore, that GaAs and InP (which makes up for In lying in a lower row than Ga, by P lying in a higher row than As) are the III–V compounds most like Si. We note in passing that if we extend our choice diagonally down to the left from Si through Ga, we reach Cd, and if we extend our choice diagonally down to the right from Si through As, we reach Te; we recognize the fact that CdTe is the major II–VI material of interest for solar cells, as discussed in Chapter 5.

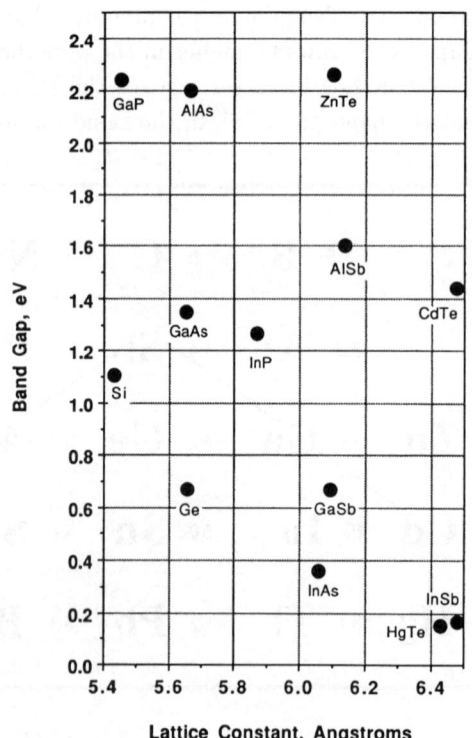

Fig. 4.2. Summary of band gaps and lattice constants for a number of zincblende III–V compounds, compared with diamond structure Si and Ge, and zincblende ZnTe, CdTe, and HgTe.

The direct band gap of GaAs is near the optimum for solar conversion, as shown in Fig. 1.2, and leads to 97% absorption of AM1 radiation in a thickness of about 2 μm. Its higher carrier mobilities than Si allow the fabrication of high-frequency devices, and it forms a variety of lattice-matched ternary compounds allowing for controlled variations of properties. GaAs solar cells should be operable at higher temperatures than silicon cells, and since they may be very thin, GaAs solar cells are expected to be radiation resistant. The band gap versus lattice constant relationship is given in Fig. 4.2 for eight III–V compounds with zincblende structure, as well as for a few other semiconductors for reference such as Si and Ge with the diamond structure, and ZnTe, CdTe and HgTe with zincblende structure. For example, Fig. 4.2 shows the close lattice match between AlAs, GaAs and Ge for three materials with a wide range of band gaps, as well as for AlSb, GaSb and InAs.

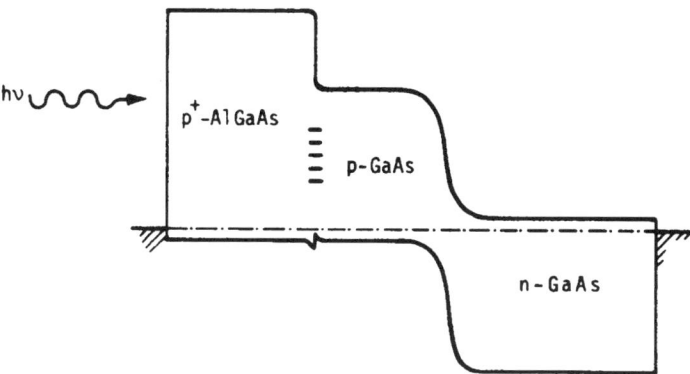

Fig. 4.3. Typical heteroface or buried homojunction p^+-AlGaAs/p-GaAs/n-GaAs solar cell band structure.

As a simple p–n junction material, GaAs is handicapped first by the fact that its high absorption coefficient leads to high carrier photoexcitation rates near the front surface, which is characterized by high surface recombination losses. Second, its direct band gap leads to a short bulk lifetime because of intrinsic recombination processes. The formation of a p-AlGaAs/p-GaAs/n-GaAs heteroface (or buried homojunction) structure with the larger band gap AlGaAs alloy, as shown in Fig. 4.3, however, greatly decreases the recombination at the front surface of the p-GaAs because of the excellent lattice match between GaAlAs and GaAs (only 0.16% mismatch between AlAs and GaAs).

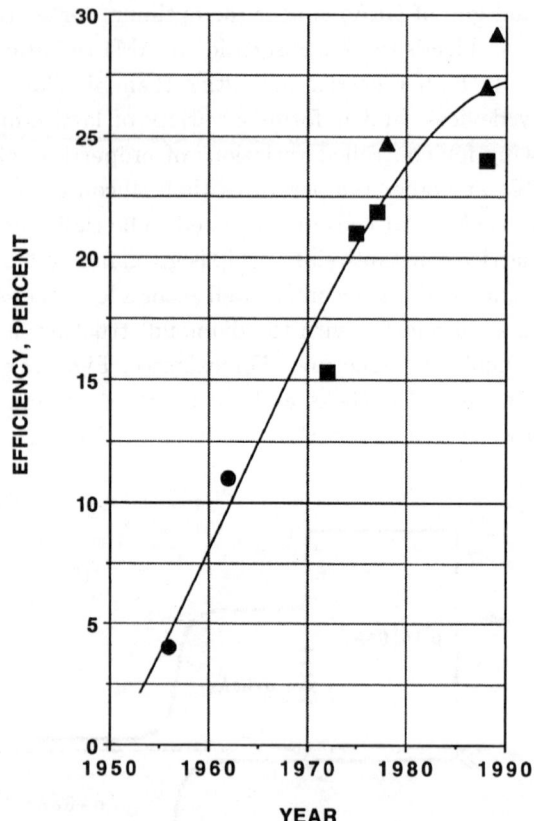

Fig. 4.4. Representative values of GaAs-based solar cell efficiency as a function of year. Three types of cells are included: (•) homojunction, (■) heteroface (buried homojunction) without concentration, and (▲) heteroface with concentration. Details are discussed further in Sec. 4.3.

Efficiencies of GaAs-based solar cells are among the highest of any solar cells, lying between 20% and 30%. Representative values of efficiency as a function of year and type of cell are plotted in Fig. 4.4 and are discussed further in Sec. 4.3. Because of the relatively high cost of the material and of cell fabrication, the development of GaAs-based (and InP-based) cells has been strongly directed toward concentrator use and space applications. General issues related to GaAs-based solar cells are discussed in several publications (Fahrenbruch and Bube, 1983; Fraas, 1985; Bube, 1990, 1993; Boes and Luque, 1993).

4.2. GaAs Fabrication Techniques

Single crystals of GaAs are usually grown by the Bridgman technique, although the Czochralski method can also be used. A high overpressure of As must be used in synthesis from the elements because the vapor pressure of As is so much larger than the vapor pressure of Ga. Common dopants include S, Se, Te, Sn, Si, C, and Ge as shallow donors, and Zn, Be, Mg, Cd, Si, Ge , and C as shallow acceptors. The dopants from column IV — C, Si, Ge and Sn — are amphoteric in GaAs, behaving as donors or acceptors depending on whether they go onto the Ga or As sublattice, and their electrical behavior depends on the conditions of growth.

GaAs-based solar cells involve layered structures that are deposited by liquid phase epitaxy (LPE), chemical vapor deposition (CVD), metallorganic chemical vapor deposition (MOCVD), or molecular beam epitaxy (MBE). Basic fabrication details are described in Fahrenbruch and Bube (1983).

A review of LPE is given by Dawson (1972). In the case of GaAs-based materials, LPE depends on the solubility of As in liquid Ga and AlGa alloys and the subsequent precipitation of $Al_x Ga_{1-x} As$ layers onto crystalline substrates, which determine the crystallographic orientation of the layers. LPE growth occurs in an atmosphere of high-purity H_2 in a growth apparatus fabricated from high-purity carbon and quartz. The apparatus consists of two portions, a top one that slides over a bottom one. For the growth of GaAs, for example, the top portion containing the Ga melt slides over the bottom, which contains a GaAs source at one location and a substrate at another. After the melt has been saturated at a temperature of about 900°C over the GaAs source, the slider carries the Ga melt over the substrate and the temperature is lowered at a rate of 0.1–0.5°C min^{-1}. During deposition, the relatively large volume of Ga acts as a sink for segregated impurities, and since the growth proceeds near thermal equilibrium, the resulting epitaxial layers are of very high quality. Variations to achieve gettering of impurities from the substrate during growth, an "etchback-regrowth" (ER) method to achieve both gettering and the growth of a compositionally graded $Al_x Ga_{1-x}$ layer, and other optimization techniques have been developed to increase cell efficiency (Woodall and Hovel, 1975, 1977; Bett *et al.*, 1991; Welter *et al.*, 1991; Habermann *et al.*, 1992).

Probably the most widely used technique in recent years is MOCVD, a variation of the CVD technique. CVD growth of $Al_x Ga_{1-x} As$ films proceeds by synthesis from gaseous Ga and As compounds, e.g. by reaction of $GaCl_3$ and AsH_3 on a substrate at 600–800°C (Johnston and Callahan, 1976). MOCVD uses a pyrolysis reaction of an organometallic Ga compound (such as trimethyl-gallium) and AsH_3 at an rf-heated substrate (Dupuis *et al.*, 1977; Dapkus *et al.*, 1978). In this technique growth is done in a cold-wall reactor and material is deposited only on heated areas. Both processes use H_2 as a carrier gas at atmospheric pressure. MOCVD is currently the growth method of choice for large scale, commercial production of III–V solar cells.

Although the large difference in vapor pressures of the elements make achievement of stoichiometry impossible in standard vacuum evaporation of GaAs from the compound, physical vapor deposition is possible due to the ability to dissociate As_4. High quality layers can also be grown by molecular beam epitaxy (MBE) (Casey and Panish, 1978).

4.3. GaAs-Based Solar Cell Development

The highest efficiencies using any materials (with Si a close second) have been achieved with solar cells based on GaAs and its solid solutions. GaAs has the optimum band gap for a single-junction solar cell, a high absorption coefficient, and the highest theoretical efficiency (about 39% for single-junction cells under 1000 suns). It can also be used in alloy form with other related materials, such as AlGaAs and InGaAs. A history of GaAs solar cell development is given by Anspaugh (1996).

In the development of GaAs-based solar cells the p^+-AlGaAs/p-GaAs/n-GaAs heteroface junction (or buried homojunction) of Fig. 4.3 plays a dominant role. In this section we summarize some of the main developments in GaAs solar cells of this type, as well as a few others of interest. For more details see Fahrenbruch and Bube (1983). In Sec. 4.4 we present a similar discussion for InP-based solar cells.

Heteroface Single Junction Cells

Except for the early report of a 4% p–n homojunction solar cell (Jenny *et al.*, 1956), the effective development of GaAs-based solar cells has been considerably more recent. Gobat *et al.* (1962) reported a p–n junction cell with an efficiency of 11%, and in 1970 Alferov *et al.* (1971) reported the first p-AlGaAs/n-GaAs heterojunction structure with an AM0 efficiency of 10–11%. In 1972 the heteroface structure shown in Fig. 4.3 was developed and an efficiency of 15.3% at AM1 and 19.1% at AM2 was reported with a p-$Al_{0.7}Ga_{0.3}As$/p-GaAs/n-GaAs cell fabricated by LPE (Woodall and Hovel, 1972). The intermediate p-type layer was formed by diffusion of Zn during the AlGaAs deposition.

Subsequently heteroface structure cells were reported with efficiencies of 24.7% at AM1 (concentration 180×) with a graded AlGaAs layer (Sahai *et al.*, 1978), 21% at AM1.4 without concentration (James and Moon, 1975), and 21.9% at AM1 without concentration (Woodall and Hovel, 1977). More recently records for highest solar-cell efficiency were held by the following GaAs-based solar cells: 22.4% for a thin-film cell, 24.3% for a single junction without concentration, and 29.2% for a single junction with concentration (Hubbard, 1989; Hubbard and Cook, 1989; Gale *et al.*, 1988, 1989; Vernon *et al.*, 1989).

The general concept of the cell structure shown in Fig. 4.3 has been changed little in recent years except for variations in thickness and composition of the layers. Although a large fraction of the carriers photogenerated at the light-incident surface of the p^+-AlGaAs is lost by surface recombination, only a small fraction of the total light current is generated there because of the large band gap of the AlGaAs. Good lattice matching and fabrication processes that yield a clean interface result in AlGaAs/GaAs interface recombination with an interface recombination velocity less than 10^4 cm sec^{-1}. The dark current-voltage characteristics of GaAs-based solar cells correspond remarkably well to either the injection model (see Eqs. 1.17 and 1.18) with $A = 1$, or the recombination model (see Eqs. 1.19 and 1.20) with $A = 2$; for high concentration light levels most cells show $A = 1$ with correspondingly low values of I_o.

Optimization of the cell performance calls for (a) minimization of the width of the p^+-AlGaAs layer consistent with a sufficiently low spreading resistance, (b) adjustment of the p-GaAs layer thickness to maximize current generation and collection, with a thickness of about 1/4 of the electron diffusion length (5–8 μm) indicated (Van der Plas *et al.*, 1978), and (c) control of the doping level in the p-GaAs and n-GaAs to obtain a high V_{oc} while retaining long carrier diffusion lengths.

One of the major advantages of the GaAs-based solar cell is its insensitivity to an increase in temperature; most measurements indicate a monotonic decrease in efficiency of 0.033 percentage points per °C, i.e. a decrease from 20.00% to 19.67% for a 10°C rise in temperature (Stuerke, 1978). An assessment of the high temperature stability of GaAs solar cells with high temperature contacts indicated minimal electrical degradation for 5 min at 550°C or for 15 min at 490°C; by elimination of processing defects and improved surface passivation, stability up to 600°C should be achievable (Tobin *et al.*, 1988).

As for other solar cells, the efficiency increases with concentration of the light for AlGaAs/GaAs solar cells. This is shown in Fig. 4.5, which also shows the effect of series resistance (Sahai *et al.*, 1978). If the series resistance can

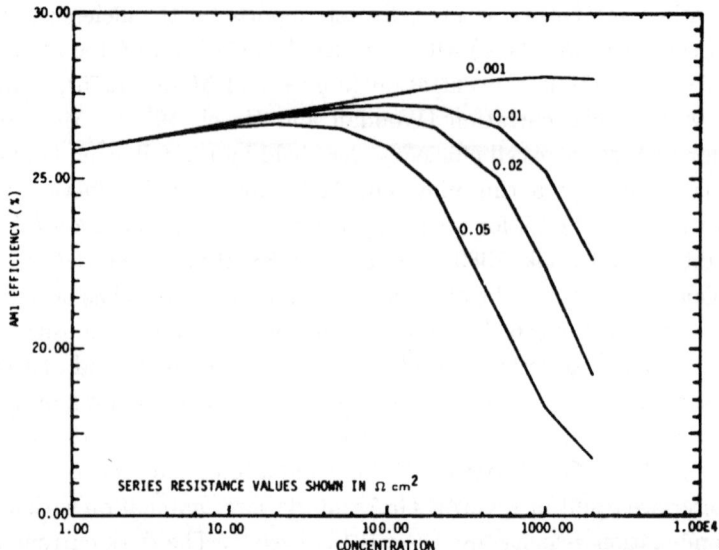

Fig. 4.5. Calculated solar efficiency as a function of concentration ratio for an AlGaAs/GaAs LPE cell for various values of lumped series resistance. (Reprinted with permission from R. Sahai *et al.*, *13th IEEE Photovoltaic Specialists Conference*, p. 946, (1978). Copyright 1978 IEEE, NY.)

be kept to less than 10^{-3} ohm-cm^2, then concentration ratios of 10^3 and above are desirable; in one design a p^+-GaAs layer is inserted between the contact metallization and the p^+-AlGaAs window to reduce contact resistivity and an overall value of series resistance was obtained between 4×10^{-5} and 4×10^{-4} ohm-cm^2.

Heteroface solar concentrator cells with efficiencies above 27% at solar concentrations over 400 suns in both n–p and p–n configurations have been investigated (MacMillan *et al.*, 1988a; Kaminar *et al.*, 1988). The higher efficiencies were obtained through improved control of the MOCVD growth conditions and improvements in cell gridline definition and edge passivation. Further optimization may be capable of increasing the efficiency to about 30% for concentrations of 500–1000 suns. A similar technique has been used to produce a GaAs solar cell with 24.0% efficiency without concentration, corresponding to the cell structure shown in Fig. 4.6 (Bertness *et al.*, 1988).

A variation on the procedure for AlGaAs/GaAs solar cells is the deposition on an inactive Ge substrate (Datum and Billets, 1991). Major advantages of the use of the Ge substrate include lower wafer cost, fracture toughness

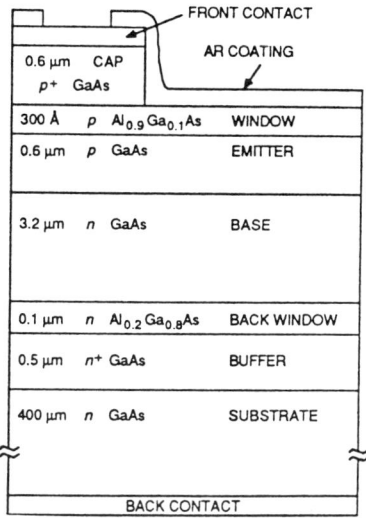

Fig. 4.6. Solar cell structure for a 24.0% efficient GaAs solar cell. At 1 sun, $\phi_{oc} = 1.05$ V, $J_{sc} = 27.1$ mA/cm^2, fill factor = 0.845. (Reprinted with permission from K. A. Bertness *et al., 20th IEEE Photovoltaic Specialists Conference*, p. 769 (1988). Copyright 1988 IEEE, NY.)

twice that of GaAs, and a higher tolerance to reverse current requirements. This work can be considered as the first step in the development of a poly-crystalline GaAs solar cell deposited on a polycrystalline Ge substrate with efficiency comparable to state-of-the-art polycrystalline cells such as CdTe and CuInSe$_2$ (Venkatasubramanian *et al.*, 1993, 1995, 1996). Solar cell parameters for AM1.5 radiation were $\phi_{oc} = 0.99$ V, $J_{sc} = 23.1$ mA/cm^2, fill factor of 0.79, and efficiency of 18.2% for a 4 cm^2 cell at AM1.5.

Related research concerns the fabrication of GaAs/AlGaAs thin film solar cells using the epitaxial liftoff (ELO) technique, which allows the transfer of these cells onto non-absorbing glass substrates (Hageman *et al.*, 1996; Lee *et al.*, 1996). Also as part of a program to fabricate GaAs cells from film structures grown on Si substrates coated with a GaP film, the properties of GaAs solar cells grown on (100) GaP by MOCVD have been investigated, with the best efficiency to date being 10.2% (Olsen *et al.*, 1996).

A method for the hybrid MBE/LPE growth of GaAs(AlGaAs) layers on Si substrates has been developed (Bett *et al.*, 1992; Baldus *et al.*, 1994a). Low-temperature ($T < 400°C$) LPE from a Sn-based melt, followed at higher temperatures by LPE from Ga-based melts, were used for the growth of GaAs

(AlGaAs) layers on Si substrates previously covered by MBE epitaxially grown GaAs, to achieve planar growth without microcracks or dissolution defects. Bismuth was investigated as an alternative to Ga and Sn solvents for the growth of GaAs layers by LPE on GaAs (MBE or MOCVD)/Si substrates at low temperatures ($T \leq 600°C$) with Sn as an n-type dopant at these temperatures (Baldus *et al.*, 1995). Attempts to use Zn p-type doping in GaAs grown from Bi-based melts were not reproducible.

The influence of substrate position (vertical or horizontal) during LPE-ER processing was investigated (Baldus *et al.*, 1994b, 1994c). A Zn post-diffusion for GaAs LPE-ER concentrator solar cells was developed to reduce the contact resistivity to $< 10^{-4}$ Ω-cm^2 (Blug *et al.*, 1995).

Transient photoluminescence decay, also known as time-resolved photoluminescence, has been used for the evaluation of GaAs solar cell materials and structures (Ahrenkiel *et al.*, 1990; Ehrhardt *et al.*, 1991; Ahrenkiel, 1992; Ahrenkiel *et al.*, 1993). This is a quick and contactless technique that directly measures the excess minority-carrier density. High resolution is obtained

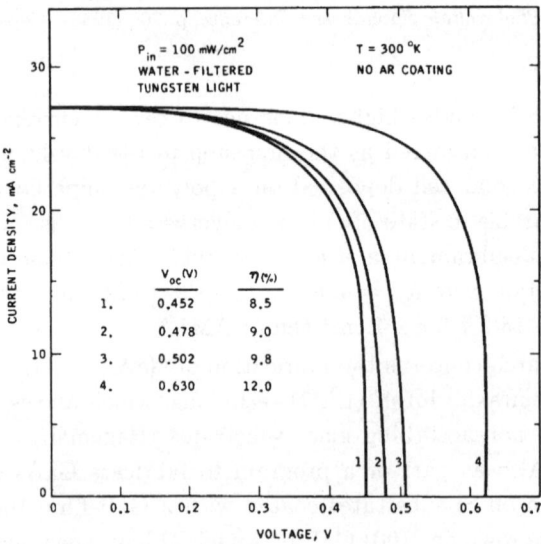

Fig. 4.7. Light current/voltage curves for Au/GaAs Schottky barriers for various treatments of the GaAs surface before application of the metal barrier contact. No antireflection coating is present. (1) "Clean" interface, (2) GaAs surface exposed to air at 300 K for 4 hr, (3) exposed to air at 300 K for 95 hr, and (4) exposed to air at 403 K for 70 hr. (Reprinted with permission from R. J. Stirn and Y. C. M. Yeh, *Appl. Phys. Lett.* **27**, 95 (1975). Copyright 1975 American Institute of Physics.)

through the use of a focused laser beam as the excitation source. Using a theoretical model to interpret measured photoluminescence decay transients, minority carrier properties in GaAs solar cells can be determined at different stages of processing. The measurements also provide minority carrier parameters, such as lifetime and interface recombination velocity, useful for device modeling.

MIS Cells

The performance of Schottky barriers on GaAs can be considerably enhanced by the inclusion of an interfacial insulator layer between the metal and the GaAs to form an MIS device. Figure 4.7 shows the striking dependence of open-circuit voltage in Au/oxide/GaAs MIS junctions on the introduction of an insulating oxide layer (Stirn and Yeh, 1975, 1977; Stirn *et al.*, 1977). High-efficiency cells can be made either by water-vapor thermal oxide growth or by deposition of Sb_2O_3 onto suitably etched GaAs surfaces. These were called AMOS cells (antireflection-coated metal oxide semiconductor) and gave efficiencies up to 17%.

It is the increase in diode factor A caused by the oxide layer that is the primary contributor to the increased open-circuit voltage. For AMOS cells on (100) GaAs surfaces, oxidation causes an increase in A from 1.0 to 1.2 without an increase in J_o. On (111) GaAs surfaces, on the other hand, J_o increases on oxidation by about an order of magnitude, and diode factors of 1.4 to 1.6 are found.

An n–i–p structure for a GaInP solar cell with an efficiency of 15.7% under AM0, among the highest efficiencies of GaInP single junction cells, was achieved by introducing an undoped, intrinsic GaInP layer at the p–n junction to reduce the interdiffusion of Zn into the emitter, and separating the p- and n-regions (Tsai *et al.*, 1995).

CLEFT Films

GaAs is the only material of current interest for solar cells that can be fabricated in the form of a single-crystal thin film solar cell by a desirable process (Gale *et al.*, 1988, 1989). This possibility is made economically attractive by the fact that the film can be grown on a reusable, single-crystal GaAs substrate by the technique known as "cleavage of lateral epitaxial films for transfer" (CLEFT). The CLEFT process does away with the necessity of using thick, non-reusable, expensive GaAs substrates.

The various steps in CLEFT fabrication are as follows (Fraas, 1985): (a) the single-crystal substrate is coated with carbon before epitaxial layer growth, (b) narrow slits are left in the carbon where the single crystal is exposed, (c) epitaxy is nucleated in the slits and the similarly oriented crystallites grow laterally rapidly across the carbon until they join to form a continuous film, (d) the thin-film solar-cell layers are then grown, (e) contacts are added and the film is glued to a glass superstrate, (f) the master substrate is removed by cleaving and can be reused.

GaAs Alloys

$In_x Ga_{1-x} As$ concentrator cells have been prepared using MOCVD on GaAs substrates, with a band gap of 1.15 eV for $x = 0.25$, and 1.35 eV for $x = 0.07$ (Werthen *et al.*, 1988). With a possible long range application to multijunction cells, these single-junction cells showed efficiencies greater than 24% at 400 suns. The cell with $x = 0.25$ has about the same band gap as Si, but has a direct band gap, and yields $\phi_{oc} = 0.80$ V compared with 0.67 V for Si. Improvements in current collection should lead to considerably higher efficiencies for the InGaAs cells.

Multijunction or Tandem Cells

Multijunction cells consist of devices involving more than one solar cell in such a way that the higher-energy solar radiation is absorbed by a larger band-gap solar cell, and the residual lower-energy solar radiation is absorbed by a smaller band-gap solar cell. One way to achieve this is to use an optical beam-splitter that reflects different portions of the solar spectrum to different cells. Such a beam splitter, for example, was built to separate the solar spectrum into two portions matching the band gaps of AlGaAs and Si (Borden *et al.*, 1981).

An operationally preferable way to achieve the same goal is to stack two or more solar cells so that the solar radiation passes through the larger band-gap material first, and then the residual radiation passes through the smaller band-gap material. This goal can be achieved in one of two ways: (a) by mechanically stacking different solar cells (MSMJ), with examples given in Sec. 4.5, and (b) by fabricating a monolithic multijunction structure using a process such as MOCVD.

One of the first successful attempts to produce a monolithic tandem solar cell (sometimes called a "cascade structure") involved the use of six or seven layers of GaAs and AlGaAs materials on a GaAs substrate to form a monolithic two-junction structure, internally connected in series by a transparent, low-resistance p^+–n^+ junction. Test cell efficiencies were about 9% (Bedair *et al.*, 1978, 1979).

Multijunction cells have been developed using different alloys of GaAs (Lewis *et al.*, 1988; MacMillan *et al.*, 1988b, 1989; Virshup *et al.*, 1988). Achievements in this effort include a monolithic, two-terminal, two-junction AlGaAs/GaAs cell with efficiency of 27.6% without concentration (MacMillan *et al.*, 1989), a 16.5% three-terminal, two junction 1.72 eV AlGaAs/1.15 eV GaInAs multijunction (Lewis *et al.*, 1988), and a 23.9% monolithically grown, two junction 1.93 eV/$Al_{0.35}Ga_{0.65}As$/1.42 eV GaAs multijunction cell (Virshup *et al.*, 1988), which from top metal to back contact includes eleven separately deposited layers using MOCVD, as shown in Fig. 4.8. A goal of the work was to produce a monolithic, three-junction device based on AlGaAs/GaAs/InGaAs grown by MOCVD.

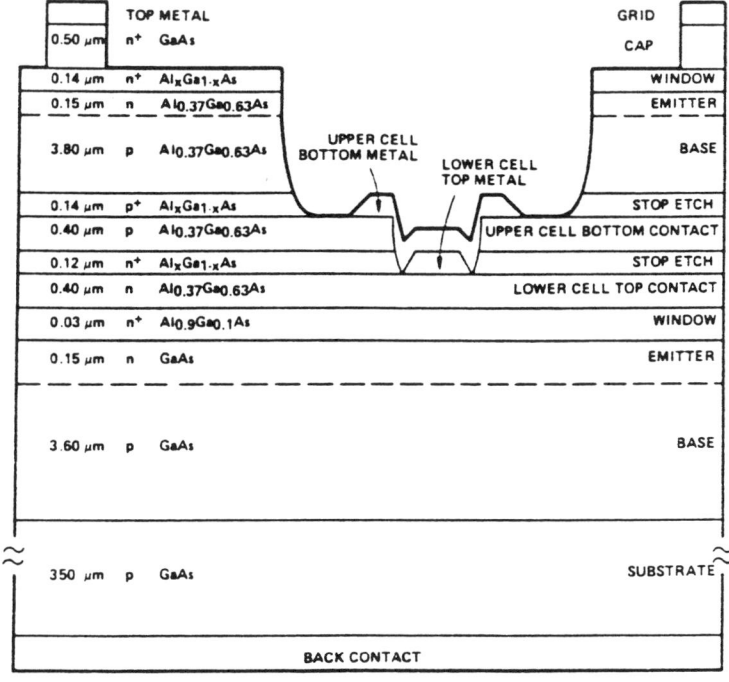

Fig. 4.8. Solar cell structure for a 23.9% efficient monolithically grown, two-junction solar cell with an $Al_{0.37}Ga_{0.63}As$ top cell and a GaAs bottom cell. At 1 sun, $\phi_{oc} = 2.41$ V, $J_{sc} = 14.9$ mA/cm^2 and fill factor = 0.84. (Reprinted with permission from G. F. Virshup *et al.*, *20th IEEE Photovoltaic Specialists Conference*, p. 441 (1988). Copyright 1988, IEEE, NY.)

GaInAsP alloys provide a range of possibilities for use in multijunction solar cells. Two compositions of GaInAsP have been shown capable of producing solar cells with an efficiency greater than 21% (Sharps *et al.*, 1993). These cells were grown in an organometallic vapor phase epitaxy (OMVPE) reactor at 675°C. Precursors were trimethylgallium, ethyldimethylindium, trimethylaluminum, arsine, and phosphine. Diethylzinc was used to supply the *p*-type dopant and a 50 ppm H_2Se in H_2 mixture was used to supply the *n*-type dopant. For terrestrial applications $Ga_{0.84}In_{0.16}As_{0.68}P_{0.32}$ (band gap = 1.55 eV), grown on *p*-GaAs, forms a cell with $\phi_{oc} = 1.047$ V, $J_{sc} = 22.5$ mA/cm^2, fill factor = 0.849, and efficiency = 21.8% for AM1.5 radiation and 23.4% for irradiation by 9.73 suns. Plans are underway to fabricate a monolithic multijunction cell with the $Ga_{0.84}In_{0.16}As_{0.68}P_{0.32}$ cell on top, then a $Ga_{0.84}In_{0.16}As_{0.68}P_{0.32}$ tunnel diode, and a Ge cell on the bottom, on which

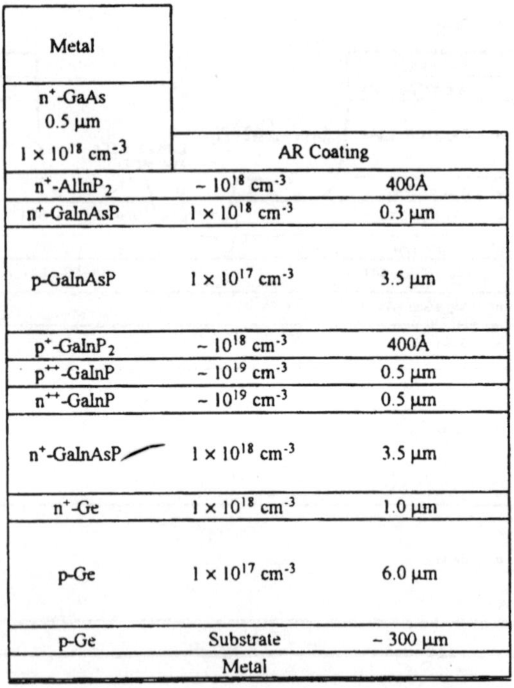

Metal		
n^+-GaAs 0.5 μm 1×10^{18} cm^{-3}	AR Coating	
n^+-AlInP$_2$	$\sim 10^{18}$ cm^{-3}	400Å
n^+-GaInAsP	1×10^{18} cm^{-3}	0.3 μm
p-GaInAsP	1×10^{17} cm^{-3}	3.5 μm
p^+-GaInP$_2$	$\sim 10^{18}$ cm^{-3}	400Å
p^{++}-GaInP	$\sim 10^{19}$ cm^{-3}	0.5 μm
n^{++}-GaInP	$\sim 10^{19}$ cm^{-3}	0.5 μm
n^+-GaInAsP	1×10^{18} cm^{-3}	3.5 μm
n^+-Ge	1×10^{18} cm^{-3}	1.0 μm
p-Ge	1×10^{17} cm^{-3}	6.0 μm
p-Ge	Substrate	~ 300 μm
Metal		

Fig. 4.9. Solar cell structure for a possible $Ga_{0.84}In_{0.16}As_{0.68}P_{0.32}$/Ge monolithic multijunction cell, involving a $Ga_{0.84}In_{0.16}As_{0.68}P_{0.32}$ tunnel diode. (Reprinted with permission from P. R. Sharps *et al.*, *23rd IEEE Photovoltaic Specialists Conference*, p. 633 (1993). Copyright 1993 IEEE, NY.)

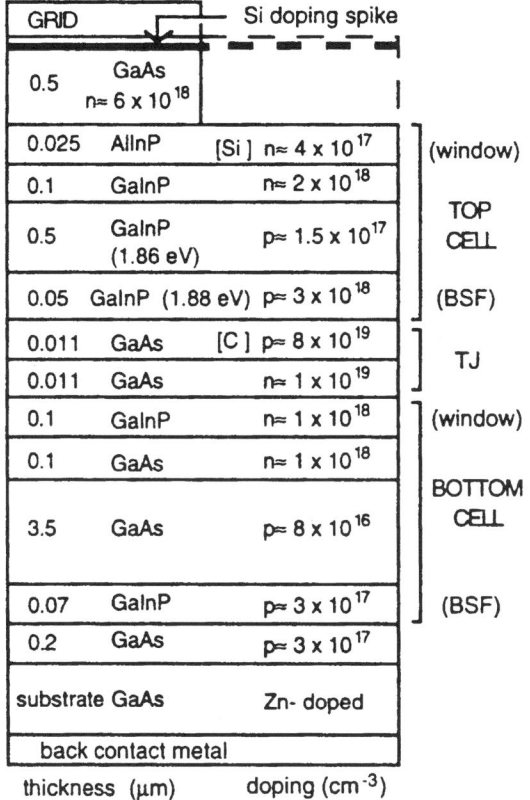

Fig. 4.10. Solar cell structure for AM0 standard solar spectrum for a GaInP/GaAs mono-lithic multijunction. (Reprinted with permission from K. A. Bertness *et al.*, *24th IEEE Photovoltaic Specialists Conference*, p. 1671 (1994). Copyright 1994, IEEE, NY.)

the $Ga_{0.84}In_{0.16}As_{0.68}P_{0.32}$ can be grown directly without the buffer layer normally needed to avoid the propagation of defects from the substrate. A schematic of this proposed cell is given in Fig. 4.9. For space applications $Ga_{0.68}In_{0.32}As_{0.34}P_{0.66}$ (band gap = 1.7 eV) has demonstrated good radiation resistance, with ϕ_{oc} = 1.161 V, J_{sc} = 28.9 mA/cm^2, fill factor = 0.86 and efficiency = 21.4% for AM0 illumination.

Record-breaking efficiencies have been reported for GaInP/GaAs two-terminal multijunction cells: 29.5% for one-sun AM1.5, 30.2% for 160-suns AM1.5, 25.7% for one-sun AM0, and 19.6% at AM0 after 10^{15} cm^{-2} 1-MeV electron irradiation (Bertness *et al.*, 1994). Efficiency advances were attained

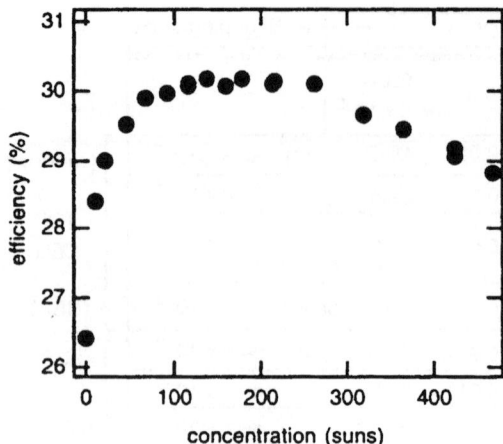

Fig. 4.11. Efficiency of a GaInP/GaAs multijunction under AM1.5 radiation as a function of the concentration. (Reprinted with permission from K. A. Bertness *et al.*, *24th IEEE Photovoltaic Specialists Conference*, p. 1671 (1994). Copyright 1994, IEEE, NY.)

primarily by improving interface passivation layers in the cells and by reducing grid coverage. The solar ·cell structure for a GaInP/GaAs multijunction cell is shown in Fig. 4.10, and the dependence of efficiency on solar concentration is given in Fig. 4.11.

A 25.8% efficient mechanical stack GaAs/Si concentrator tandem-solar cell has been fabricated using an LPE-grown GaAs top-cell (Blieske *et al.*, 1994).

4.4. InP-Based Solar Cells

InP, like GaAs, is a direct band-gap material with a band gap of 1.34 eV, close to the maximum for solar energy conversion. Also like GaAs, homojunction cells are limited by surface recombination at the incident surface (Galavanov *et al.*, 1967), but heterojunctions with good lattice matching, and heteroface junctions (buried homojunctions), are much more efficient. InP crystals are grown by the Czochralski method at high pressures or by using a liquid encapsulation technique to preserve stoichiometry.

Heterojunction Cells

Highly efficient CdS/InP heterojunction cells have been fabricated by vacuum evaporation, chemical vapor deposition, and close-spaced vapor transport (CSVT). For more details see Fahrenbruch and Bube (1983).

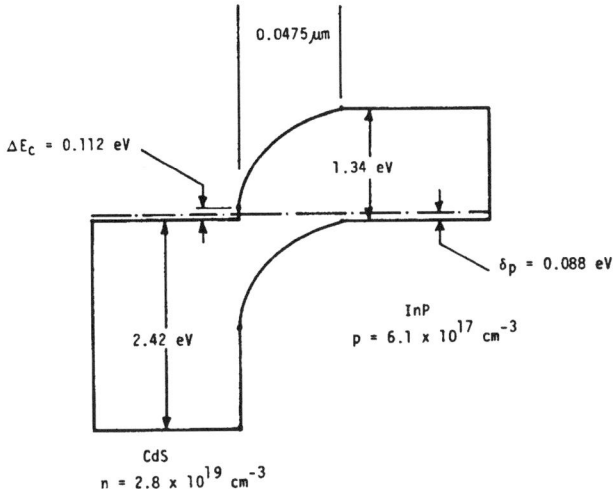

Fig. 4.12. Energy band diagram for a CdS/InP heterojunction prepared by close-spaced vapor transport of a CdS film onto single crystal InP. (Reprinted from A. Yoshikawa and Y. Sakai, *Solid State Electronics* **20**, 133 (1977). Copyright 1977, with kind permission from Elsevier Science Ltd., The Boulevard, Langford Lane, Kidlington 0X5 1GB, UK.)

An efficiency of 12.5% for an n-CdS/p-InP heterojunction was reported by Wagner *et al.* (1975) for which the CdS layer was formed by vacuum deposition on InP single crystals doped with Cd. A schematic energy band diagram is given in Fig. 4.12. CdS is an excellent choice for such a heterojunction since there is only a 0.32% lattice mismatch between the (111) plane of zincblende InP and the basal plane of wurtzite CdS. Also the tetrahedral atomic distance is 25.33 nm in InP and 25.32 nm in CdS.

Heteroface Cells

A heat treatment at about 600°C in a non-oxidizing atmosphere improves the CdS/InP cell described above, giving an efficiency of 14% (Shay *et al.*, 1976), and suggesting the formation of a heteroface junction, n-CdS/n-InP/p-InP.

The efficiency of a CdS/InP cell was further increased to 15% by using chemical vapor deposition of CdS on InP using an open-tube H_2S/H_2 flow system (Shay *et al.*, 1977; Bettini *et al.*, 1977, 1978). Once again the properties of the cell suggest that it is an InP buried homojunction. Solar cell parameters for AM2 illumination were $\phi_{oc} = 0.79$ V, $J_{sc} = 18.7$ mA/cm^2, $ff = 0.735$.

In view of the importance of lattice matching for high efficiency that underlay the above discussion of CdS/InP solar cells, it is perhaps surprising to

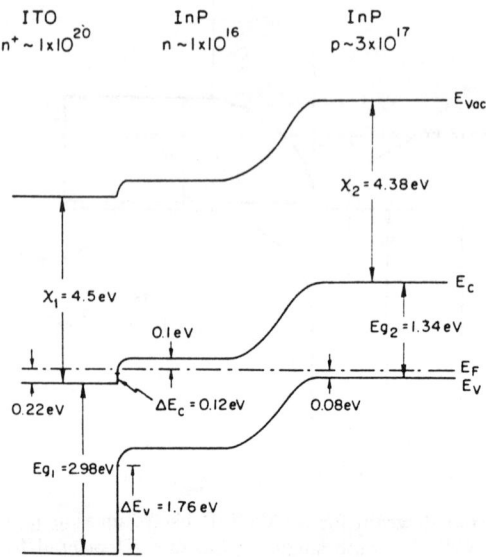

Fig. 4.13. Energy band diagram for an n-ITO/n-InP/p-InP buried homojunction produced by rf sputtering of the ITO onto p-InP. (Reprinted with permission from M. J. Tsai *et al.*, *J. Appl. Phys.* **51**, 2696 (1980). Copyright 1980 American Institute of Physics.)

find that cells with efficiency greater than 14% can be made from junctions between indium-tin-oxide (ITO) and p-type InP for which the lattice mismatch is large (Sree Harsha *et al.*, 1977). Subsequent investigation (Bachmann *et al.*, 1979; Tsai *et al.*, 1980) has shown that the high efficiency of the ITO/InP junctions is the result of the formation of a buried homojunction in the InP by the action of sputter deposition, with the resulting band diagram shown in Fig. 4.13.

Epitaxial Film Cells

Single-crystal p-type InP homoepitaxial layers were grown by a chemical vapor deposition method on single crystal p^+-InP substrates (Manasevit 1978). CdS/InP junction cells made by vacuum evaporation of CdS onto these layers showed efficiencies between 8 and 12%. Similar results were reported for sputtered ITO layers on homoepitaxial p-InP layers.

The electrical properties were determined (Tsai and Bube, 1978) for a series of epitaxial n-type undoped InP films deposited on high-resistivity GaAs:Cr or InP:Fe substrates by MOCVD, which showed that epitaxial undoped n-type

InP films can be deposited by MOCVD on InP substrates to produce material with mobility close to that of single crystal InP. The same is true for GaAs substrates only for values of electron densities greater than 10^{16} cm^{-3}.

Homojunction Cells

A large scale commercial process has been described for the fabrication of n^+–p homojunction InP solar cells for space power generation, involving the diffusion of sulfur into p-type InP substrates in a sealed quartz ampoule (Okazaki *et al.*, 1988). The best 2×2 cm^2 cell had $\phi_{oc} = 0.828$ V, $J_{sc} = 33.7$ mA/cm^2, fill factor $= 0.816$ and efficiency $= 16.6\%$ for AM0.

In spite of the front surface recombination problems, and in the absence of a suitable lattice-matched, wide-bandgap material that could be used as a window layer for passivation of the front surface, a high efficiency n^+–p junction has been prepared by MOCVD with a graded-junction, or front-surface-field structure, to decrease front surface recombination losses (Keavny *et al.*, 1990). The cell consists of a p-type heavily-doped buffer layer, a lightly-doped p-type base layer, an n-type emitter, and an n-type In$_{0.53}$Ga$_{0.47}$As cap layer. The n-type emitter is made up of two layers: A moderately heavily doped region (3×10^{18} cm^{-3}) near the junction, and a very heavily doped region (3×10^{19} cm^{-3}) at the surface. After metallization with Cr-Au-Ag, the cells were mesa etched, and the InGaAs layer was removed from the active area with a selective etch. Finally the cells had a ZnS/MgF$_2$ two-layer antireflection coating. A 4 cm^2 cell for AM0 had $\phi_{oc} = 0.876$ V, $J_{sc} = 36.34$ mA/cm^2, fill factor $= 0.824$ and efficiency $= 19.1\%$. The cells showed a degradation of 4.7% after exposure to 10^{14} cm^{-2} 1 MeV electrons.

Hydrogenation of n^+–p InP solar cells has been shown to increase the efficiency from 14.8% to 17.5% at AM0 because of a reduction in carrier concentration in the near-surface layer due to the formation of an acceptor-hydrogen complex (Min *et al.*, 1993).

MIS Cells

MIS cells with efficiency of 14.5% have been fabricated on InP, similar to those described above for GaAs (Kamimura *et al.*, 1981).

Polycrystalline InP Cells

Because of the expense of InP, it has been attempted to use thin films of polycrystalline InP less than 2 μm thick with grain size about the same as

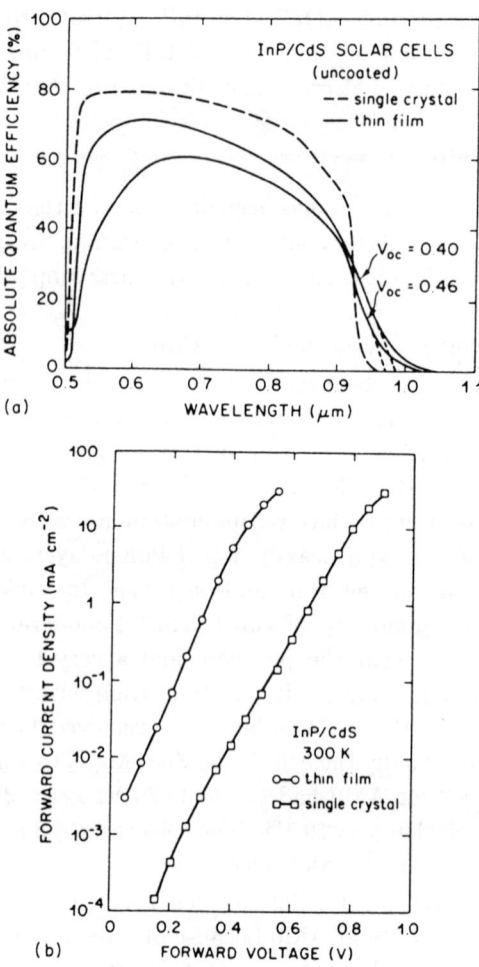

Fig. 4.14. Comparisons of CdS/InP photovoltaic junction properties for single-crystal InP and polycrystalline InP cells. (a) Wavelength dependence of the quantum efficiency. (b) Forward-bias current/voltage curves. (Reprinted with permission from J. L. Shay *et al.*, *IEEE Transactions Electron Devices* **ED-24**, 483 (1977). Copyright 1977, IEEE, NY.)

the thickness. CdS/InP junctions have been prepared on polycrystalline InP deposited by PCl$_3$ CVD on carbon substrates (Bachmann *et al.*, 1976; Shay *et al.*, 1977). Values of J_{sc} are high, but so are values of J_o because of the polycrystalline nature of the material, and ϕ_{oc} is considerably reduced; the highest efficiency values were 5.7%.

Figure 4.14 compares the properties of CdS/InP cells prepared on single-crystal InP with those prepared on polycrystalline InP (Shay *et al.*, 1977). The quantum efficiency of the polycrystalline InP cells is only slightly less than that of the single-crystal cells, but the junction characteristics show both a hundred fold increase in J_o and a decrease in A from 2.0 to 1.7, resulting in a decrease in ϕ_{oc} to values of about 0.46 V.

4.5. Hybrid Multijunction Solar Cells

A number of multijunction combinations have been investigated, using a GaAs-based solar cell as one member.

GaAs/Si

A record-breaking 31% efficient GaAs/Si mechanically stacked multijunction (MSMJ) concentrator solar cell has been developed (Gee and Virshup, 1988). Special care was needed to insure that the GaAs cell had maximum transparency for sub-band gap radiation. InGaAs is a possible substitute for the Si cell.

GaAs/CuInSe₂

An MSMJ four-terminal device, consisting of a GaAs thin film top cell made by the CLEFT process and a ZnCdS/CuInSe$_2$ thin bottom cell, gave an efficiency of 21.3% (19.5% from the GaAs cell, and 2.8% from the CuInSe$_2$ cell) (Stanbery *et al.*, 1987; Kim *et al.*, 1988). Good radiation resistance makes these cells candidates for flat-plate space applications.

GaInP₂/GaAs

Efficiencies of about 25% have been reported for monolithic GaInP$_2$/GaAs multijunction cells with a tunnel-junction interconnect (Olson *et al.*, 1989; Kurtz *et al.*, 1990), and plans for a GaInP$_2$/GaAs/Ge triple junction have been described (Chiang *et al.*, 1994). Comparable efficiencies are obtained for *p*–on–*n* GaInP$_2$/GaAs two-terminal, monolithic multijunction cells (Sharps *et al.*, 1994).

A single junction InGaP solar cell with an n–p–p^+ structure and a BSF achieved an efficiency of 18.48%, and InGaP/GaAs monolithic multijunctions were prepared with an efficiency of 27.3% (Takamoto *et al.*, 1994; Kurita *et al.*, 1995).

GaAs/GaSb

An efficiency of 34% has been achieved in an MSMJ GaAs/GaSb cell (Fraas *et al.*, 1990, 1991). Cells based on *p-n*-InGaSbAs/*n*-GaSb lattice matched heterostructures have also been fabricated by LPE with the long-wavelength edge of photosensitivity of the cells at about 2.15–2.2 μm (Andreev *et al.*, 1996).

InP/GaInAs

An efficiency of 22.2% has been achieved with a monolithic multijunction concentrator cell based on InP/GaInAs, and research has been carried out on other variations in this system (Wanlass *et al.*, 1991, 1994).

4.6. Radiation Resistance

Since III–V solar cells have considerably better radiation resistance than Si solar cells (the shorter intrinsic carrier lifetime in III–V materials due to the direct band gap makes it less sensitive to the creation of imperfection recombination centers), they are especially favored for space applications where the power-to-weight ratio and the resistance to radiation damage are of prime importance. InP-based cells show better radiation damage resistance than GaAs-based cells. A review of GaAs solar cell properties with special reference to radiation effects is given by Anspaugh (1996).

Fabrication procedures can also be varied to increase resistance to radiation damage. One such procedure involves band-gap grading via compositional grading to achieve high electric fields near the incident surface to reflect carriers away from the surface and reduce surface recombination loss, or to achieve high electric fields throughout the generation region to increase the effective diffusion length there (Tauc, 1957; Hutchby and Fudurich, 1976; Woodall and Hovel, 1977; Kordos *et al.*, 1979; Kordos and Pearson, 1980).

The influence of 20.6 MeV proton and 1 MeV electron radiation on CdS/InP solar cells has been investigated (Botnaryuk *et al.*, 1990). It was found that CdS/InP solar cells possess a higher resistance to radiation than either Si or AlGaAs/GaAs cells. Since indium is an expensive metal with a rather small natural abundance, InP cannot compete economically in most applications with GaAs and other solar cells, but since it does have a better radiation tolerance than GaAs, development for such applications continues (Gessert *et al.*, 1989).

Radiation effects due to protons and electrons have been investigated with n^+–p homojunction InP solar cells made by closed-ampoule diffusion of In_2S_3

into a *p*-type substrate (Takamoto *et al.*, 1990). The greatest damage is done by low-energy protons stopped in the active region. A 50 μm coverglass shields the cell from low-energy protons, which is effective because InP solar cells are highly resistant to electron and high-energy proton irradiation. Analysis of space flight data on InP solar cells indicates that InP cells are suitable for space applications (Takahashi *et al.*, 1991; Yamaguchi *et al.*, 1990, 1991), and that the development of thin film InP cells is highly desirable.

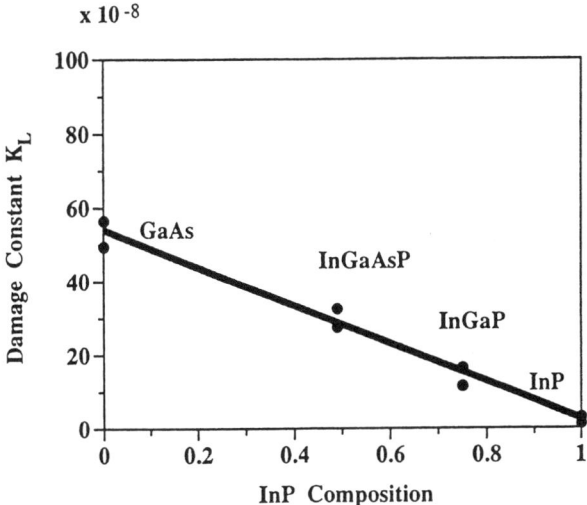

Fig. 4.15. Changes in damage constant K_L for minority-carrier diffusion length for InP-related materials and GaAs as a function of InP composition. (Reprinted with permission from M. Yamaguchi *et al.*, *Japanese J. Appl. Phys.* **34**, 6222 (1995). Copyright 1995, *Japanese Journal of Applied Physics*.)

The radiation resistance of a number of different compound semiconductor solar cells to 1 MeV electrons has been recently investigated (Yamaguchi, 1995), including InP, InGaP, InGaAsP, GaAs, AlGaAs, InGaAs, Si, Ge, and CuInSe$_2$. The major conclusions of the study are: (a) InP and CuInSe$_2$ have the highest radiation resistance; (b) differences of radiation resistance between GaAs-based solar cells such as GaAs, AlGaAs, and InGaAs are due to band-gap energy effects with the resistance to damage increasing with band gap; (c) the superior radiation resistance of CuInSe$_2$ cells is associated with the higher optical absorption coefficient of CuInSe$_2$ compared to other materials; and (d) the better radiation resistance of InP-based solar cells, such as InP, InGaP, and InGaAsP, is due to a lower defect introduction rate compared to other materials. Figure 4.15 shows the dependence of the damage constant K_L for InGaAsP alloys as a function of the composition (Yamaguchi *et al.*, 1995). K_L is defined by $K_L F = (1/L_F^2) - (1/L_o^2)$, where F is the integrated fluence, L_F is the minority-carrier diffusion length after irradiation, and L_o is the minority carrier diffusion length before irradiation. The value of K_L decreases linearly with the fractional InP composition of the materials.

CHAPTER 5

CADMIUM TELLURIDE AND OTHER II–VI MATERIALS

5.1. Overview

When work on thin-film Cu_xS/CdS cells came to an end because of the unavoidable degradation associated with the effects of Cu diffusion, the question naturally arose as to what other p-type materials could form useful heterojunctions with CdS. Two such materials are InP, discussed in Chapter 4, and $CuInSe_2$ discussed in Chapter 6. It would also be desirable to find suitable p-type materials among other II–VI compounds, capable of being deposited in thin film polycrystalline form in ways analogous to those used for CdS. Of the six II–VI chalcogenides (ZnS, ZnSe, ZnTe, CdS, CdSe, and CdTe) only ZnSe and CdTe can be prepared in both n-type and p-type form, and of the rest only ZnTe can be prepared in high-conductivity p-type form. For the preparation of p–n heterojunctions from II–VI chalcogenides, therefore, the choice of the p-type member is essentially restricted to ZnTe or CdTe, since p-type ZnSe has yet to be demonstrated in polycrystalline form. This leaves the nine potential possibilities summarized in Table 5.1, together with their alloys (Bube, 1976). The II–VI compounds n-type CdS and n-type ZnO have been used for the high-conductivity window-layer in several heterojunction cells and examples are given in several places in this book.

The two most likely candidates for efficient solar cells are n-ZnCdS/p-CdTe and n-ZnSSe/p-CdTe. A plot of band gap versus lattice constant for the II–VI chalcogenides and a few other materials of interest are shown in Fig. 5.1 (Bube and Mitchell, 1993). All of the II–VI heterojunctions show appreciable lattice mismatch except n-CdSe/p-ZnTe, but such mismatch by itself does not appear to be a serious problem for CdTe-based cells. The stoichiometry limits of existence are shown in Fig. 5.1, although several of the compounds have been prepared outside these limits in metastable form.

In all of the different ways in which solar cells have been fabricated based on junctions involving p-CdTe crystals and polycrystalline thin films, three problems appear to be dominant: recombination losses associated with the

Table 5.1. Energy Relations at II–VI Binary Heterojunction Interfaces

Larger Band gap Material (E_G, eV)	Smaller Band gap Material (E_G, eV)	Diffusion Voltage, ϕ_D^a	Lattice Mismatch %
n-CdSe (1.70)	p-CdTe (1.44)	0.57	6.3
p-ZnTe (2.26)	n-CdSe (1.70)	0.61	0.5
n-CdS (2.42)	p-CdTe (1.44)	1.02	9.7
n-CdS (2.42)	p-ZnTe (2.26)	1.06	3.9
p-ZnTe (2.26)	n-CdTe (1.44)	1.28	5.8
n-ZnSe (2.69)	p-CdTe (1.44)	1.43	12.5
n-ZnSe (2.69)	p-ZnTe (2.26)	1.47	7.1
n-ZnS (3.70)	p-CdTe (1.44)	1.62	15.3
n-ZnS (3.70)	p-ZnTe (2.26)	1.66	10.5

[a] Based on values of electron affinity given in A. G. Milnes and D. L. Feucht (1992)

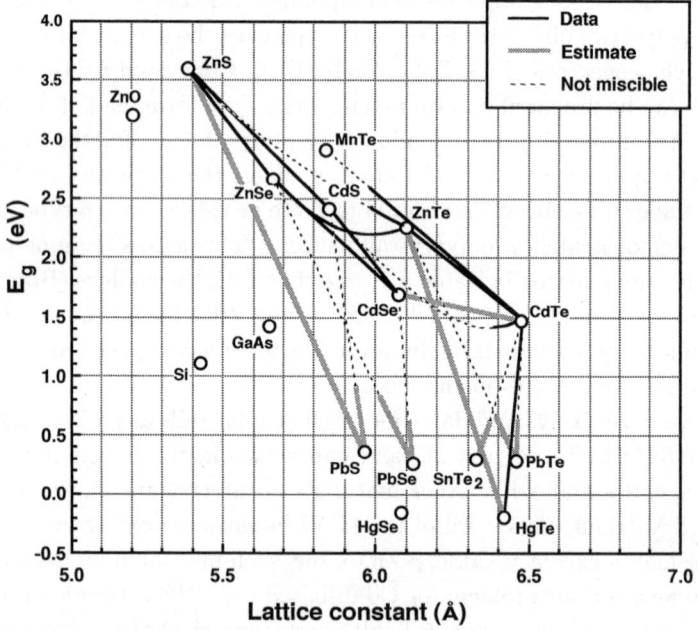

Fig. 5.1. Band gap versus lattice constant. For hexagonal structures, $a^*(2)^{1/2}$ is used. Only ternary tie lines are shown. "Estimate" indicates that a linear estimate was used in lieu of lacking data. (Data compiled by A. L. Fahrenbruch, 1993. Reprinted with permission from R. H. Bube and K. W. Mitchell, *J. Electronic Materials* **22**, 17 (1993), a publication of the Minerals, Metals & Materials Society, Warrendale, Pennsylvania 15086.)

junction interface, difficulty in doping the p-CdTe, and difficulty in obtaining low-resistance contacts to the p-CdTe. These major areas and related topics are discussed in this chapter.

By 1976 efficiencies between 6 and 7% had been found for (a) evaporated n-CdS on single crystal p-CdTe with or without an indium-tin oxide (ITO) transparent, conducting coating on the CdS, and (b) a solution-sprayed n-CdS layer on single crystal p-CdTe (Bube, 1976). Since then a rather wide variety of film deposition techniques have been used in the fabrication of CdTe-based solar cells, both in the deposition of window materials on p-type single crystal or polycrystalline CdTe, and in the deposition of polycrystalline p-CdTe layers themselves. These include vacuum evaporation (VE), hot-wall vacuum evap-oration (HWVE), close-spaced vapor transport (CSVT) — sometimes called close-spaced sublimation (CSS), electron-beam evaporation (EBE), chemical vapor deposition (CVD) or metalorganic chemical vapor deposition (MOCVD), screen printing (ScP), electrodeposition (ED), spray pyrolysis (SP), and sput-tering (ST). These techniques are described in this chapter as related to the development of CdTe polycrystalline thin films for solar cells. A summary of the general properties of CdTe is available in Zanio (1978).

Attention must also be paid to the potential health and safety issues that arise because of the toxic nature of cadmium, both in connection with the manufacturing of CdTe devices, and also during field use (Moskowitz *et al.*, 1990; Moskowitz and Zweibel, 1990; Moskowitz, 1994). Given the environ-mental concerns, the development of responsible means for disposing of CdTe solar cells is essential. The technology is straight-forward (Doty and Meyers, 1988; Patterson *et al.*, 1994) and new science is not required (Birkmire and Meyers, 1994). Techniques for dealing with off-specification modules and pro-cess by-products, using acid leaching of the CdTe from the glass substrates and recovery by means of an ion exchange process, have also been discussed (Sasala *et al.*, 1994).

A summary of representative solar cell results are given in Table 5.2 (Bube, 1988), showing that CdTe solar cells with an efficiency approaching 10% or higher have been made as heterojunctions, homojunctions, buried homojunc-tions and MIS junctions, using single crystal CdTe or thin film polycrystalline CdTe, and a variety of deposition techniques. A plot of optimum CdTe-based solar cell efficiencies as a function of year is given in Fig. 5.2, where compara-ble values for the major competing $CuInSe_2$-based polycrystalline solar cells, described in Chapter 6, are also included.

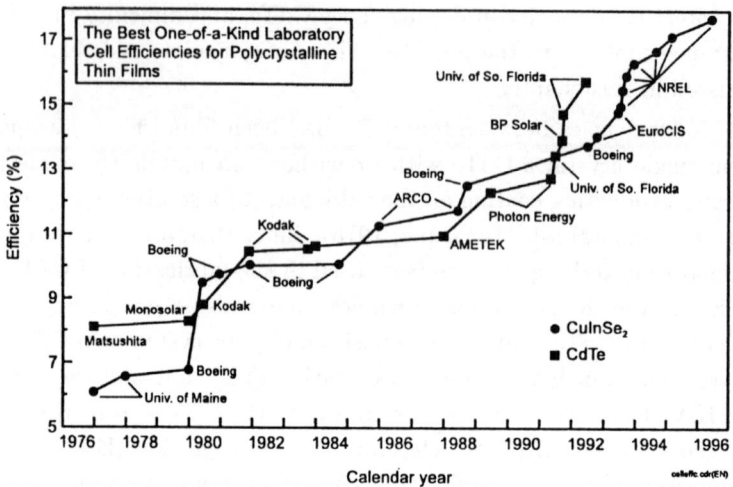

Fig. 5.2. Best one-of-a-kind laboratory cell efficiencies for polycrystalline thin-film cells based on CdTe or CuInSe₂. (Reprinted with permission from K. Zweibel *et al., 25th IEEE Photovoltaic Specialists Conference*, p. 745 (1996). Copyright 1996, IEEE, NY.)

Table 5.2. Representative High Efficiency CdTe Solar Cells

Cell	Open-Circuit Voltage, ϕ_{oc}	Short-Circuit Current, J_{sc} mA/cm^2	Solar Efficiency, %	Reference
n-CdS/p-CdTe heterojunction CdTe by CSVT, CdS by CBD (all thin-film cell)	0.84	25.1	15.8	Ferekides *et al.*, 1993
Au/n-CdTe Schottky barrier SX n-CdTe	0.81	21.6	15.1	Raychaudhuri *et al.*, 1987
n-CdS/p-CdTe heterojunction CdTe by CSVT CdS by CVD (all thin-film cell)	0.82	25.4	15.05	Nishio *et al.*, 1996

Table 5.2. (*Continued*)

Cell	Open-Circuit Voltage, ϕ_{oc}	Short-Circuit Current, J_{sc} mA/cm^2	Solar Efficiency, %	Reference
n-CdS/p-CdTe heterojunction both layers by CSVT (all thin-film cell)	0.85	21.7	14.2	Ferekides *et al.*, 1996
n-ITO/p-CdTe buried homojunction ITO react. dep. on SX p-CdTe	0.89	20	13.4	Nakazawa *et al.*, 1987
n-CdS/p-CdTe buried homojunction both layers by screen printing (all thin-film cell)	0.75	22	12.8	Uda *et al.*, 1982 Nakayama *et al.*, 1980 Suyama *et al.*, 1987 Matsumoto *et al.*, 1984
n-CdS/p-CdTe buried homojunction CdS by CVD on SX p-CdTe	0.67	20	11.7	Yamaguchi *et al.*, 1976, 1977
n-CdS/p-CdTe heterojunction both layers by electrodeposition (all thin-film cell)	0.76	22.0	11.0	Kim *et al.*, 1994
CdTe homojunction p-CdTe by CSVT on large-grain n-CdTe	0.82	21	10.7	Mimila-Arroyo *et al.*, 1979 Barbe *et al.*, 1982
n-CdS/p-Cd$_{0.9}$Hg$_{0.1}$Te heterojunction both layers by electrodeposition (all thin-film cell)	0.62	27	10.6	Basol *et al.*, 1986

Table 5.2. (*Continued*)

Cell	Open-Circuit Voltage, ϕ_{oc}	Short-Circuit Current, J_{sc} mA/cm^2	Solar Efficiency, %	Reference
n-ITO/*p*-CdTe heterojunction *e*-beam evap. ITO on SX *p*-CdTe	0.81	20	10.5	Werthen *et al.*, 1983b
n-CdS/*p*-CdTe heterojunction both layers by CSVT (all thin-film cell)	0.75	17	10.5	Tyan *et al.*, 1982
n-CdS/*p*-CdTe heterojunction *p*-CdTe by CSVT on vac. evap. *n*-CdS (all thin-film cell)	0.75	22	10.5	Chu *et al.*, 1987
n-CdS/*p*-CdTe heterojunction *p*-CdTe electrodep. on SP *n*-CdS (all thin-film cell)	0.74	22	10.4	Meyers, 1987
n-CdS/CdTe/*p*-ZnTe *n–i–p* cell SP CdS, electrodep. CdTe, vac. evap. ZnTe (all thin-film cell)	0.69	22	9.4	Meyers, 1987
n-ZnO/*p*-CdTe heterojunction *n*-ZnO by SP on SX *p*-CdTe	0.54	19.5	8.8	Aranovich, 1980
Au/*n*-CdTe MIS cell by electrodep. (all thin-film cell)	0.72	19	8.7	Fulop *et al.*, 1982
n-ITO/*p*-CdTe buried homojunction *n*-ITO sputtered on SX *p*-CdTe	0.82	14.5	8.0	Courreges *et al.*, 1980

5.2. Surface and Contact Properties

Surface Properties

The effects of CdTe surface preparation on the subsequent properties of Cr/CdTe junctions, and of ITO/CdTe and CdS/CdTe heterojunction solar cells have been extensively investigated with single-crystal p-type CdTe (Patterson and Williams, 1978; Werthen *et al.*, 1983d).

In a study of cleaved, oxidized, etched and heat-treated p-type CdTe surfaces, the chemical compounds present on the surface of CdTe have been investigated by x-ray photoelectron spectroscopy (XPS) after different surface preparations and correlated with the properties of junctions formed by Cr on CdTe (Haering *et al.*, 1983).

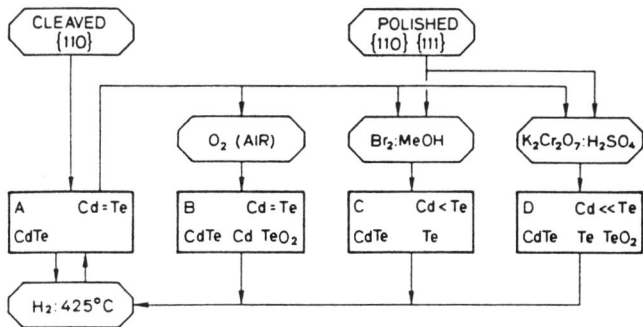

Fig. 5.3. Flowchart summarizing the various CdTe surface treatments and their effects. (Reprinted with permission from J.-P. Haering *et al.*, *J. Vacuum Science and Technology* **A1**, 1469 (1983). Copyright 1983, American Institute of Physics.)

A flowchart summarizing the various CdTe surface treatments and their effects is given in Fig. 5.3. Results can be summarized as follows: (1) cleaving in nitrogen gives a stoichiometric CdTe surface, (2) a 2% Br_2-in-methanol etch (Br:MeOH), or a chromate etch with $K_2Cr_2O_7$:H_2SO_4 (Chr) of {110} or {111} CdTe surfaces results in excess Te, (3) hydrogen heat treatment of Br:MeOH-etched surfaces results in stoichiometric surfaces by removing excess Te, (4) oxidation of the {110} CdTe surfaces is much more rapid than the {111} surfaces, (5) the presence of Te on the Br:MeOH etched surfaces results in larger J_o and reduced ϕ_{oc}, (6) stoichiometric surfaces obtained either by cleaving or by hydrogen heat treatment result in junctions characterized by barrier heights of 0.99 V, and (7) $\phi_{oc} = 0.5$ V is obtainable with Cr/CdTe junctions formed on stoichiometric CdTe surfaces. Figure 5.4(a) shows the

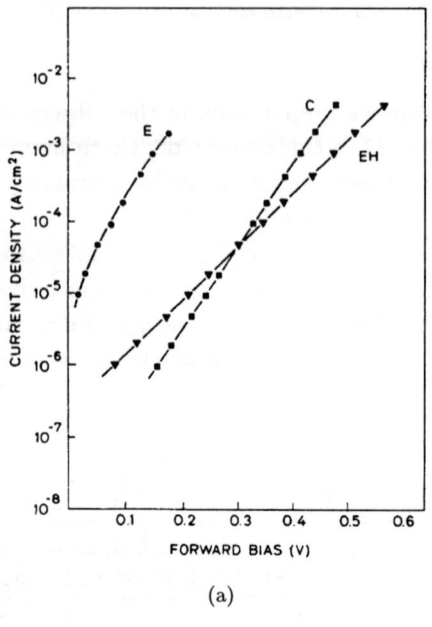

(a)

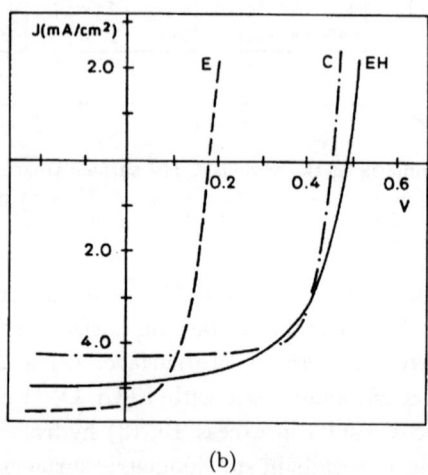

(b)

Fig. 5.4. (a) Dark J–V characteristics of Cr/CdTe junctions formed on {110} surfaces: cleaved (C), Br:MeOH-etched (E), and Br:MeOH-etched plus hydrogen heat treated (EH). (b) Light J–V characteristics of Cr/CdTe junctions formed on {110} surfaces: cleaved (C), Br:MeOH-etched (E), and Br:MeOH-etched plus hydrogen heat treated (EH). (Reprinted with permission from J.-P. Haering *et al.*, *J. Vacuum Science and Technology* **A1**, 1469 (1983). Copyright 1983, American Institute of Physics.)

dark current-voltage characteristics of Cr/CdTe junctions formed on {110} surfaces, and Fig. 5.4(b) shows the light current-voltage characteristics for the same junctions. Silberman *et al.* (1983) showed that CdTe cleaved in vacuum does not oxidize on exposure to dry oxygen unless light or electrons are present.

Surface photovoltage measurements of Br:MeOH-etched single-crystal surfaces indicated that the Fermi level was pinned due to the presence of the etch-induced excess Te layer (Werthen *et al.*, 1983c). Metal/CdTe junctions formed on such surfaces show low barriers and no dependence on metal work function. Stoichiometric surfaces obtained either by cleaving or by hydrogen heat treatment after etching result in metal/CdTe junctions characterized by large barriers that depend on metal work function. The dependence of light current-voltage curves on Cr/CdTe junctions on etching, heat treatment after etching, and crystal orientation is illustrated in Fig. 5.5.

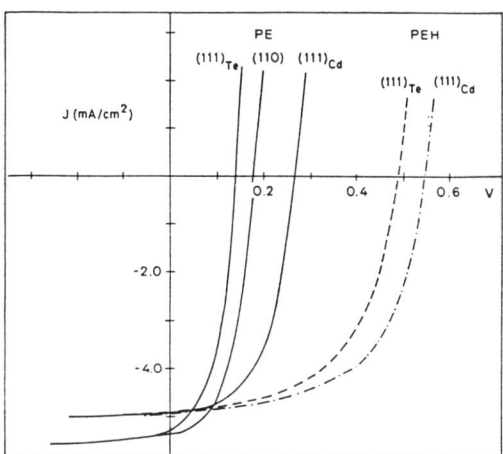

Fig. 5.5. Light J–V characteristics of Cr/CdTe junctions formed on polished and Br:MeOH etched (PE), and polished, Br:MeOH etched and heat treated (PEH) surfaces of different orientations. (Reprinted with permission from J. G. Werthen *et al.*, *J. Appl. Phys.* **54**, 5982 (1983). Copyright 1983 American Institute of Physics.)

These surface properties were investigated further through the characterization of ITO/CdTe and CdS/CdTe heterojunction solar cells formed by electron-beam evaporation (EBE) of ITO and CdS onto single crystal p-CdTe (Werthen *et al.*, 1983b). ITO/CdTe cells prepared on polished and Br:MeOH etched surfaces showed solar efficiencies of about 10%, with major losses attributable to reduced fill factors and ϕ_{oc} because of the non-ideal nature of the interface.

ITO/CdTe junctions prepared on cleaved as well as heat-treated surfaces show poor ϕ_{oc} because of large J_o values induced by a necessary air-heat treatment for ITO transparency. CdS/CdTe heterojunctions also show a strong dependence on CdTe surface condition, but with less influence by the junction formation process, with solar efficiencies of 7.5% being achieved on a Br:MeOH etched and heat-treated surface. Consistent with the results obtained from metal/CdTe junctions, polished and Br:MeOH etched surfaces result in poor ϕ_{oc} values, whereas cleaved, as well as polished, Br:MeOH etched, and heat-treated surfaces result in larger ϕ_{oc} values. It was found that ϕ_{oc} increased with decreasing hole density in the CdTe, and values of ϕ_{oc} as high as 0.81 V were observed for a junction prepared on CdTe with $p = 1 \times 10^{15}$ cm^{-3}, together with $J_{sc} = 19$ mA/cm^2 and a fill factor of 0.55, to give an efficiency of 10.5%, with no anti-reflection coating.

Heat treatment of p-type CdTe:P single crystals in hydrogen at temperatures between 310°C and 475°C decreases the hole density near the CdTe surface by as much as a factor of 25, apparently due to the out-diffusion of acceptors, and produces an approximately exponential acceptor density profile (Nishimura and Bube, 1985).

Surface Oxidation

A high-resolution electron microscopy study of a CdTe single crystal, after being stored in air at room temperature under room illumination for a period of abut 30 days, showed the existence of a 60 nm thick TeO$_2$ layer on the {110} surface (Ponce et al., 1981).

The effect of this surface oxide layer on the properties of Cr/CdTe junctions formed on cleaved surfaces exposed to air for different lengths of time was observed (Werthen et al., 1983a). An increase in open-circuit voltage observed with length of time the cleaved surface was exposed to air was attributed to the presence of the TeO$_2$ on the air-exposed surface.

Thermal oxidation of single-crystal p-type CdTe was carried out in dry and wet oxygen at temperatures between 350°C and 500°C (Wang et al., 1987). The growth rate was increased by about a factor of 2 by changing from dry to wet oxygen, and was slightly higher on (111) Cd and (111) Te faces than on (110) faces. The thickness of the oxide varied as the square-root of the oxidation time, implying a diffusion controlled process with an activation energy of 1.2 eV for thermal oxidation. Examination of the oxide by Auger electron spectroscopy, x-ray photoelectron spectroscopy, and transmission electron microscopy showed that the composition of thick layers is CdTeO$_3$. A related study of the solid-state quaternary phase equilibirum diagram for the Hg-Cd-Te-O system, including ternary diagrams for Cd-Te-O and Hg-Te-O, indicated that $CdTeO_3$ is the first oxide to form and remains stable (Rhiger and Kvaas, 1983).

Cr/CdTe and $Cr/CdTeO_3/CdTe$ (MIS) junctions, involving the thermal oxide $CdTeO_3$, were fabricated on both p- and n-type CdTe single crystals by oxidation of single crystal CdTe after polishing, Br:MeOH etching, and hydrogen heat treatment (Wang *et al.*, 1989). For n-type CdTe the oxide increases the open-circuit voltage (to 0.71 V for a 1 nm thick oxide layer) over that of the Schottky barrier (0.39 V), whereas for p-type CdTe the oxide decreases the open-circuit voltage (from 0.47 to 0.42 V). Postulated energy-band diagrams for the two types of junction are given in Fig. 5.6(a) for the

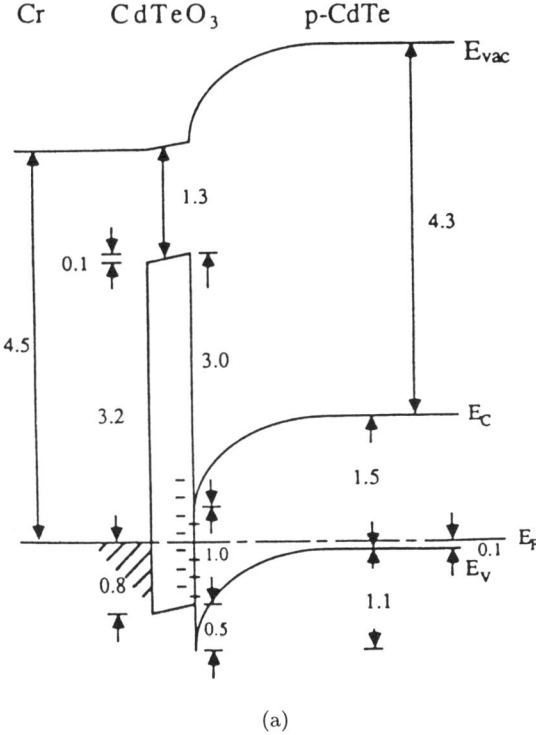

(a)

Fig. 5.6. (a) Energy band diagram for a Cr MIS Junction on p-CdTe crystal with native $CdTeO_3$ oxide as the insulator. (b) Energy band diagram for a Au MIS junction on n-CdTe with native $CdTeO_3$ oxide as the insulator. (Reprinted with permission from F. F. Wang *et al.*, *J. Appl. Phys.* **65**, 3552 (1989). Copyright 1989 American Institute of Physics.)

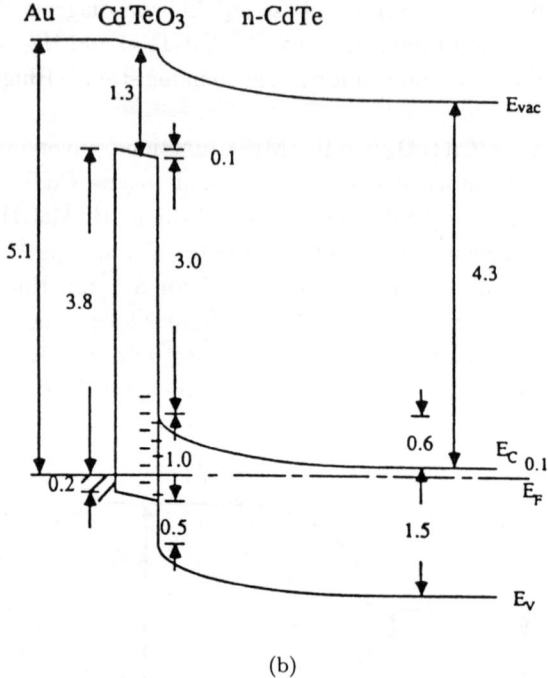

(b)

Fig. 5.6. (*Continued*)

p-CdTe and 5.6(b) for the n-CdTe. In p-CdTe photogenerated electrons are partially blocked at the CdTeO$_3$/p-CdTe interface, become trapped to decrease the positive charge in the oxide, and decrease the barrier height. In n-CdTe the oxide is transparent to photogenerated holes, and larger values of open-circuit voltage than for an MS junction are obtained because majority electrons must tunnel through the oxide.

Contacts

The work function of p-type CdTe is sufficiently large (of the order of 5.7 eV) that no metal has a large enough work function to make an ideal ohmic contact. In addition, the Fermi energy at the surface can be partially pinned by surface states, depending on the surface treatment. The effort to obtain a low-resistance contact with ohmic behavior has therefore focused on obtaining a contact to highly-surface-doped p-CdTe suitable for tunneling transport. Low-resistance contacts to p-CdTe have been reviewed by Ponpon (1985) and Fahrenbruch (1987).

Early work (Gu *et al.*, 1975; Jager and Seipp, 1981) indicated that etching the surface with $K_2Cr_2O_7:H_2SO_4$ provided the basis for a low-resistance contact using Au or Ni-Au. Related results were obtained by Anthony *et al.* (1982) who found that the best contact was a CuAu alloy, prepared by co-evaporating Cu and Au from the same boat with Cu comprising approximately 12 at.% of the evaporant, on a *p*-type CdTe surface etched with $K_2Cr_2O_7:H_2SO_4$ (a CuAu/Chr contact). Auger profiling measurements indicated that there is a marked enhancement of the Te/Cd ratio over a region of about 48 nm from the surface. The presence of Cu in the contact increases the effective doping density in the depletion region. The contact resistivity at room temperature depends on the CdTe bulk resistivity, as shown in Fig. 5.7, but it increases rapidly upon cooling, to a value greater than 10^6 ohm-cm^2 at 125 K. The CuAu/Chr contacts were found to be essentially stable at room temperature, but increased in resistivity by a factor of about four when annealed between 100°C and 200°C. These contacts could be described by a thermally assisted tunneling model (Padovani and Stratton, 1966).

Contact resistivities comparable to those of CuAu/Chr contacts can also be obtained by the use of a material like *p*-ZnTe as the contact material to polycrystalline CdTe in the formation of *p–i–n* junctions as described below

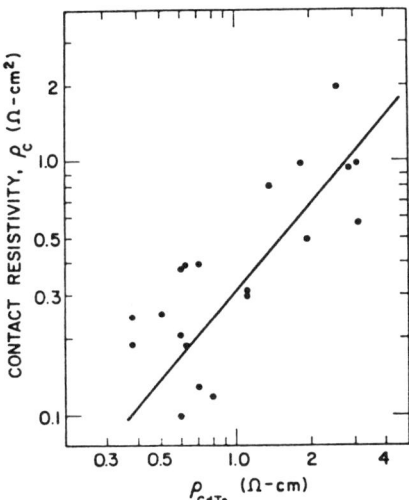

Fig. 5.7. Dependence of contact resistivity on bulk resistivity for CuAu/Chr contacts on *p*-CdTe at room temperature. (Reprinted with permission from T. C. Anthony *et al.*, *J. Electronic Materials*, **11**, 89 (1982), a publication of the Minerals, Metals & Materials Society, Warrendale, Pennsylvania 15086.)

(Meyers, 1989; Nouhi *et al.*, 1989), or of *p*-HgTe as the contact material because of its larger work function than *p*-CdTe (Chu *et al.*, 1988). Contact resistances of the order of 0.4–0.8 ohm-cm^2 were obtained for deposition of HgTe, either by direct combination of the elemental vapors in a gas flow system or by CSVT, onto large-grain polycrystalline CdTe with resistivity of 200–400 ohm-cm. Thin (< 100 nm) ZnTe:Cu films were deposited by an electrochemical method to make contact to the polycrystalline CdTe in a CdTe/CdS/ITO solar cell, giving higher optical transmission than similar vacuum evaporated ZnTe:Cu contacts and leading to an 8.7% efficient cell (Mondal *et al.*, 1991, 1992). Cells made with electrodeposited CdTe and ZnTe:Cu vacuum-evaporated contacts to the CdTe have been reported with an efficiency of 12.9% (Tang *et al.*, 1996).

Reproducible low-resistance contacts of the order of 0.01 ohm-cm^2 at 30 K and 0.15 ohm-cm^2 at 77 K were achieved on Li-diffused single-crystal CdTe, using *n*-butyllithium, with Au as the contact metal (Lee and Bube, 1983). Li doping achieved a p^+ surface layer with $p = 10^{19}$ cm^{-3} that made tunneling currents possible to produce the low contact resistivities observed (from Au to p^+ surface), independent of the bulk resistivity or even the bulk conductivity type. Unfortunately the contact resistivity changes with time, first decreasing and then later increasing, and with temperature, probably involving diffusion of Li acceptors away from the surface and loss of Li acceptors by interaction with lattice defects or by precipitation. Even so, contact resistivities less than 0.03 ohm-cm^2 can be achieved at 300 K for periods of several months.

Subsequent research on these contacts showed that values of contact resistivity as low as 2.1×10^{-3} ohm cm^2 at room temperature and 3.7×10^{-2} ohm cm^2 at liquid N$_2$ temperature can be achieved on Li-diffused CdTe with Au as the contact metal (Lee and Bube, 1985). An MIS tunnel diode model was proposed to account for the observed current-voltage characteristics of the contacts, in view of a surface analysis that indicated that the crystalline structure of the CdTe surface had been greatly disturbed and a Te-rich surface layer produced, and the degradation of the contacts with time is probably caused both by diffusion of Li and by a decrease of interface state density through out-diffusion of excess Te.

A contact process for making low-resistance contacts to polycrystalline *p*-CdTe has been claimed to be compatible with module fabrication needs (McCandless *et al.*, 1994). After CdCl$_2$ processing, the CdTe surface is coated with a thin layer of Cu by electron beam evaporation that is then heated and reacted, diffusing Cu into the CdTe to dope it *p*-type. The reaction removes elemental Cu from the surface, that might lead to chemically-induced instability and produces a conductive surface that can be contacted using different materials. Using CdTe films from a variety of sources, ϕ_{oc} varied from 0.78 to 0.86 V, the fill factor from 0.71 to 0.77, and the efficiency from 11.0 to 14.1%. A related process involving the deposition of a very thin $Cu_x Te$ layer on the CdTe surface has been described by Florez *et al.* (1996).

Although most polycrystalline-based devices are made with contacts that involve Cu in some way, there are a number of alternatives. Ghosh *et al.* (1995) reported making contacts with resistivities of about 0.1 ohm-cm^2 by depositing electroless Ni containing P onto a previously prepared surface.

5.3. Doping of *p*-type CdTe

A number of different techniques have been applied to the specific problem of increasing and controlling the bulk *p*-type conductivity of single crystal CdTe, which are primarily the subject of this section. We consider examples of ion implantation doping, CVD and MOCVD doping, photon-assisted doping, doping from a Cd_3As_2 source, and ion-assisted doping. In polycrystalline materials there is considerable compensation of incorporated dopants by grain boundary states, so that the effective carrier densities can be orders of magnitude lower then the doping density.

Ion Implantation Doping

Doping of *n*-type CdTe single crystals with conversion from *n*- to *p*-type by standard ion-implantation of As$^+$ and P$^+$ has been investigated in some detail (Chu *et al.*, 1980). Photovoltaic CdTe homojunctions were made by implantation of As$^+$ ions into *n*-type undoped Cd-annealed CdTe crystals (Chu *et al.*, 1978). Controlled annealing of the implanted samples with Cd overpressure improved the doping efficiency. High values of $\phi_{oc} = 0.84$ V were achieved, but with low quantum efficiency resulting in a cell efficiency of only 3.0%, probably associated with low electron diffusion lengths in the *p*-type implanted layer.

CVD and MOCVD Doping

Chu *et al.* (1985) reported doping of *p*-type CdTe multicrystalline films deposited on non-crystalline substrates from the direct combination of the elements in a CVD system using PH$_3$ or AsH$_3$, yielding a maximum hole density of about 6×10^{15} cm^{-3}. Ghandhi *et al.* (1987) obtained hole densities of 2×10^{17} cm^{-3} in epitaxial films grown on single-crystal CdTe substrates at 350°C using metalorganic chemical vapor deposition (MOCVD) growth with AsH$_3$ as the dopant.

Photon-Assisted Doping (PAD)

Beginning in 1986, Schetzina and his co-workers pioneered a new technique that they called Photo-Assisted Molecular Beam Epitaxy (PAMBE) in which controlled doping of single crystal CdTe grown by MBE was obtained by illumination of the growing films ($\lambda = 0.5$ μm at 100 mW/cm^2) and co-evaporation of a dopant (Bicknell *et al.*, 1986, 1987, Hwang *et al.*, 1988, Fahrenbruch *et al.*, 1992). They reported highly doped n- and p-type CdTe films, using In for n-type and Sb or As for p-type. The highest hole density reported was for As-doped films with $p = 6.2 \times 10^{18}$ cm^{-3}. Films grown by this technique were also reported to have better crystalline quality than those grown without illumination. The general assumption was that illumination supplies additional kinetic energy to the growth surface and enables enhancement of specific chemical reactions and surface reactions (Benz *et al.*, 1990). In subsequent years the only research groups that reported successful PAD results with CdTe are Schetzina's group at North Carolina State University, and two groups at other laboratories which include his former students (Myers, 1989; Harris *et al.*, 1990, 1991; Bicknell-Tassius *et al.*, 1989, 1990). There appears to be agreement among other workers that the use of the PAMBE technique to obtain CdTe and HgCdTe layers yields an improvement in the structural quality of the films with illumination during growth, but has little if any influence on doping (Koestner *et al.*, 1989; Arias *et al.*, 1990; Harris *et al.*, 1990; Tiwari *et al.*, 1991).

Several results indicate that Cd/Te flux ratios greater than unity during CdTe film growth facilitate dopant incorporation and activation (Harper *et al.*, 1989; Bicknell-Tassius *et al.*, 1990; Wu *et al.*, 1991; Arias *et al.*, 1990).

A consensus of workers in the field seems to favor the opinion that the effect of photons is to increase the desorption rate of Te (thus creating Te vacancies) and to enhance the surface mobility of atoms at the growing surface (Fahrenbruch *et al.*, 1992). Arsenic is the most widely used dopant for growth of p-CdTe by PAMBE.

Doping from a Cd_3As_2 Source

Co-evaporation of Cd_3As_2 during epitaxial film growth by vacuum evaporation on single crystal substrates enabled incorporation of electrically active As into CdTe films, giving hole densities up to 2×10^{16} cm^{-3}, whereas co-evaporation of elemental Cd or As$_4$ did not (Fahrenbruch *et al.*, 1992). In general, doping p-CdTe and p-ZnTe films using compound dopants sources such as Zn$_3$As$_2$

and Cd_3As_2 has produced some interesting results. Incorporation of As into MBE-grown ZnSe films using Zn_3As_2 as the As source (Shibli *et al.*, 1990; Turco-Sandroff *et al.*, 1991), and deposition of a mixture of Cd_3As_2 + As onto the surface of a CdTe film grown by CSVT followed by indiffusion has been used to fabricate *p–n* junction solar cells (Cohen-Solal, 1985).

Ion-Assisted Doping (IAD)

In the IAD process ionized and accelerated dopant beams (usually of energy less than 100 eV) are directed onto a film as it is being grown from the vapor. The technique has been successfully used to control the dopant incorporation in MBE-grown Si and GaAs (Greene *et al.*, 1985). IAD with phosphorus impurity to produce *p*-type epitaxial CdTe films by vacuum evaporation on single crystal substrates has been investigated as a function of ion dose, ion energy, growth temperature, and growth rate (Sharps *et al.*, 1990), and an overview given by Fahrenbruch *et al.* (1992). A diagram of the apparatus used is given in Fig. 5.8. The system was designed so that the CdTe effusion cell and the ion source were simultaneously directed onto the substrate.

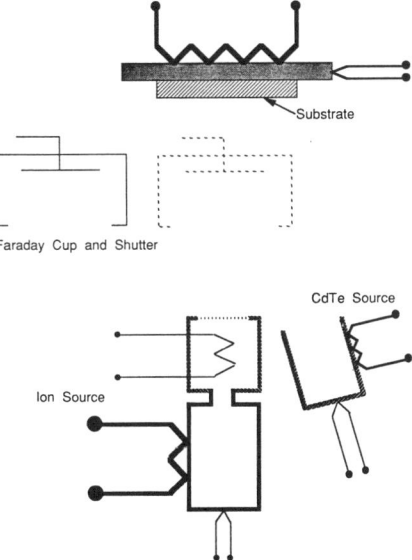

Fig. 5.8. Schematic diagram of the ion-assisted doping system. (Reprinted with permission from P. Sharps *et al.*, *J. Appl. Phys.* **68**, 6406 (1990). Copyright 1990 American Institute of Physics.)

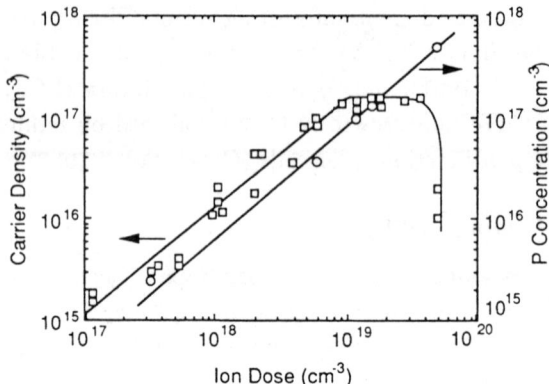

Fig. 5.9. Comparison of hole density induced in CdTe by IAD, and SIMS measurements on incorporated phosphorus, as a function of ion dose. Squares refer to the hole density data, and circles refer to the SIMS data. (Reprinted with permission from P. Sharps *et al., J. Appl. Phys.* **68**, 6406 (1990). Copyright 1990 American Institute of Physics.)

The variation of hole density and the density of incorporated P by SIMS measurements are compared in Fig. 5.9 as a function of ion dose. The growth temperature was 400°C, the ion energy was 60 eV, the growth rate was 10 μm/h, the growth time was 1 hour, and the substrates were single crystal p-type CdTe. A linear dependence of hole density on ion dose is observed until saturation occurs at a value of $p = 2 \times 10^{17}$ cm^{-3}; for higher ion doses, the hole density remains independent of ion dose until a dose of about 4×10^{19} cm^{-3}, after which a further increase in ion dose leads to a decrease in the measured hole density. The maximum hole density corresponds well with the maximum hole density in p-type CdTe single crystals observed by Selim and Kroger (1977), for example, who found that introducing P at levels of up to 5×10^{19} cm^{-3} into single-crystal CdTe produced a hole density of only 1×10^{17} cm^{-3}. The parallel shift of the curves in Fig. 5.9 is probably due to an overestimate of the electrical activity of the P in the "standard" Bridgman-grown single-crystal sample of CdTe:P. Overall the data suggest that the electrical activity of the incorporated P is approximately unity for the positively sloped portion of the curve.

The decrease in hole density for the highest ion doses is possibly due to more compensating damage occurring than can be annealed out at the growth temperature. An investigation of the effects of ion bombardment of p-CdTe with Ar ions showed that the threshold for damage occurs at an accelerating voltage of slightly less than 100 V, and that bombardment by Ar ions at higher voltages results in the formation of an *n*-type surface layer on the *p*-type CdTe crystals (Chien *et al.*, 1988).

An investigation of the effects of growth variables indicated that for the growth conditions considered, the hole density was independent of growth temperature between 350°C and 500°C, independent of the ion energy over the range from 60–300 eV, and decreased with decreasing growth rate at a given ion dose (Sharps *et al.*, 1990). The films showed the same hole mobilities as for high quality Bridgman-grown single crystals doped with P to about the same level, and the hole lifetime for the films was about an order of magnitude larger than that obtained for a P-doped single crystal. Unfortunately, however, the electron minority-carrier lifetime was between three and ten times smaller than the minority carrier lifetime in single crystals. When the ion beam current was increased, the minority carrier lifetime decreased by another order of magnitude, indicating that the IAD process itself is responsible for an increased density of recombination centers. If these layers were to be used in a CdS/CdTe solar cell, it would be necessary to have a two-layer structure at the CdTe surface, with an IAD CdTe contact layer below an undoped light-absorbing CdTe layer making contact with the CdS top layer. Such heterojunctions did not exhibit the same decrease in J_{sc} seen with a heterojunction prepared with doped CdTe directly at the junction interface with CdS.

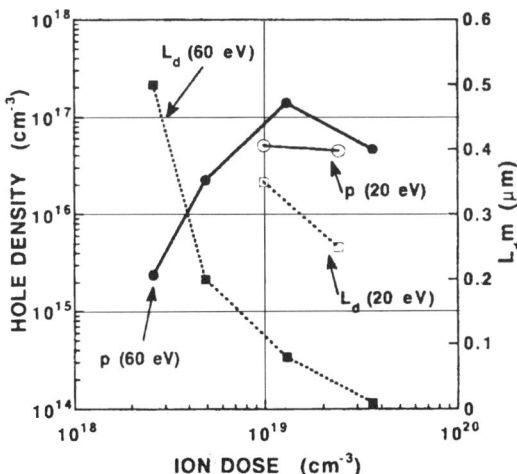

Fig. 5.10. Hole density and minority carrier diffusion length versus ion dose for IAD *p*-CdTe:P at 60 eV, compared to 20 eV with electron irradiation during growth. (Reprinted with permission from D. Kim *et al.*, *J. Appl. Phys.* **75**, 2673 (1994). Copyright 1994 American Institute of Physics.)

An investigation of the nature of the ion damage associated with IAD on single crystals revealed the existence of such damage through measurements of transmission electron microscopy, etch-pit density, and minority-carrier diffusion length, and showed that the ion damage depends on both the ion dose and the ion energy (Kim *et al.*, 1994). Although reducing the ion energy below 60 eV results by itself in lower doping densities, the use of simultaneous electron irradiation and Cd overpressure during deposition make it possible to achieve equivalent doping levels for 20 eV ions while reducing the ion damage. Figure 5.10 shows the dependence of hole density and minority carrier diffusion length on ion dose for both an ion energy of 60 eV, and for an ion energy of 20 eV with simultaneous electron irradiation and Cd overpressure. It is proposed that simultaneous electron irradiation makes the surface Cd-stabilized (Singh and Arias, 1989; Wu *et al.*, 1991), and creates free electron-hole pairs at the growing surface that change the Cd–Cd bond strength and enhance Cd kinetics at the surface (Arias *et al.*, 1990). Figure 5.10 shows that the minority-carrier diffusion length is three times larger when using 20 eV ions combined with electron irradiation, compared to using 60 eV ions alone, in achieving approximately the same hole density.

A second way to control the microstoichiometry of the CdTe surface is to vary the Cd/Te flux ratio by applying a Cd overpressure. By using excess Cd corresponding to Cd/Te = 1.002, together with 20 eV P ions and electron irradiation, p-CdTe:P films with $p = 1 \times 10^{17}$ cm^{-3} and $L_d = 0.35$ μm were obtained (Kim *et al.*, 1994). A CdS/CdTe heterojunction with a 20 eV IAD CdTe film showed an efficiency almost three times greater than a heterojunction made with a 60 eV IAD CdTe film.

Transport properties of IAD CdTe films deposited with either 60 eV or 20 eV were measured using the Hall effect between 8 K and 400 K, and were compared with single crystal values (Moesslein *et al.*, 1993). The hole mobility in the films was essentially the same as in single crystals, and showed the same type of temperature dependence corresponding to ionized impurity scattering at lower temperatures and polar mode scattering at higher temperatures.

5.4. Post-Deposition Processing and Interface Properties

One of the keys to optimizing solar cells based on polycrystalline CdTe is a postdeposition heat treatment to improve the properties of the junction interface (Basol *et al.*, 1985; Meyers, 1989; Basol, 1992; McCandless and Birkmire, 1991; Birkmire *et al.*, 1992; Birkmire and Meyers, 1994; Ferekides *et al.*, 1994). Such heat treatment is usually carried out at temperatures greater than 400°C and in a medium including $CdCl_2$ and/or O_2. The heat treatment causes CdTe grains to regrow and become larger, and decreases the defect density at the junction interface. The p-type character of CdTe films appears to be enhanced upon annealing in O_2 at elevated temperatures, although the exact mechanism is not yet understood. The importance of this processing step led to the general recipe: "The golden rule for making CdTe devices is to deposit cadmium and tellurium as inexpensively as possible and then to heat-treat the devices to activate them." (Zweibel and Barnett, 1993). The equivalent sentiment is pictured cartoon-style in Fig. 5.11. In some cases these treatments can be made an integral part of the film growth process itself so that a post-deposition step is not necessary (Mitchell *et al.*, 1985b; Basol, 1992).

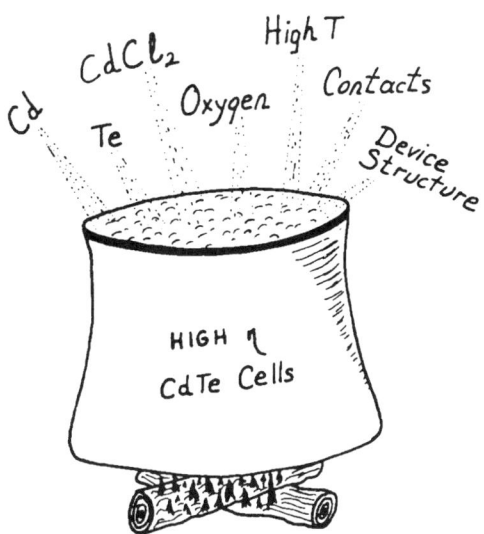

Fig. 5.11. A general recipe for the preparation of high efficiency thin film CdTe solar cells. (Reprinted with permission from B. M. Basol, *International Journal of Solar Energy*, **12**, 25 (1992). Copyright 1992, Harwood Academic Publisher, Langhorne, PA.)

The following design rules for preparing efficient CdS/CdTe solar cells can be summarized (Basol, 1992). (1) The CdTe films should be deposited by any one of several suitable procedures onto a transparent CdS/TCO/glass substrate. Here TCO stands for transparent conducting coating, e.g. SnO_2 or ITO. (2) The thickness of the CdS layer is made as thin as possible to improve the short wavelength cell response. The thickness can easily be reduced to below 0.2 μm using CdS deposition processes such as solution growth, electroplating, and CSVT. (3) The CdTe layer is chemically treated to produce a low-resistivity p-type Te-rich surface suitable for a tunneling contact with typical contact materials graphite or Ni. A p-type dopant such as Cu is usually introduced at the contact/CdTe interface with an annealing step before the contact deposition. (4) Finally a heat treatment involving O_2 and $CdCl_2$ is an important ingredient in the fabrication process, promoting crystal growth and the p-type character of the CdTe film, as mentioned above.

The process by which an n-type, as-deposited CdTe film is converted into p-type by heat treatment in air has been named the type-conversion/junction-formation (TCJF) process. The application of the TCJF process to CdTe and HgCdTe films produced by electrodeposition has also been described (Basol, 1984, 1988).

A two-stage deposition process has also been proposed as a possible method for CdTe/CdS solar cell fabrication (Basol and Kapur, 1990; Bhatti *et al.*, 1992; Basol, 1992). This involves CdTe film growth (1) by deposition of alternating Cd and Te layers onto a suitable substrate by such techniques as electrodeposition, vacuum evaporation, or sputtering, followed by (2) reaction of these stacked layers to form the compound after introducing varying amounts of $CdCl_2$ into the unreacted stacked layers to promote grain growth.

Several investigations have led to the conclusion that a $CdS_{1-x}Te_x$ alloy is formed at the CdTe/CdS interface due to interdiffusion that takes place during fabrication, a process that is enhanced by the presence of O_2 (Clemminck *et al.*, 1991; Birkmire *et al.*, 1992; Mao *et al.*, 1995; Ferekides *et al.*, 1996). It is believed that the formation of this interfacial layer reduces structural defects due to lattice mismatch between the semiconductors and leads to a lower defect density in the critical region near the interface. The extent of interdiffusion and the value of x obtained in a particular case, depend on the deposition temperature of CdTe and the $CdCl_2$ present during subsequent heat treatment conditions. The tendency of O_2 during fabrication to produce small-grain CdS films is consistent with an enhancement of Te diffusion into CdS. The net effect is to produce CdTe/CdS cells with higher ϕ_{oc}'s, fill factors, and higher cell efficiencies.

The effect of reducing the thickness of the CdS film in polycrystalline CdS/CdTe solar cells fabricated by vacuum evaporation, with subsequent treatment with $CdCl_2$ and a heat treatment in air, has been investigated (McCandless and Hegedus, 1991; Granata *et al.*, 1996). Spectral response measurements indicate that CdS and CdTe interdiffusion, corresponding to the formation of both S-rich and Te-rich $CdTe_{1-x}S_x$ regions, increases as the CdS thickness decreases from 1.5 to 0.12 μm. When the CdS thickness is < 0.1 μm, the spectral response indicates the complete disappearance of a distinct CdS layer, and the subsequent formation of a $TCO/CdTe_{1-x}S_x$ junction. In general, the TCO/CdS_xTe_{1-x} junction is inferior to that involving CdS, so shunt currents are increased as CdS thickness is decreased. There are some exceptions (Mitchell *et al.*, 1985b; Nakazawa *et al.*, 1987), suggesting that with proper treatment of the TCO (e.g. formation of an insulating layer of TCO at the interface) ϕ_{oc} can be maintained for very thin or non-existent CdS. A related investigation of the effect of CdS thickness on CdS/CdTe quantum efficiency suggests that the difference between thick and thin CdS might well be as much as 12% versus 15% in device efficiency (Granata *et al.*, 1996).

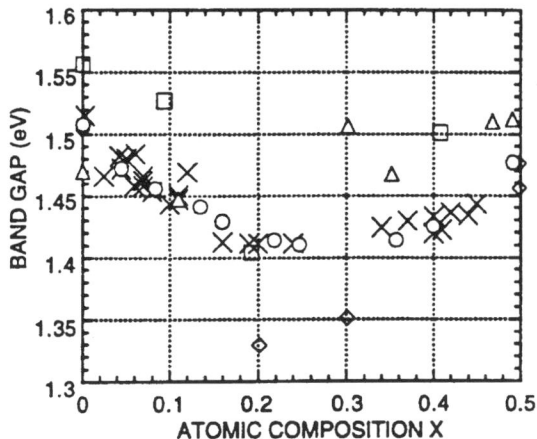

Fig. 5.12. Measured band gaps of as-deposited $CdTe_{1-x}S_x$ thin films as a function of x, as reported in this work (X), Bonnet 1970 (triangles), Hill and Richardson 1973 (diamonds), Ohata *et al.*, 1973 (circles), and Al-Ani *et al.*, 1993 (squares). (Reprinted with permission from D. G. Jensen *et al.*, in *Thin Films for Photovoltaic and Related Device Applications*, eds. D. Ginley, A. Catalano, H. W. Schock, C. Eberspacher, T. M. Peterson and T. Wada, Materials Research Society Proc. **426**, Pittsburgh, PA (1996a), p. 327.)

Predominantly single-phase $CdTe_{1-x}S_x$ films with $0 \leq x \leq 0.45$ were grown by vacuum co-evaporation of CdS and CdTe, with the dependence of band gap on x shown in Fig. 5.12 (Jensen *et al.*, 1996a,b). Phase segregation was promoted by heat treatment of the films at 415°C in the presence of $CdCl_2$, and the solubility limit of S in CdTe was determined to be 5.8%. $CdTe_{1-x}S_x/CdS$ solar cells made with compositionally uniform $CdTe_{1-x}S_x$ layers with x near the solubility limit, gave an efficiency of 10.8% and showed cell characteristics very much like those of a conventional CdTe/CdS cell where the alloy is formed by cell processing.

Transmission electron microscopy (TEM) was used to investigate the microstructures and interfaces of thin film CdTe/CdS/ITO/glass photovoltaic devices, particularly as a result of annealing in air after prior exposure to $CdCl_2$ (Loginov *et al.*, 1996). The cells used were prepared by vacuum evaporation of CdS onto an ITO/glass substrate, followed by CSVT deposition of CdTe. It is concluded that a $CdCl_2$ treatment reduces the influence of interfacial states on the current transport mechanism (changing from tunneling to thermally activated), causes grain growth in the CdS and CdTe, and decreases the density of stacking faults and dislocations in the CdTe. It also introduces Cl-rich precipitates into the lattice, but these have not been observed in the region near the interface, and their effect on device performance remains unknown.

TEM indicates that the CdTe grains grow epitaxially on the CdS grains during deposition and that heat treatment causes substantial grain growth in the CdTe, such that the epitaxial relationship is lost and one CdTe grain may bridge several differently oriented CdS grains (Al-Jassim *et al.*, 1993).

An investigation of the micro through nanostructure properties of polycrystalline CdTe using atomic force microscopy (AFM) has shown correlations with processing and electronic structures for several standard deposition techniques such as vacuum evaporation, CSVT, and sputtering (Levi *et al.*, 1994). Correlations between nanostructure and optical properties indicate that $CdCl_2$ heat treatment passivates defects. Except for CSVT-deposited films, the $CdCl_2$ treatment also enhances grain size. Diffusion of sulfur across the CdTe/CdS interface during heat treatment has been verified.

The $CdCl_2$ anneal of CdTe/CdS solar cells usually consists of coating the CdTe surface with a saturated solution of $CdCl_2$ in methanol and heat treating the film in air at 400°C to 450°C for 10 to 30 minutes. An alternate approach using a $CdCl_2$ vapor treatment has been developed that results in uniform large grain films, preferred (111) crystallographic texture, less chloride surface residue, less S diffusion into the CdTe, and more reproducible current-voltage performance than is obtained with other $CdCl_2$ treatments (Zhou *et al.*, 1994; Birkmire *et al.*, 1995; McCandless *et al.*, 1996). In the vapor treatment, samples were annealed at 425°C for 20 minutes in air with a partial pressure of $CdCl_2$ of 0.04 Torr. Vapor treated films do not require a rinse step prior to device completion, unlike samples that are $CdCl_2$:methanol treated. Solar cells prepared with the $CdCl_2$ vapor treatment were comparable to the best CdTe cells: $\phi_{oc} = 0.83$ V, $J_{sc} = 19.6$ mA/cm^2, and $ff = 0.65$. Similar microstructural changes were observed for HCl vapor treatment of films (Qu *et al.*, 1996).

Interdiffusion between CdS and CdTe during cell preparation has been deliberately limited by a process in which the CdS film on a glass substrate is treated with ZnCl$_2$ and CdCl$_2$ to reduce the reactivity of the CdS when CdTe is subsequently deposited on top of the CdS by close-spaced sublimation (Romeo *et al.*, 1997). Resulting cells showed $\phi_{oc} = 0.845$ V, $J_{sc} = 23$ mA/cm^2, $ff = 0.71$, and efficiency of 13.8%.

The need to evaluate the uniformity of the microstructure of CdS/CdTe large-area solar cells has led to application of the non-destructive testing method of time-resolved photoluminescence to map recombination lifetimes in such cells (Ahrenkiel, 1992; Ahrenkiel *et al.*, 1994). Typical lifetime profiles indicated spatial variation by factors of two to three across 1 cm dimensions, and that the open-circuit voltage of the cells was controlled by the area with the minimum lifetime.

5.5. SIS Junctions

One possible solution to the problems posed by difficulties with doping and contacts on p-CdTe has been proposed by Meyers (1989). The solar cell consists of a p–i–n heterojunction made up of n-CdS/i-CdTe/p-ZnTe/Ni, as shown in Fig. 5.13. The n-CdS layer is deposited on SnO$_2$-coated glass, the CdTe

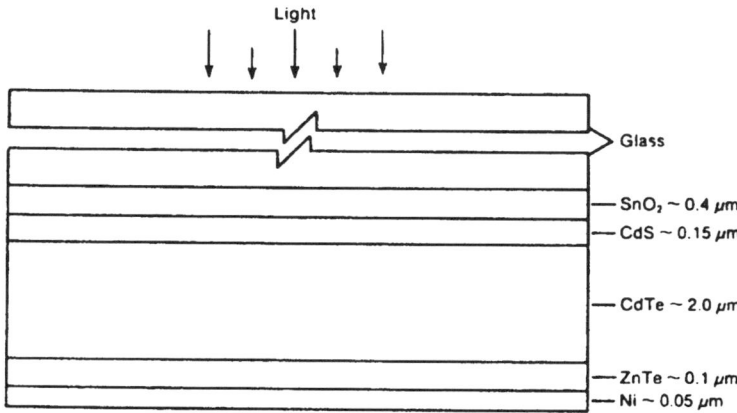

Fig. 5.13. Solar cell structure for an 11% CdTe n–i–p solar cell. At 1 sun, $\phi_{oc} = 0.68$ V, $J_{sc} = 20.5$ mA/cm^2, fill factor = 0.625. (Reprinted with permission from P. V. Meyers, *Photovoltaics Advanced Research and Development* (1989), *9th Annual Review Meeting*, SERI/CP-213-3495, p. 19. Copyright 1989, SERI, Golden, CO.)

layer is deposited by electrodeposition on the CdS, and the final ZnTe layer is deposited on the CdTe. Work on this kind of cell has been extended by Nouhi *et al.* (1989) with the use of CdTe films deposited by MOCVD to achieve small-area cells with an efficiency over 9%.

SIS-type cells have also been prepared with polycrystalline ZnCdTe or CdMnTe (Rohatgi *et al.*, 1988, 1989) as the *i*-region material; these devices have a lower ϕ_{oc} than similar devices using CdTe alone.

5.6. Grain Boundary Effects

The effect of grain boundaries on the electrical transport properties of poly-crystalline *n*-type CdTe films doped by coevaporation of In was examined by comparing the properties of polycrystalline films evaporated by HWVE on glass, with those of epitaxial films evaporated by HWVE on BaF$_2$ single-crystal

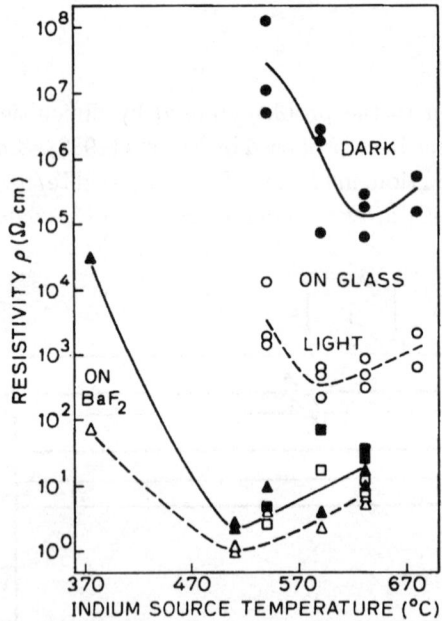

Fig. 5.14. Dark and light resistivity of CdTe:In films grown on glass and on single crystal BaF$_2$ by HWVE, as a function of the indium source temperature. Films on glass in the dark (•) and in the light (○); films on BaF$_2$ grown with $T_{\mathrm{sub}} = 480°$C in the dark (▲) and in the light (△); films on BaF$_2$ grown with $T_{\mathrm{sub}} = 450°$C in the dark (■) and in the light (□). (Reprinted with permission from W. Huber *et al.*, *J. Appl. Phys.* **54**, 4038 (1983). Copyright 1983 American Institute of Physics.)

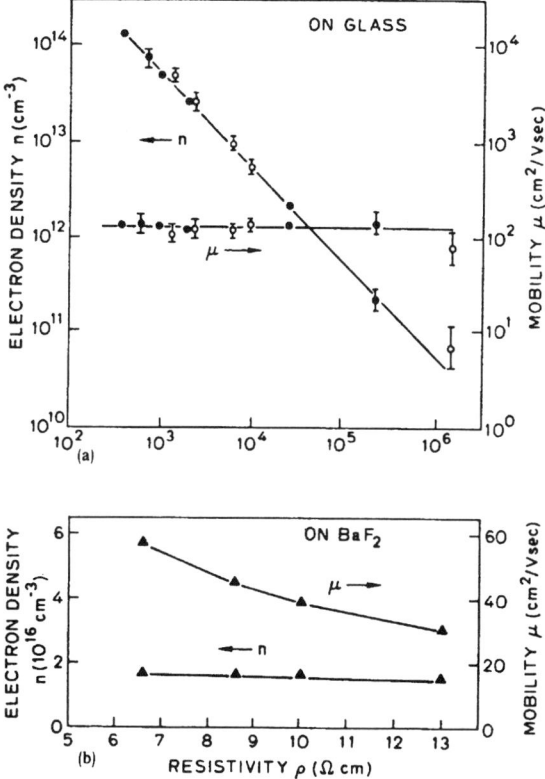

Fig. 5.15. Variation of the electron density and mobility under optical excitation as a function of the resistivity in the light for CdTe films deposited by HWVE (a) on glass substrates, and (b) on a BaF$_2$ substrate. (Reprinted with permission from W. Huber *et al., J. Appl. Phys.* **54**, 4038 (1983). Copyright 1983 American Institute of Physics.)

substrates (Huber *et al.*, 1983). Resistivity and Hall effect measurements could be successfully interpreted in terms of the grain boundary model first proposed by Seto (1975), which is a one-dimensional model including the contribution of the bulk resistivity of the undepleted portion of the grain, and carrier transport across the barrier described by thermionic-diffusion theory (Sze, 1969).

The dependence of the dark and light resistivity of the two types of films on the temperature of an indium source used for doping is shown in Fig. 5.14, and the variation of the electron density and mobility under photoexcitation, measured by photo-Hall effect, with the light resistivity for the two types of films is shown in Fig. 5.15. The dark resistivity of CdTe films deposited on glass is uniformly high. Photoexcitation causes the resistivity to decrease by orders of magnitude by a process in which the electron density is increased but the mobility is unchanged. These effects are consistent with the existence of fully depleted grains in these films (grain size about 1 μm) in the dark, as described by the models of Card and Yang (1977) and Seager (1981).

The dark resistivity of CdTe films deposited on BaF$_2$ single crystal substrates under the same HWVE deposition conditions is orders of magnitude smaller than that for films deposited on glass. Photoexcitation produces a change in electron mobility but not in electron density. These effects are consistent with the existence of partially depleted grains in these much larger-grain films.

The optical and electrical properties of grain boundaries in n- and p-type CdTe have been investigated by analysis of single boundaries within bicrystals and of polycrystalline thin films (Thorpe et al., 1986). Figure 5.16 shows the temperature dependence for conductivity in dark and light of a single grain boundary in a p-type CdTe bicrystal. Grain boundaries control the magnitude

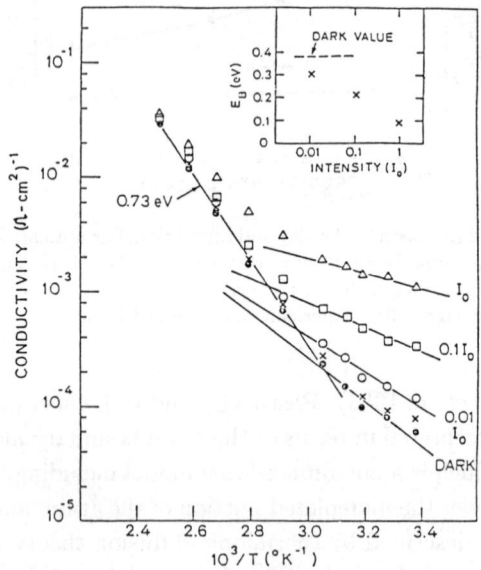

Fig. 5.16. Grain-boundary conductivity as a function of temperature in the dark and for several different illumination intensities ($I_0 = 60$ mW/cm^2 white light) for the grain boundary in a p-type CdTe bicrystal. The inset shows the variation of the activation energy with light intensity at low temperatures. (Reprinted with permission from T. P. Thorpe et al., $J.$ $Appl.$ $Phys.$ **60**, 3622 (1986). Copyright 1986 American Institute of Physics.)

of the mobility of majority carries in polycrystalline CdTe, with resistance values associated with grain boundaries being 3 to 5 orders of magnitude larger than expected for bulk material. At higher temperatures the grain boundary conductivity is not affected by illumination, but at lower temperatures the grain boundary conductivity is thermally activated with an activation energy that decreases with illumination. The results appear to be consistent with an interpretation that proposes thermally-assisted tunneling in the low-temperature range and thermal excitation over the boundary barrier at higher temperatures, as earlier proposed for the temperature dependence of the mobility in PbS (Espevik *et al.*, 1971) and CdS (Wu and Bube, 1974) due to grain boundary effects.

The energy dependence of the grain-boundary state densities was determined for CdTe bicrystals by boundary J–ϕ dependence following the method of Pike and Seager (1979) for states above the valence band, and by spectral response measurement of boundary photocapacitance and/or photoconductivity using the method developed by Werner *et al.* (1982) for states below the

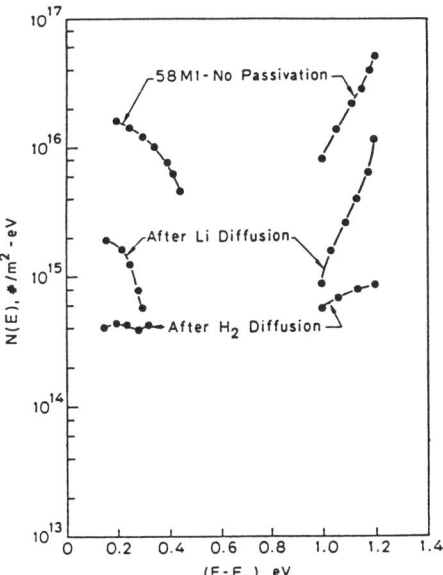

Fig. 5.17. Density of grain-boundary interface states as a function of energy in the forbidden gap of CdTe, for a particular p-type CdTe bicrystal, before any passivation, after passivation by Li diffusion and after passivation by hydrogen heat treatment. (Reprinted with permission from T. P. Thorpe *et al.*, *J. Appl. Phys.* **60**, 3622 (1986). Copyright 1986 American Institute of Physics.)

conduction band (Thorpe *et al.* 1986). Figure 5.17 shows a typical variation of grain boundary state density without any passivation, and the effects of Li diffusion and H_2 heat-treatment passivation. The passivation effects are striking, but unfortunately are not stable for more than a few weeks. Similar experiments with grain boundaries in n-type CdTe showed that in this case passivation could be temporarily achieved by heating in air, not in H_2.

5.7. Film Deposition by Vacuum Evaporation

An n-CdS/p-CdTe solar cell with an efficiency of 7.9% ($\phi_{oc} = 0.63$ V, $J_{sc} = 16.1$ mA/cm^2, $ff = 0.66$, $J_o = 1.7 \times 10^{-8}$ A/cm^2) was prepared by vacuum evaporation (VE) of an n-CdS film on prepared p-CdTe single-crystal substrates (Mitchell *et al.*, 1977b). Optical absorption coefficients of CdTe greater than 10^4 cm^{-1} were calculated from the photovoltaic spectral-response curve

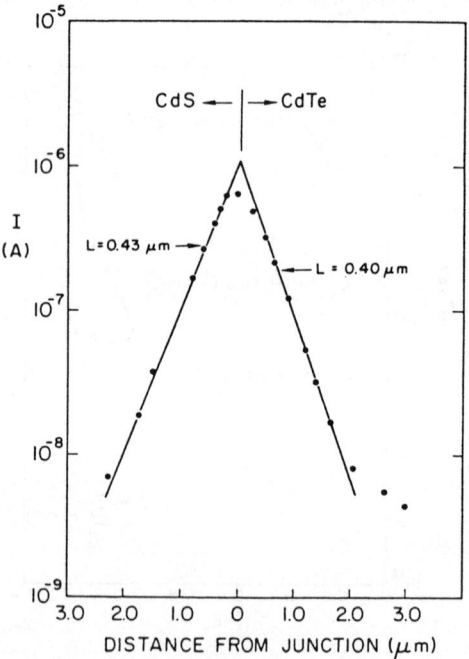

Fig. 5.18. Electron-beam-induced current as a function of beam position for the 7.9% n-CdS/p-CdTe cell. The calculated values of the minority carrier diffusion lengths are 0.40 μm in p-CdTe, and 0.43 μm for n-CdS. (Reprinted with permission from K. W. Mitchell *et al.*, *J. Appl. Phys.* **48**, 4365 (1977). Copyright 1977 American Institute of Physics.)

(Mitchell *et al.*, 1977c), and an analysis of the fill-factor in such cells was carried out (Mitchell *et al.*, 1977a). Sputtered indium-tin-oxide (ITO) was used as a transparent electrode to the CdS, and glycerol was used as an anti-reflection coating. Two CdTe surface treatments were compared (mechanical polishing, and mechanical polishing followed by a bromine-in-methanol (MB) etch), and the effects of different CdTe and CdS resistivity were investigated. The MB-etched cells were significantly better than the mechanically polished cells, giving a saturation current J_o some one to two orders of magnitude smaller, and a quantum efficiency significantly larger. A typical spectral response of the cells is shown in Fig. 1.13 in Chapter 1 of this book, and a plot of electron-beam-induced current (EBIC) for the 7.9% cell is given in Fig. 5.18.

A variation of vacuum evaporation called hot-wall vacuum evaporation (HWVE) is illustrated in Fig. 5.19, similar to that used by Lopez-Otero (1978) and Lopez-Otero and Huber (1979). The apparatus fits inside a standard vacuum system, and consists of a heated quartz liner with three independent furnaces that allow an independent control of the CdTe, dopant, and Cd or Te source materials. The substrate is placed in a fourth furnace on top of the quartz liner. The use of a heated wall confining the region of evaporation almost eliminates waste of material and provides an environment in which the deposited films are closer to equilibrium conditions. Many of the phenomena encountered in depositing CdTe films are similar using CSVT or HWVE. Investigation of grain boundary effects using CdTe films deposited by HWVE is described above (Huber *et al.*, 1983).

Polycrystalline CdTe films were deposited by HWVE on graphite substrates, and single-crystal CdTe films were deposited by HWVE on single-crystal *p*-CdTe:P substrates (Fortmann *et al.*, 1987). CdS/CdTe heterojunctions were formed by the vacuum evaporation of CdS. It was observed that the ϕ_{oc} was a maximum at about 0.80 V for both types of cells when the carrier densities in the *n*-CdS and *p*-CdTe were approximately equal. The effects were successfully modeled in terms of an unidentified level lying 0.45 eV below the bottom of the conduction band in CdTe.

Difficulties in doping CdTe films using a doped source, such as are described below for CSVT film deposition (Anthony *et al.*, 1985), were also encountered with attempts to dope CdTe using HWVE (Huber *et al.*, 1983; Anthony *et al.*, 1984; Bube *et al.*, 1984). Attempts were made to dope CdTe films by co-evaporation of such dopants as As, Sb, Na, and Ag, but all attempts to dope *p*-type CdTe films during HWVE deposition were unsuccessful. Likely causes include: (1) small sticking coefficients for the dopants on the growing surface

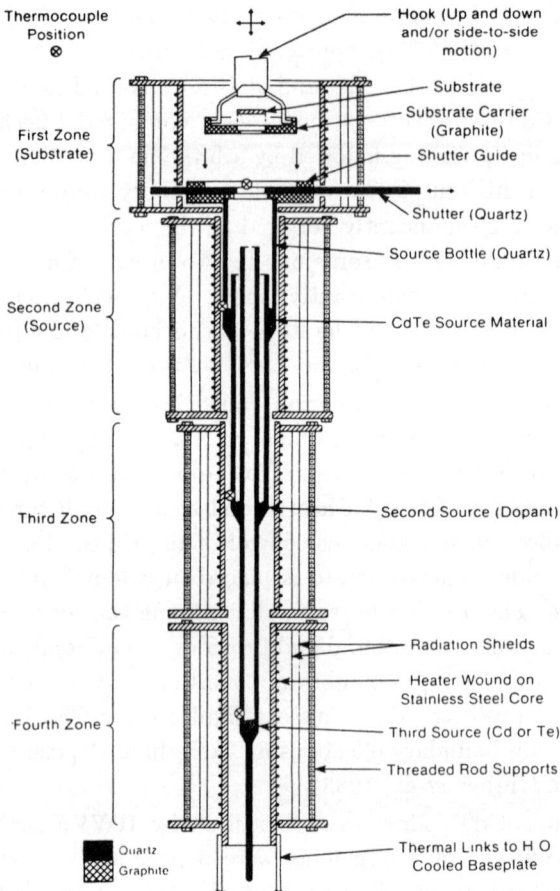

Fig. 5.19. Schematic diagram of the hot-wall vacuum evaporation (HWVE) apparatus used to grow CdTe thin films. The chamber pictured is enclosed in a vacuum system. (Reprinted with permission from R. H. Bube *et al., IEEE Transactions on Electron Devices* **ED-31**, 528 (1984). Copyright 1984, IEEE, NY.)

of CdTe at elevated temperatures, (2) evaporation of many of these dopants as molecular species such as As_4 or Sb_2 with high dissociation energies, (3) self-compensation in CdTe, and (4) impurities being incorporated on more than one site, giving rise to amphoteric behavior.

As mentioned in Table 5.2, an ITO/CdTe heterojunction fabricated by e-beam evaporation of ITO on single-crystal p-CdTe gave $\phi_{oc} = 0.81$ V, $J_{sc} = 20$ mA/cm^2, and an efficiency of 10.5% (Werthen *et al.*, 1983b). Nakazawa *et al.*, (1987) fabricated In_2O_3/CdTe cells on single-crystal CdTe with $\phi_{oc} = 0.89$ V, $J_{sc} = 20.1$ mA/cm^2 and efficiency of 14.4%. The In_2O_3 layer was vacuum deposited directly on the CdTe single crystal by reactive vacuum evaporation of In in O_2 and ultrasonic cleaning was required during etching for the highest ϕ_{oc} values (to 0.91 V). These are the highest ϕ_{oc} values observed to date for CdTe-based cells.

5.8. Film Deposition by Close-Spaced Vapor Transport (CSVT) [or Close-Spaced Sublimation (CSS)]

In the CSVT method, sometimes called close-spaced sublimation (CSS), material is transported from a source wafer to a substrate through a gas at atmospheric pressure. The source and substrate are in contact with carbon blocks that are heated by infrared lamps. The source and substrate are generally held about 1.5 mm apart by quartz spacers.

CSVT was first used by Nicoll (1963) for the heteroepitaxial growth of GaAs on Ge. Since then it has been used to grow a wide variety of semiconductors, as summarized by Anthony *et al.* (1984) in their investigation of the growth process itself.

Early investigations of *p*-CdTe films deposited on single crystal *n*-CdS by the technique of close-spaced vapor transport are described by Saraie *et al.* (1972), Fahrenbruch *et al.* (1974), and Buch *et al.* (1977). One of the striking

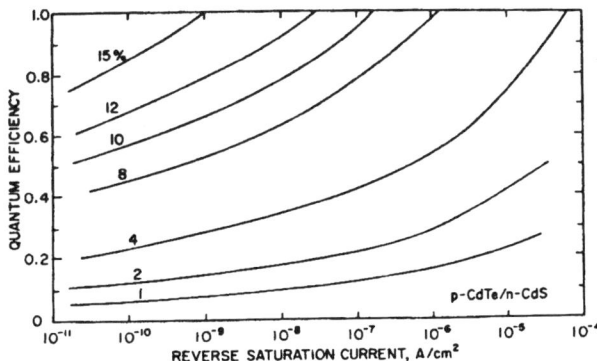

Fig. 5.20. Equiefficiency curves for the *p*-CdTe/*n*-CdS heterojunction system as a function of quantum efficiency and reverse saturation current density. Calculated for negligible reflection, series resistance and area utilization losses. (Reprinted with permission from A. L. Fahrenbruch *et al.*, *Appl. Phys. Lett.* **25**, 605 (1974). Copyright 1974 American Institute of Physics.)

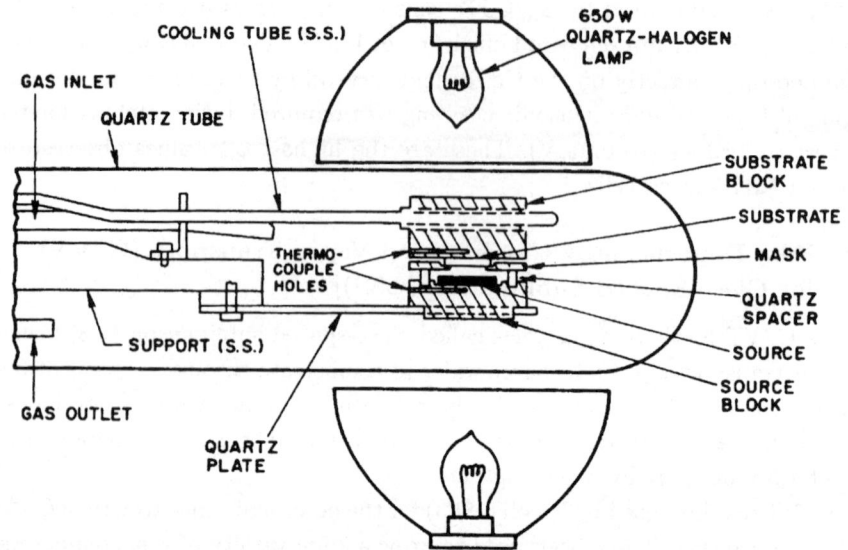

Fig. 5.21. Basic features of the close-spaced vapor transport (CSVT) system used to grow CdTe thin films. (Reprinted with permission from T. C. Anthony *et al.*, *J. Vac. Sci. Technology*, **A2**, 1296 (1984). Copyright 1984, American Institute of Physics.)

consequences of this early research was the realization that the reverse saturation current density J_o was about 10^3 times larger than the lowest values observed for junctions prepared by other means, resulting in values of open-circuit voltage between only 0.5 and 0.6 V instead of the highest observed value of about 0.9 V. Figure 5.20 shows the dependence of cell efficiency on quantum efficiency and J_o for p-CdTe/n-CdS heterojunctions; for these early cells prepared by CSVT, the quantum efficiency was about 0.85 and J_o was about 3×10^{-7} A/cm^2. Use of CSVT for the growth of CdTe films has also been reported by Mimila-Arroyo *et al.* (1977, 1979) and Tyan and Perez-Albuerne (1982).

A diagram of the CSVT apparatus used by Anthony *et al.* (1984) is given in Fig. 5.21. The source and substrate susceptors, as well as the mask, are machined from purified, high-density graphite. Ceramic-sheathed chromel-alumel thermocouples monitor the temperatures of the graphite blocks that are heated independently by two 650 W quartz-halogen lamps. A stainless-steel cooling tube positioned within the substrate block enables lower substrate temperatures to be achieved and permits rapid cooling of the substrate. In addition, the spring action of the cooling tube ensures good thermal contact between the substrate and the substrate block. A perforated, cylindrical quartz spacer maintains a separation of about 1 mm between the source and substrate. The graphite source block is supported by a quartz plate and the entire chamber is enclosed within a quartz tube. It was concluded that the growth of CdTe films by CSVT in inert ambients at atmospheric pressure is diffusion limited. Important parameters in the diffusion theory are the spacing, ambient gas, ambient gas pressure, and source temperature, each offering independent control over the growth rate. Approximately equal growth rates in He and in H_2 indicated the relative ineffectiveness of H_2 as a transport agent for CdTe.

In a subsequent investigation of the electrical properties of CdTe films and junctions, polycrystalline films of p-CdTe were deposited by CSVT on graphite

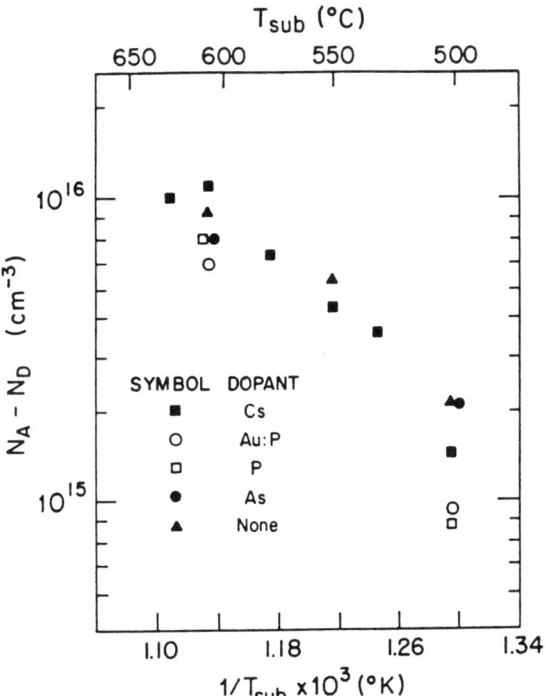

Fig. 5.22. Uncompensated ionized acceptor density in CdTe films deposited by CSVT as a function of the graphite substrate temperature, using a source temperature of 653°C in hydrogen at 1 atm, for several different sources doped as indicated. (Reprinted with permission from T. C. Anthony *et al.*, *J. Appl. Phys.* **57**, 400 (1985). Copyright 1985 American Institute of Physics.)

substrates, and homoepitaxial films on single-crystal CdTe:P substrates, with hole densities as high as 1.5×10^{16} cm^{-3} without intentional doping of the films (Anthony *et al.*, 1985). Investigation showed that this doping in the films was controlled neither by dopant transport from the source nor by impurities in the CSVT system. The first of these conclusions is dramatically illustrated in Fig. 5.22 showing the uncompensated acceptor density as a function of substrate temperature for a variety of sources containing Cs, Au:P, P, As or no dopant. The acceptor density increases with substrate temperature, but is independent of the doping of the source or the growth rate. The results can be consistently interpreted in terms of the creation of intrinsic acceptor defects significantly affected by self-compensation by other intrinsic defects (with no incorporation of the dopant species).

CdS/CdTe/graphite solar cells were prepared using these CdTe films, with the most efficient having $\phi_{oc} = 0.52$ V, $J_{sc} = 17.4$ mA/cm^2, fill factor = 0.64, and solar efficiency = 6.4%. The effects of the three problems with CdTe cells enumerated at the beginning of Sec. 5.2 are all clearly illustrated by the properties of these particular cells. The fill factor decreases as the hole density in the CdTe decreases, because of the effects of series resistance associated both with the grain-boundary resistance of the CdTe film and the back contact CdTe/graphite resistance. Analysis of current-voltage dependence indicates that the transport process is controlled by recombination at interface states above room temperature, and by tunneling below room temperature.

As early as 1982 Tyan and Perez-Albuerne (1982) showed that the CSVT technique could be used to produce efficient CdTe solar cells. Both CdS and CdTe layers were deposited by CSVT to form a glass/In$_2$O$_3$/CdS/CdTe/Au cell with $\phi_{oc} = 0.75$ V, $J_{sc} = 17$ mA/cm^2, and an efficiency of 10.5%. In one of the first mentions of the apparent importance of oxygen in CdS/CdTe solar cell preparation, it was reported that a small concentration of oxygen introduced during the deposition enhanced the *p*-type character of the CdTe film and ensured a shallow junction between the CdS and the CdTe.

Chu *et al.* (1987) used the CSVT technique to produce higher efficiency CdS/CdTe heterojunction cells with the structure glass/SnO$_2$/CdS/CdTe/ graphite:Cu with the parameters given in Table 5.2. An important step used by Chu *et al.* in their preparation technique was to include an *in situ* cleaning step for the CdS/SnO$_2$/glass structure in the CSVT apparatus by heating in a hydrogen atmosphere before depositing the CdTe; at room temperature, values of J_o were two orders of magnitude smaller for cells with *in situ* cleaning than for cells without such a cleaning step. In both the work of Tyan and Albuerne

(1982), and of Chu *et al.* (1987), it appears that the inverted configuration (CdTe on CdS, rather than CdS on CdTe) allows fine tuning of the junction parameters to achieve higher cell performance.

Although almost all workers have found the CdS layer to be necessary for high ϕ_{oc}, it is possible to make cells without it. Mitchell *et al.* (1985b) deposited polycrystalline CdTe directly on tin-oxide coated glass by CSVT. These 4 cm^2 cells produced $\phi_{oc} = 0.66$ V, $J_{sc} = 28.1$ mA/cm^2, and an efficiency of 10.5%.

The effect of various deposition parameters on CdS films deposited by CSVT on Corning 7059 glass, and the subsequently formed CdTe/CdS junction, finally resulting in an efficiency of 15.8%, has been investigated in some detail using measurements of SEM, photoluminescence, current versus voltage versus temperature, spectral response, and capacitance versus voltage measurements (Ferekides *et al.*, 1993, 1996). It was found that the presence of O$_2$ during deposition of CdS produced smaller and denser grains in the CSVT-deposited CdS films, resulted in a considerable decrease in the deposition rate that could be compensated for by raising the source temperature, and resulted in improved solar cell performance. Measurements indicate that this is the result of the addition of O$_2$ forming pinhole free, thin (70–80 nm) CdS films, eliminating the possibility of the subsequently deposited CdTe coming into direct contact with the SnO$_2$/glass substrate. The inclusion of O$_2$ in the deposition also appears to enhance the formation of a CdS$_{1-x}$T$_x$ region due to Te diffusion into the CdS films, which shifts the electrical junction away from the metallurgical interface, as discussed above in Sec. 5.2.

There still appears to be incomplete agreement on the necessity for O$_2$, however. CdS/CdTe solar cells have been prepared by CSVT with an efficiency of 13% without the inclusion of any oxygen in a He-only ambient by using thicker CdTe films to minimize pinholes, and a stronger CdCl$_2$ anneal to increase interdiffusion at the CdS/CdTe interface. (Rose *et al.*, 1996a,b).

By modifying the source and substrate temperature profiles used during CSVT deposition, it has been shown that it is possible to grow large-grain material, fill pinholes, improve device ϕ_{oc}, and achieve an efficiency of 12% (Li *et al.*, 1996).

5.9. Film Deposition by Spray Pyrolysis (SP)

The technique of spray pyrolysis (Chamberlin and Skarman, 1966) was used to prepare n-CdS/p-CdTe heterojunction solar cells by SP deposition of CdS films on single-crystal CdTe to yield solar efficiencies greater than 6% without

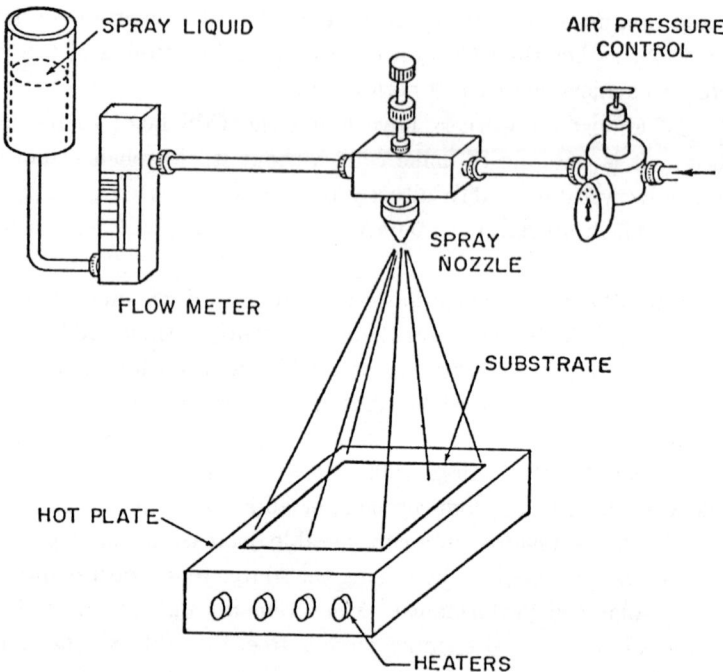

Fig. 5.23. Schematic diagram of a spray-pyrolysis (SP) apparatus (Reprinted with permission from Y. Ma, Ph.D. Dissertation, Stanford University (1977).)

optimization (Ma *et al.*, 1977a). A schematic diagram of the apparatus is given in Fig. 5.23. The reacting species are carried to the substrate in droplets of liquid sprayed on a hot substrate. The liquid is often an organic complex containing the material to be deposited, diluted in water and/or alcohol; in the case of CdS for example, the organic used to supply the sulfur is thiourea, and $CdCl_2$ is used to supply the cadmium. The major advantage of spray pyrolysis is its ability to produce thin films with good uniformity and adherence over large areas economically.

A detailed investigation of the properties of CdS films deposited by SP was carried out by showing that the optical and electrical properties of these films are strongly influenced by substrate temperature, cooling rate, spraying rate, and post-deposition heat-treatments (Ma and Bube, 1977b). It was possible to prepare a wide variety of films with room temperature conductivity between 10^{-4} and 10^2 (ohm-cm)$^{-1}$, corresponding to effective electron mobilities between 10^{-3} and 10^2 cm^2/V-sec.

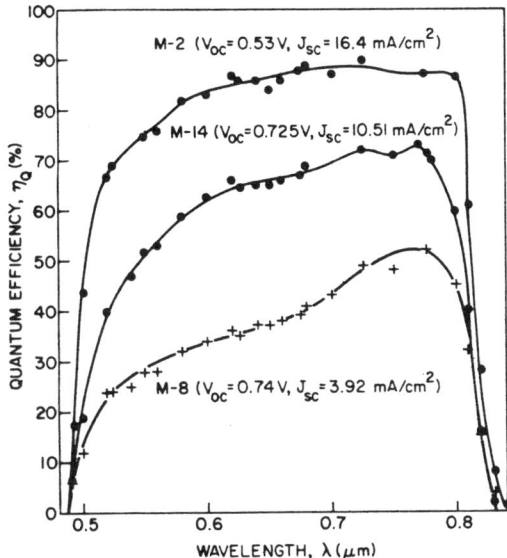

Fig. 5.24. Spectral response of quantum efficiency for three cells of n-CdS/p-CdTe prepared by SP deposition of CdS onto single-crystal p-CdTe. Sample M-2 used one CdTe boule, samples M-8 and M-14 used a second boule. CdTe substrate temperature was 450°C for M-2 and 400°C for M-8 and M-14. H$_2$ heat treatment was for 5 min at 400°C for M-2 and M-8, and for 5 min at 420°C for M-14. Solar efficiency of M-2 was 6.5% (uncorrected for reflection losses). (Reprinted with permission from Y. Y. Ma *et al., Appl. Phys. Lett.* **30**, 422 (1977). Copyright 1977 American Institute of Physics.)

Significant variables in the deposition are the temperature of the substrate and the effects of a subsequent heat treatment in H$_2$, which reduces the CdS resistivity and improves the solar efficiency. Typical variations in spectral response for three cells are shown in Fig. 5.24. All show response typical of a buried homojunction with very long diffusion length (sharp high shoulder on the long wavelength side). They show progressively more loss due to front surface recombination.

One of the advantages of the SP method is the relative ease with which solid-solution films can be prepared by a simple mixture of the components in the appropriate ratios. A complete range of solid-solution films was prepared by SP for CdS$_x$Se$_{1-x}$ and Zn$_x$Cd$_{1-x}$S (Feigelson *et al.*, 1977). In both the CdSSe and ZnCdS systems, complete solid solutions are formed over the entire composition range and the lattice parameters follow Vegard's law. In the CdS$_x$Se$_{1-x}$ system, the film resistivity is controlled by chlorine donors and does not vary appreciably with x; in the $Zn_xCd_{1-x}S$ system, however, the resistivity increases almost exponentially from about 10 ohm-cm for $x = 0$ to 10^{11} ohm-cm for $x = 1.0$, due both to an increase in the ionization energy of chlorine donors and to compensation of these donors with increase in x (Chynoweth and Bube, 1980).

A series of photovoltaic heterojunctions of the type n-$Zn_{0.1}Cd_{0.9}S$/p-CdTe have been prepared by SP of ZnCdS films on single-crystal CdTe with variation of the CdTe substrate temperature and of the postdeposition heat-treatment temperature in hydrogen (S.-Y. Yin *et al.*, 1978). Results indicated that the heterojunction system itself could provide $\phi_{oc} = 0.8$ V with an efficiency of 8% without an antireflection coating if the technological problems associated with the contacts to the ZnCdS film could be resolved.

A detailed investigation was made of the electrical and photovoltaic properties of heterojunctions prepared by SP deposition of thin ZnO films on single-crystal p-type CdTe:P (Aranovich *et al.*, 1979, 1980), using the substrate temperature and the postdeposition temperature for annealing in H_2 as the principal experimental variables. In spite of a large 28% lattice mismatch between ZnO and CdTe, high quantum efficiencies were achieved with values greater than 90% if a reverse bias was applied, the spectral response is that expected for a heterojunction, and cells with efficiency up to 8.8% were obtained referred to the active area of the cell. A thermally-assisted tunneling model (Padovani and Stratton, 1966) in which bulk and interface deep traps control the forward characteristics provides a good correlation with experimental data.

Spray pyrolysis has been used for the production of relatively large area SnO_x/CdS/CdTe/Ni/Sn solar cells, with small area samples showing efficiencies greater than 12% (Albright *et al.*, 1988, 1989; Jordan *et al.*, 1988).

It has been shown that CdTe films and a CdS/CdTe heterojunction can be deposited on stainless-steel substrates using spray pyrolysis, although considerable refinements would be required for a viable solar cell (Berry *et al.*, 1990).

It has been proposed that high-quality CdTe films can be produced by SP using nanoparticle precursors making possible materials with improved smoothness, density, and a lower processing temperature (Schulz *et al.*, 1996). CdI_2 was reacted with Na_2Te in a methanol solvent, resulting in the formation of soluble NaI and insoluble CdTe nanoparticles. CdTe thin films were produced by spray depositing the nanoparticle colloids (2.5–7.5 nm diameter) onto substrates at 280–440°C with no further thermal treatment. A layer-by-layer film growth mechanism was proposed for such a one-step spray deposition of nanoparticle precursors.

5.10. Other Film Deposition Methods

Chemical Vapor Deposition (CVD)

The first CdS/CdTe heterojunction solar cell with an efficiency of 5–6% was made by Bonnet and Rabinhorst (1972) by CVD techniques.

The possibility of fabricating n-ZnO/p-CdTe solar cells by the deposition of n-ZnO single crystal films on p-CdTe single crystal substrates by chemical vapor deposition (CVD) has been investigated by Kimata *et al.* (1984). The (111) Cd surface was used as the growth surface to avoid growing ZnTe at the interface. The efficiency was limited to 2.2% because of the effect of interface states and the high electrode resistance to the p-CdTe.

Nishimura and Bube (1985) showed that CVD deposition of CdS on single-crystal CdTe in hydrogen, under conditions in which there is out-diffusion of acceptors and large decrease in hole density near the surface for the uncovered CdTe surface, reduced the loss of acceptors by serving as a diffusion barrier. CdS/CdTe and CdS:In/CdS/CdTe heterojunctions formed in this way showed solar efficiencies between 8 and 10%. In both n- and p-type CdTe, heat treatment results in a reduction of carrier density near the surface. In n-type CdTe this effect is attributed to the creation and inward diffusion of Cd vacancies, but in p-type CdTe, Kim *et al.* (1988) have shown that the effect in P-doped p-type CdTe is associated with the formation of compensating P_{Cd} anti-site donor defects.

Screen Printing (ScP)

Reasonably efficient solar cells have been prepared from CdS and CdTe by simple screen printing techniques (Nakayama *et al.*, 1976, 1980; Uda *et al.*, 1982). A 20 μm layer of 0.2 ohm-cm CdS was screen printed onto an In_2O_3-coated glass substrate, using a CdS paste obtained by mixing CdS powder with 9% by weight $CdCl_2$ powder to serve as flux and propylene glycol to serve as binder. This was followed by the screen printing of a 10 μm layer of 0.1–1 ohm-cm, n-type CdTe, using a paste of (Cd + Te) prepared by mixing the powdered elements and adding 0.5% by weight of $CdCl_2$ powder and propylene glycol. The layers were sintered in close-fitting (but not leak-tight) boxes in air. A layer of Cu_xTe was formed on the n-CdTe by the dipping process and Ag-paint electrodes were applied to the Cu_xTe. An efficiency of 8.1% was reported for a cell for which an EBIC scan indicated that the electrical junction did not coincide with the CdS/CdTe interface. The observation that the efficiency is sharply peaked for a sintering temperature of 620°C suggests that the need to optimize the thickness of a CdS_xTe_{1-x} layer at the interface may be an important factor. Small screen-printed cells of this type have been extensively used in calculators, and are the only CdTe cells that are sold commercially at this writing. Developments in this area have been described by Ikegami (1988) and Kim *et al.* (1994). Screen printing techniques have also been used to investigate the properties of CdS_xTe_{1-x} solid solutions (de Melo *et al.*, 1994).

Electrodeposition (ED)

It was discovered early that reasonably efficient CdTe solar cells could be fabricated using the relatively simple techniques of electrodeposition (Basol, 1988; Meyers, 1988). Panicker *et al.* (1978) laid the foundation for this by showing the feasibility of electroplating CdTe thin films. Fulop *et al.* (1982) produced an MIS cell involving an n-CdTe film deposited by cathodic electrodeposition from an acidic aqueous solution of Cd- and Te-containing ions, and an insulating oxide layer obtained by annealing in air after deposition. The electrodeposition technique has been developed by Basol *et al.* (1985), Meyers (1989), Das and Morris (1992), Woodcock *et al.* (1993, 1994), and Paulson *et al.* (1994). Lincot *et al.* (1995) have investigated the epitaxial electrodeposition of CdTe films on InP; the presence of a thin layer of CdS, formed by chemical bath deposition on the InP markedly improved the epitaxial growth of the CdTe.

An investigation of the formation of CdS/CdTe solar cells using the techniques of electrodeposition resulted in a cell with an efficiency of 11.0% (Kim *et al.*, 1994a). CdS layers were deposited by chemical bath deposition on SnO_2-coated glass substrates, using cadmium acetate, ammonium acetate and ammonium hydroxide as described by Chu *et al.* (1992). A $CdCl_2$ coating was applied to the surface of the CdS, the sample was annealed at 450°C for 50 min in N_2, and then was rinsed in hot DI water and dried in N_2. The CdTe deposition was done in an electroplating system with Cd and Te anodes, a Ag/AgCl reference electrode, and the sample substrate as the cathode; after deposition the CdTe samples were rinsed in hot DI water and dried in N_2. A $CdCl_2$ coating was applied to the CdTe surfaces, samples were annealed at 410°C for 45 min in air, cooled, rinsed and dried. Contacts to the CdTe were made by evaporating Au through masks to form dot cells (0.033 cm^2) after etching with Br-MeOH. Contacts to the CdS were made by indium solder and silver paint. EBIC measurements indicated a maximum inside the CdTe layer, suggesting a buried homojunction.

Kim *et al.* (1994b) also investigated the effect of annealing on the microstructure and stress state of electrodeposited CdTe, involving recrystallization and grain growth. Additional discussion of stress testing of cells has been given by Meyers and Phillips (1996).

Using a similar approach, Kampmann *et al.* (1995) carried out a detailed investigation of the influence of the electrodeposition potential on the optical, photoelectrochemical, and structural properties of CdTe. An efficiency of over 12% has been reported for thin film cells fabricated using electrodeposition (Song *et al.*, 1996).

CdS films of device quality were deposited by electrodeposition by electroreduction of Cd and S at the substrate in a bath of $CdCl_2$ and sodium thiosulfate (McCandless *et al.*, 1995). These films were similar to those deposited by vacuum evaporation, respond similarly to air heat treatments, and can be used in the fabrication of CdTe/CdS solar cells with efficiencies comparable to those made with evaporated CdS. Electrodeposition provides a fabrication method using a low temperature ($85°C$), very low deposition current densities, and high utilization of constituent materials.

ZnO has been of interest as a transparent n-type window in solar cells. ZnO films of good quality have been prepared by cathodic electrodeposition from aqueous solutions at low deposition temperatures ($\leq 80°C$) (Peulon and Lincot, 1995, 1996). These films have high transparency in the visible corresponding to a band gap of about 3.4 eV, and high n-type doping is possible. The structure of the films can be varied by control of the deposition conditions from dense films to open-structured layers of columnar crystals.

Sputtering (ST)

The optical and electrical properties of indium-tin oxide (ITO) sputtered on glass have been investigated to determine the conditions for maximum conductivity and transmissivity (Haines and Bube, 1978).

When an ITO/CdTe junction is formed by rf sputtering of the ITO onto single crystal CdTe, a 1-μm thick surface layer of the p-type CdTe is converted to n-type as a consequence of the changes in the defect structure induced by ion bombardment and temperature, producing a buried homojunction of the type n-ITO/n-CdTe/p-CdTe with solar efficiencies up to 8% and $\phi_{oc} = 0.82$ V (Courreges *et al.*, 1980). A proposed energy band diagram is given in Fig. 5.25. Unfortunately the n-CdTe layer is unstable and shows degradation in photovoltaic parameters as the homojunction is converted to a heterojunction at room temperature. Related investigation of the properties of In/CdTe

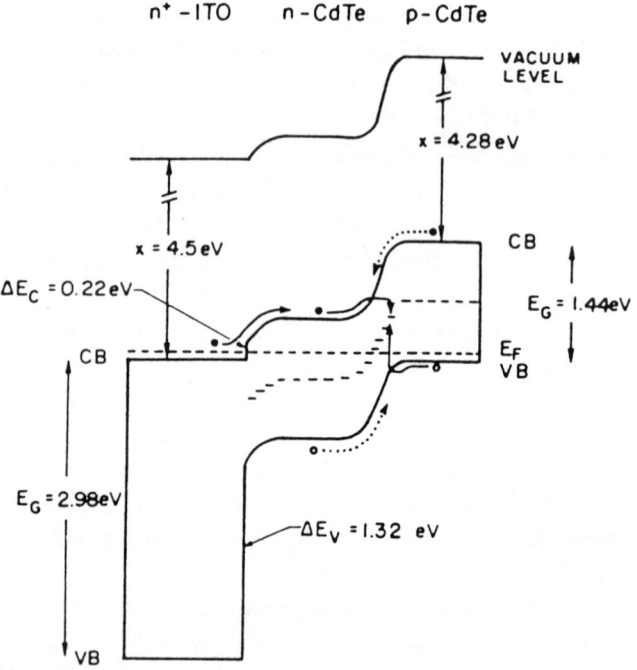

Fig. 5.25. Proposed energy-band diagram for an ITO/CdTe junction produced by rf sputtering of ITO onto the surface of a p-type CdTe:P crystal, showing the n-type CdTe layer formed by thermal effects during sputtering. (Reprinted with permission from F. G. Courreges *et al.*, *J. Appl. Phys.* **51**, 2175 (1980). Copyright 1980 American Institute of Physics.)

junctions indicated that an approximation to a normal Schottky barrier is obtained by vacuum evaporation of In onto an MB-etched CdTe surface, but that for evaporated In onto a sputter-etched surface or by sputtering In onto an MB-etched surface, the In is only a contact to a CdTe homojunction formed by the sputtering process.

It has been shown that rf planar magnetron sputtering of both CdS and CdTe can produce CdS/CdTe solar cells on soda-lime glass with efficiencies above 10% (Compaan *et al.*, 1993). The influence of substrate temperature, rf power, target erosion, and magnetic field configuration on rf sputtering of CdTe and CdS has been investigated (Compaan *et al.*, 1994). The effect of microstructure in rf sputtered CdS/CdTe cells (Shao *et al.*, 1966), and of interdiffusion of CdS and CdTe layers in laser-ablation-deposited and sputtered cells (Fischer *et al.*, 1996), have also been discussed.

All high efficiency $Cu(In,Ga)Se_2/CdS/ZnO$ solar cells have been fabricated using sputtered ZnO as the front contact. The standard two layer rf-sputtered process from ceramic targets consists of a highly resistive pure ZnO layer followed by an Al-doped ZnO. The role of the highly resistive ZnO layer deposited on top of the chemical-bath-deposited CdS buffer layer, and possible influences of sputtering pressure and O_2 addition to the Ar sputtering gas have been investigated (Ruckh *et al.*, 1996). A high rate reactive DC-magnetron sputtering process of metallic Zn + Al targets without additional substrate heating was also developed, by which suitable ZnO layers can be prepared in less than 5 minutes. All cell parameters are essentially the same for rf-sputtered and DC-magnetron reactive sputtered ZnO.

Chemical Bath Deposition

CdS films have been a key window layer in several of the leading solar cell systems, including InP (see Chapter 4), CdTe, and $CuInSe_2$ (see Chapter 6). Several different techniques have been described for the deposition of CdS films in CdS/CdTe cells in the above sections of this chapter. Another method of interest for all of these applications, not specifically discussed before this, is the method known as chemical bath deposition (CBD). In this section we summarize the principal work in this area and the application to the different solar cell systems described in Chapters 4, 5 and 6.

CBD resembles CVD by being based on the reaction between precursor materials in a metastable situation, but in CBD this occurs in a liquid phase rather than in the vapor phase as in CVD. It also has some relationship to spray pyrolysis, except that in CBD films are formed by precipitation from solution rather than by being sprayed onto a hot substrate. CdS is usually formed from the reaction between dissolved cadmium ions (e.g. cadmium tetrammine complex ions), and thiourea molecules, in ammonia solutions (Lincot and Ortega-Borges, 1992; Ortega-Borges and Lincot, 1993). It is a technique that allows the formation of films near room temperature and that is adaptable to large area processing at low cost. Historically it began with the fabrication of PbS photoconductive detectors (Kicinski, 1948) for which it was the principal fabrication technique for many years, and it has been used to prepare thin films of a variety of chalcogenides and in the fabrication of high-efficiency CdS/CdTe (Turner *et al.*, 1991) and $CdS/CuInSe_2$ solar cells (Birkmire *et al.*, 1989; Mauch *et al.*, 1991a, 1991b).

Electrochemical impedance spectroscopy and quartz crystal microbalance techniques have been applied to characterize films of CdS prepared by CBD

(Lincot and Vedel, 1991; Lincot and Ortega-Borges, 1992; Ortega-Borges and Lincot, 1993). They conclude that film growth proceeds in two consecutive steps, the formation of a dense compact inner layer followed by the formation of a porous less adherent layer. A simple columnar growth model is shown to account reasonably well for the experimental results. CBD allows the deposition of very thin films of the order of a few nanometers in the presence of excess thiourea, a possibility that can be quite important in seeking high-efficiency solar cells with CdS window layers. The process has been studied further through the use of high-resolution transmission electron microscopy, which indicates that the growth in CdS probably proceeds by way of an atomic mechanism, rather than by cluster coagulation as is indicated for CBD growth of ZnS and CdSe (Lincot *et al.*, 1993; Froment *et al.*, 1995). The growth of CBD CdS thin films on SnO_2 and glass substrates has been analyzed using atomic force microscopy (AFM) (Moutinho *et al.*, 1996). CBD CdS films have been optically and electrically characterized (Ozsan *et al.*, 1994). The band gap of these films depends strongly on the deposition conditions, increasing with an increase in deposition rate. The films are all *n*-type with doping densities aground 10^{17} cm^{-3} under illumination without intentional doping in the bath.

CBD CdS films have been recently investigated in the formation of CdS/ InP, CdS/CdTe, and CdS/CuInSe$_2$ junctions. Epitaxial growth of hex-CdS on (111)* InP single crystals has been achieved using CBD from cadmium ammonia-thiourea aqueous solutions (Lincot *et al.*, 1994; Froment *et al.*, 1995), and a solar cell based on a *p*-InP single crystal with InP/CBD-CdS/In$_2$O$_3$ structure gave an efficiency of 17.8% (Saito *et al.*, 1994). Epitaxial growth of cub-CdS on (100) InP was achieved by CBD, probably because of lattice matching between the CdS (100) plane and the InP (100) plane (Cortes *et al.*, 1996).

The interaction between CBD deposited CdS films and subsequently electrodeposited CdTe films in the formation of CdS/CdTe junctions has been investigated (Ozsan *et al.*, 1995). Results indicated that the as-deposited cub-CdS CBD films with small grain size and high density of defects are beneficial for the recrystallization in the cell structure during the subsequent heat treatment.

CBD CdS films were successfully used in the fabrication of ZnO/CdS/ CuInSe$_2$ solar cells, where it was found that the final cell performance depends on the specific Cd-salt used in the CBD CdS film deposition process: High efficiency was obtained for CdI$_2$ and CdSO$_4$ (Mauch *et al.*, 1991a,b). Further investigation indicated that the CBD reaction is not limited to the abrupt deposition of a CdS layer on CuInSe$_2$, but involves chemical modifications at the interface, with the formation of a $CuInSe_2/CdIn_xSe_y/CdS$ gradual structure (Lincot *et al.*, 1992), much like the $CdS_{1-x}Te_x$ structure in $CdS/CdTe$ junctions described in Sec. 5.4. This solid solution formation has the beneficial effects of removing superficial oxides, eliminating Cu_xSe conducting phases in excess, and forming a graded interface structure. Nucleation and growth of the CdS layer on $Cu(InGa)Se_2$ thin films is described by Friedlmeier *et al.* (1996).

The effect of impurities in CBD CdS films for use in $Cu(In,Ga)Se_2$ solar cells has been investigated by Kylner *et al.* (1996) using secondary ion mass spectroscopy, Fourier transform infrared spectroscopy, Rutherford backscattering spectrometry, and x-ray photoelectron spectroscopy. As-deposited films were subjected to air annealing between 200°C and 350°C, and storage for 1.5 months in different environments. These films were found to contain O (primarily in $CdCO_3$ and H_2O), C, N (CN bonds), and H beside the CdS matrix. Impurities were distributed uniformly and did not depend on the substrate material. The CN bonds were investigated further (Kylner *et al.*, 1997a), who found that a high degree of disorder exists around the CN bond in as-deposited CBD CdS films, and that the CN impurity is water-soluble and is effectively reduced by water at 60°C. The solar cells involving these CBD CdS films had efficiencies greater than 15%, but correlation between CN density and cell efficiency has not been determined. To investigate these impurity effects further, CBD CdS films were deposited with varying thiourea concentrations (Kylner *et al.*, 1997b). Both impurity content and deposition rate are increased with increasing thiourea concentration. CBD CdS films with thiourea concentration greater than 0.03 M showed a higher degree of disorder, a mixture of hexagonal and cubic phases, compressive stresses in the films, and about 0.1 eV higher optical bandgap. Photoconductivity measurements indicated that the CBD CdS film with the highest amount of impurities (thiourea concentration of 0.1 M) had the shortest photodecay time of all investigated films. The effects associated with higher thiourea concentrations, however, do not seem to affect the solar cell efficiency.

ZnS has a large band gap of 3.6 eV and can be used as a window layer in thin film heterojunction solar cells (Dona and Herrero, 1994). Chemical bath deposition has been shown to be a way to obtain adherent, specular, homogeneous and stoichiometric ZnS thin films from thioacetamide-based baths (Ortega Borges *et al.*, 1992; Mokili *et al.*, 1995).

5.11. CdTe Alloys

Interest in alloys related to CdTe include research on ZnTe and $Hg_{1-x}Zn_xTe$ using deposition by MOCVD (Chu *et al.*, 1989), $Zn_xCd_{1-x}Te$ using deposition by CSVT (Peters *et al.*, 1988a,b) and by MBE (Rohatgi *et al.*, 1988, 1989), $ZnSe_{1-x}Te_x$ using LPE (Koval *et al.*, 1991), and $Cd_{1-x}Mn_xTe$ using deposition by MOCVD (Nouhi *et al.*, 1989; Rohatgi *et al.*, 1988, 1989).

$Zn_xCd_{1-x}Te$ films with Zn concentrations as high as $x = 0.45$ were deposited onto graphite substrates using CSVT (Peters *et al.*, 1988a). Values of x in the deposited films equal to that in the source could be achieved either by deposition with a small temperature difference (2–8°C) between source and substrate at atmospheric pressure of an ambient gas, or by deposition at low ambient pressures (0.3 Torr) with any desired temperature difference between source and substrate. For films deposited at a substrate temperature of 610°C, $p = 5 \times 10^{15}$ cm^{-3} as determined from capacitance versus voltage measurements on In/ZnCdTe Schottky barriers. No change in this hole density was found with variations in Zn concentration.

A series of $CdS/Zn_xCd_{1-x}Te$ heterojunctions were prepared by vacuum evaporation of CdS onto single crystal p-type $Zn_xCd_{1-x}Te$ substrates with $x = 0$, 0.1, and 0.3 (Peters *et al.*, 1988b). These junctions showed a maximum ϕ_{oc} when the $Zn_xCd_{1-x}Te$ surface was made stoichiometric by heat treatment in H_2 before CdS deposition, a maximum ϕ_{oc} for any value of x for CdS deposition conditions yielding approximately equal carrier densities in CdS and $Zn_xCd_{1-x}Te$, and an increase in J_o with increasing x. The best junction for $x = 0.3$ had a ϕ_{oc} of only 0.36 V. The junction transport process resulting in relatively high values of J_o (2.6×10^{-8} A/cm^2 in the best cell with $x = 0.3$), could be modeled by thermally assisted tunneling of holes from the $Zn_{0.3}Cd_{0.7}Te$ to the interface of the junction followed by recombination there with electrons from the CdS, provided that a high density of interface states was assumed. It was suggested that the large values of J_o were associated with lattice distortions resulting from the solid solution formation.

Thin films of $Hg_xZn_{1-x}Te$ over the range from $x = 0$ to $x = 0.4$ have been deposited by electrodeposition (Thero and Singh, 1996). These films show optical absorption coefficients of the order of 10^4 cm^{-1}, indicating their possible suitability for use as absorber materials in thin film solar cells.

The processing steps involving TCJF, described in Sec. 5.4, necessary for a solar cell involving electrodeposited HgCdTe have been discussed (Basol and Tseng, 1986; Basol, 1988). A completed $CdS/Hg_{0.1}Cd_{0.9}Te$ cell was characterized by $\phi_{oc} = 0.62$ V, $J_{sc} = 27.03$ mA/cm^2, $ff = 0.63$ and efficiency = 10.6%. Compared to a similarly electrodeposited CdS/CdTe cell (band gap of CdTe = 1.47 eV, compared to that of $Hg_{0.1}Cd_{0.9}Te = 1.30$ eV) with the same efficiency, the CdS/HgCdTe device shows a 27% larger J_{sc} and a 14% smaller ϕ_{oc}.

5.12. Other II–VI Junctions

Photovoltaic properties of some 50 heterojunctions formed by CSVT in the form of five different II–VI heterojunctions were investigated by Buch *et al.* (1976, 1977). These included n-CdSe/p-ZnTe, n-CdSe/p-CdTe, n-ZnSe/p-CdTe, p-ZnTe/n-ZnSe, and n-CdTe/p-ZnTe. Single-crystal and polycrystalline ZnTe, single-crystal CdTe, and polycrystalline ZnSe substrates were used. Under CSVT conditions, at temperatures of 375°C and above, several of the II–VI heterojunctions involving four different atoms showed signs of alloying and third phase formation at the heterointerface, as shown in Fig. 5.26. Values of the reverse saturation current J_o were several orders of magnitude larger than that expected for a simple recombination process for all the junctions, suggesting the importance of recombination at the interface.

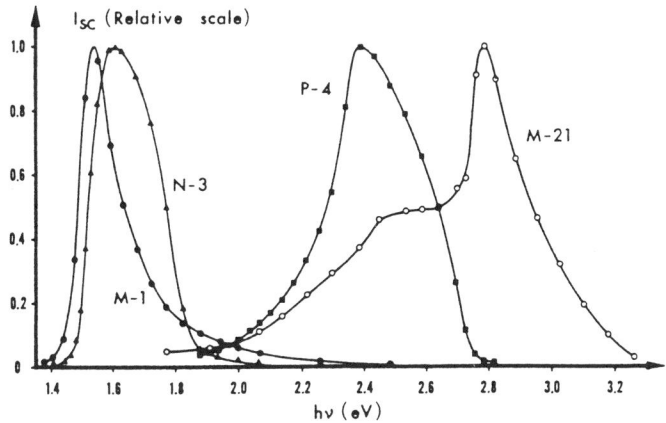

Fig. 5.26. Short-circuit spectral response curves for four II–VI compound solar cells prepared by CSVT. Both M-1 and M-21 are n-ZnSe/p-CdTe cells, but their spectral responses are quite different. The response of M-1 resembles that of n-CdSe/p-CdTe cell N-3, suggesting CdSe formation at the junction in M-1. The low-energy spectral cutoff for M-21 resembles that of p-ZnTe/n-ZnSe cell P4, suggesting that in this case ZnTe is formed at the junction. (Reprinted with permission from F. Buch *et al., J. Appl. Phys.* **48**, 1596 (1977). Copyright 1977 American Institute of Physics.)

CHAPTER 6

COPPER INDIUM DISELENIDE AND OTHER I–III–VI MATERIALS

6.1. Overview

As photovoltaic research extended beyond the simple Group IV elemental Si, to Group III–V compounds like GaAs, and then to Group II–VI compounds like CdTe, as described in the previous chapters of this book, it moved with one more step of complexity to I–III–VI$_2$ compounds like copper indium diselenide, CuInSe$_2$. We could think simplistically of CuInSe$_2$, or CIS as it has become known for the sake of brevity, as the result of substituting a Group I (Cu) and a Group III (In) element to give an average result similar to that of a Group II element in II–VI materials.

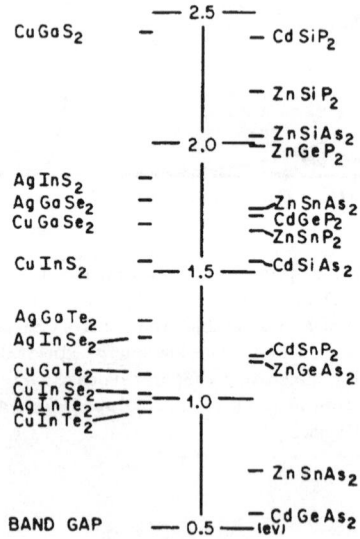

Fig. 6.1. Lowest energy gaps of I–III–VI$_2$ and II–IV–V$_2$ semiconductors in the range of interest for solar photovoltaic conversion (Reprinted from S. Wagner and P. M. Bridenbaugh, *J. Crystal Growth* **39**, 151 (1977) with kind permission of Elsevier Science — NL, Sara Burgerhartstraat 25, 1055 KV Amsterdam, The Netherlands. Copyright 1977 Elsevier Science.)

Such a ternary compound represents the first step to more complicated solid solutions with more than three different atoms, so that both the desired band gap and the proper heterojunction lattice match could in principle be obtained to provide maximum efficiency. Research on quaternary and pentenary systems has been undertaken (Antypas and Moon, 1974; Loferski *et al.*, 1981) but to date these much more complex materials have not met the stringent demands of reproducibility and operating lifetime required for a solar cell material. Some twenty-three ternary compounds of type I–III–VI$_2$ and II–IV–V$_2$ with band gaps in the range of interest for solar cells are summarized in Fig. 6.1.

In 1974 it was shown that a solar cell efficiency of 12%, a high value for cells of any type at that date, could be obtained with a CdS/CuInSe$_2$ heterojunction made by vacuum evaporation of CdS onto single crystal p-type CuInSe$_2$, previously Se-annealed to increase its p-type conductivity (Wagner *et al.*, 1974; Shay *et al.*, 1975).

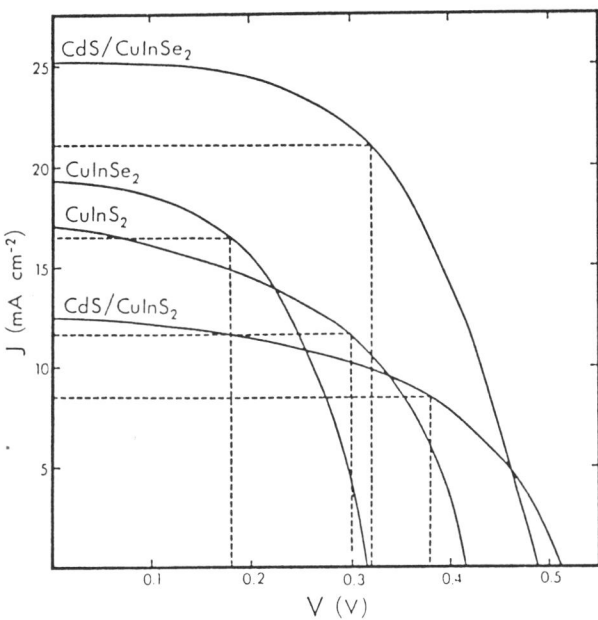

Fig. 6.2. Current versus voltage curves under illumination by 100 mW cm^{-2} by tungsten-halogen for early solar cells: all-thin-film heterojunctions of CdS/CuInSe$_2$ and CdS/CuInS$_2$, and all-thin-film homojunctions of CuInSe$_2$ and CuInS$_2$. (Reprinted with permission from L. L. Kazmerski, Univ. of Maine, Orono, Second Quarter Report to NSF-RANN and ERDA, NSF/AER 75-19576/PR/77/2, p. 13 (1977).)

These promising initial results led to a search for methods of preparing thin-film CuInSe$_2$ cells with equal or better solar efficiencies. All thin-film CdS/CuInSe$_2$ cells prepared by vacuum evaporation were fabricated with an efficiency of 6.6%, and similar all-thin-film CdS/CuInS$_2$ cells were fabricated with an efficiency of 3.2% (Kazmerski *et al.*, 1975, 1976; Kazmerski, 1977). All-thin-film homojunctions of CuInS$_2$ and CuInSe$_2$ were also prepared by varying the flux of S or Se during vacuum evaporation (Kazmerski and Sanborn, 1977). Efficiencies for the homojunctions were in the range of 3–4%. Interesting summaries of the properties of these early heterojunction and homojunction thin-film cells are given in Fig. 6.2 that compares their current-voltage curves in the light, and Fig. 6.3 that compares their spectral response of quantum efficiency.

A step forward was marked next by the announcement of the vacuum-evaporation preparation, using a three-source method, of all-thin-film CdS/CuInSe$_2$ heterojunctions with an efficiency of 10%, approaching that of the

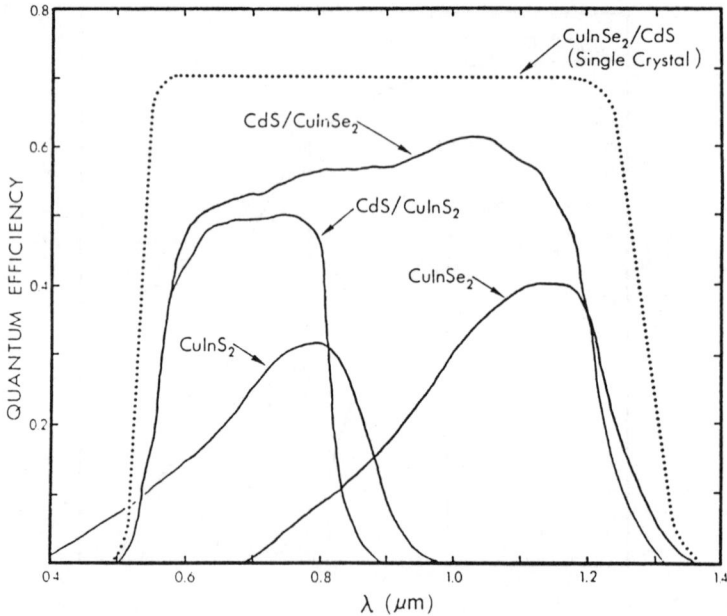

Fig. 6.3. Spectral response of quantum efficiency for early solar cells: all-thin-film heterojunctions of CdS/CuInSe$_2$ and CdS/CuInS$_2$, and all-thin-film homojunctions of CuInSe$_2$ and CuInS$_2$. (Reprinted with permission from L. L. Kazmerski, Univ. of Maine, Orono, Second Quarter Report to NSF-RANN and ERDA, NSF/AER 75-19576/PR/77/2, p. 16 (1977).)

initial single-crystal device, achieved primarily by a major increase in the short-circuit current to 39 mA/cm^2 (Mickelsen and Chen, 1981, 1982). Because of its smaller band gap, CuInSe$_2$ cells give a higher short-circuit current than other high-efficiency solar cells, but a lower open-circuit voltage ($\phi_{oc} = 0.40$ V in this case). It is noteworthy that a quantum efficiency of about 0.9 was achieved in a film with a grain size of only about 1 μm, indicating that grain boundary recombination is not important in CuInSe$_2$ films.

Kazmerski and Wagner (1985), Bube (1987), and Zweibel and Barnett (1993) summarize the properties of a number of solar cells involving thin-film

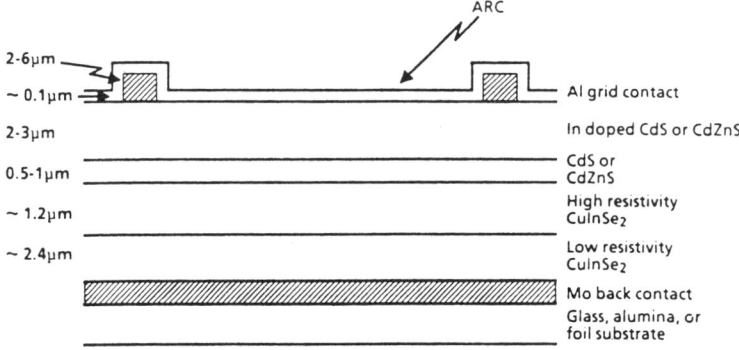

Fig. 6.4. Solar cell structure for a 9.6% large area CuInSe$_2$ solar cell. At 1 sun, $\phi_{oc} = 0.43$ V, $J_{sc} = 35.3$ mA/cm^2, and fill factor = 0.63. (Reprinted with permission from W. E. Devaney and R. A. Mickelsen, *Solar Cells* **24**, 19 (1988). Copyright 1988, Elsevier Sequoia S.A., Lausanne, Switzerland.)

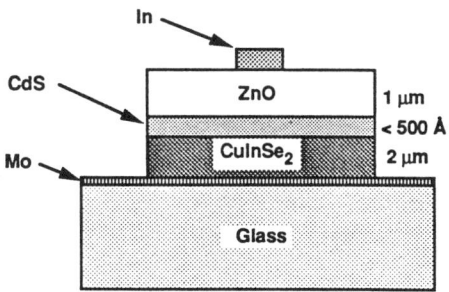

Fig. 6.5. Solar cell structure for the ARCO (now Siemens) ZnO/CdS/CuInSe$_2$ solar cell. For simulated sunlight of 100 mW/cm^2, $\phi_{oc} = 0.36$–0.42 V, $J_{sc} = 35$–40 mA/cm^2, $ff = 0.5$–0.65, and cell efficiency = 7–9%. (Reprinted with permission from J. B. Yoo, *et al.*, *J. Appl. Phys.* **68**, 4694 (1990). Copyright 1990 American Institute of Physics.)

$CuInSe_2$ as the absorbing material. Table 6.1 provides data and references for specific solar cell parameters as of 1987. A reported $CdS/CuInTe_2$ cell showed no photovoltaic effect.

Overall better performance in $CdS/CuInSe_2$ cells was achieved using the backwall configuration shown in Figs. 6.4 and 6.5. Three reasons were advanced for this: (1) the lower deposition temperature for CdS as compared to $CuInSe_2$ minimizes diffusion between the two materials, (2) the CdS or CdZnS window layer protects the $CuInSe_2$ absorber layer from interaction with the surface environment, and (3) a heat treatment in oxygen that proves critical for optimizing cell performance is effective only if the CdS or CdZnS is on top.

Table 6.1. Polycrystalline Thin-Film $CuInSe_2$-based Solar Cells (1987)

Cell	Open-Circuit Voltage, V	Short-Circuit Current, mA/cm^2	Efficiency %	Reference
A. High Efficiency $CuInSe_2$ Cells				
$ZnCdS/CuInSe_2$ (1 cm²)	0.44	39.4	11.9	Devaney et al., 1985 Mickelsen and Chen, 1986
$ZnCdS/CuInSe_2$ (4 cells; 91 cm²)	0.43	35.3	9.6	Stanbery et al., 1984 Mickelsen et al., 1987
$ZnO/CdS/CuInSe_2$	0.49	36.7	12.5	Potter et al., 1985
B. Experimental $CuInSe_2$ Cells				
$ZnO/ZnSe/CuInSe_2$	0.42	37	8.5	Stirn and Nouhi, 1986 Nouhi and Stirn, 1986 Nouhi et al., 1987
$CdS/CuInSe_2$ (sputtered)	0.31	36	5.2	Thornton et al., 1984, 1987 Lommasson et al., 1987
$CdS/CuInSe_2$ (electrodeposited)	0.30	30	4.0	Shih and Qiu, 1987 Pern et al., 1987
C. $CuIn_{1-x}Ga_xSe_2$ Cells				
$ZnCdS/$ $CuIn_{0.77}Ga_{0.23}Se_2$	0.51	30.4	10.2	Chen et al., 1987
$ZnCdS/CuGaSe_2$	0.85	11.6	5.8	Dimmler et al., 1987
$ZnCdS/$ $CuIn_{0.44}Ga_{0.56}Se_2$	0.52	15.6	2.9	Dimmler et al., 1987
$CdS/CuGaSe_2$	0.58	9.6	2.6	Noufi et al., 1986

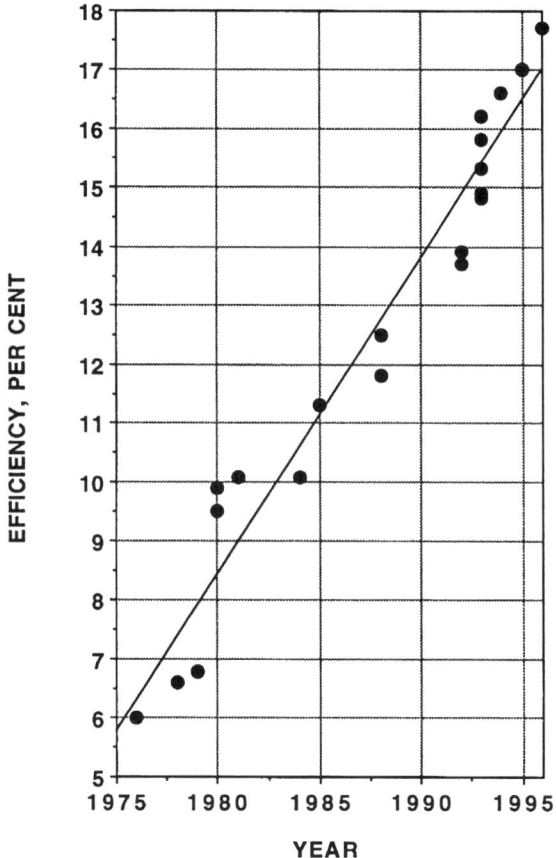

Fig. 6.6. NREL solar cell efficiency as a function of year for all thin-film $CuInSe_2$ cells. (Data from R. F. Service 1996.)

The increase in efficiency of all thin-film $CuInSe_2$ solar cells over the years up to the present is shown in Fig. 6.6. Reviews of the progress have been given by Meakin (1985), Rockett and Birkmire (1991), Basol (1992), and Tarrant and Gay (1995). By 1996, several improvements developed over the intervening years, including increasing the crystallite size by varying the concentration of copper as the film is deposited with a copper-rich layer at the bottom, and the inclusion of Ga to make a larger band gap $CuIn_{1-x}Ga_xSe_2$ (CIGS) material graded with larger values of x near the bottom of the film to aid in electron collection, have led to a 0.4 cm^2 cell with an efficiency of 17.7% (J. Tuttle *et al.*, 1996), marking the high point on Fig. 6.6. At the same time a CIGS cell with 100 times larger area has shown an efficiency of 13.9% (Schock *et al.*, 1996).

6.2. Materials Properties of CuInSe$_2$

CuInSe$_2$ has a number of properties that make it particularly suitable for use as a solar-cell absorber material: (1) a band gap of 1.04 eV, which enables it to absorb a large fraction of the solar spectrum; (2) a direct band gap, which minimizes the requirements for minority carrier diffusion length, since carriers are photoexcited close to the collecting junction; (3) the possibility of being prepared in either *n*- or *p*-type form; (4) an electron affinity such that harmful band gap spikes are not expected upon heterojunction formation with CdS, ZnCdS or ITO; (5) a lattice constant that matches well with CdS or ZnCdS, e.g., there is only a 1.2% lattice mismatch between the chalcopyrite structure of CuInSe$_2$ and the wurtzite structure of CdS; (6) the highest absorption constant $(3\text{--}6 \times 10^5 \text{ cm}^{-1})$ reported for any semiconductor (see Fig. 6.7); (7) the ability to be fabricated in thin-film form by a variety of techniques, as was the case for CdTe as described in Chapter 5; and (8) excellent stability under operating conditions (Mickelsen *et al.*, 1982; Hsiao *et al.*, 1983).

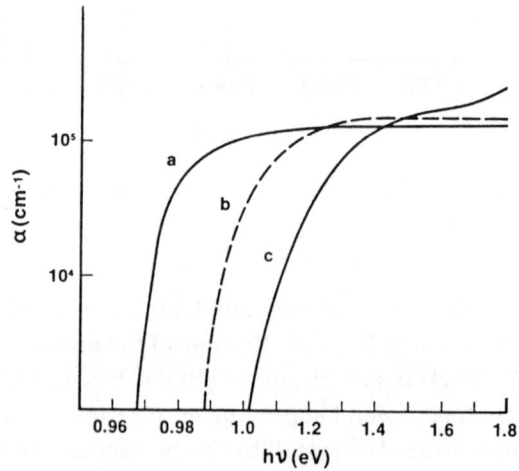

Fig. 6.7. Dependence of absorption coefficient α on photon energy for CuInSe$_2$: (a) single crystal, 300 K, (b) single crystal, 100 K, and (c) thin film, 300 K. (Reprinted with permission from L. L. Kazmerski *et al.*, *J. Vac. Sci. Technol.* **A1**, 395 (1983). Copyright 1983, American Institute of Physics.)

Crystal Structure

CuInSe$_2$ crystallizes in a diamondlike lattice with a face-centered tetragonal unit cell (Parkes *et al.*, 1973). The lattice structure and unit cell for the chalcopyrite structure are shown in Fig. 6.8. Each selenium atom serves as the center of a tetrahedron of two Cu and two In atoms, and each metallic atom is surrounded by a tetrahedron of selenium atoms. The lattice parameters are $a = 0.5784$ nm, and $c = 1.1614$ nm. CuInS$_2$ and CuInTe$_2$ have similar structures.

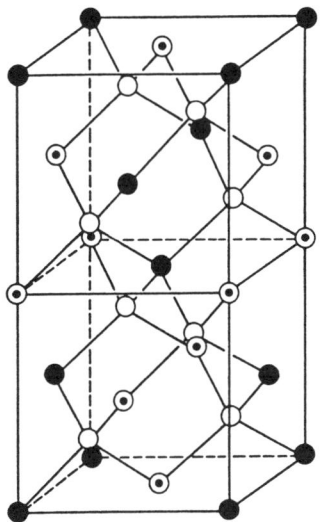

Fig. 6.8. Chalcopyrite unit cell for CuInSe$_2$. (●) Cu, (⊙) In, (○) Se. (Reprinted with permission from L. Kazmerski and S. Wagner, in *Current Topics in Photovoltaics*, ed. T. J. Coutts and J. D. Meakin, Academic Press, London, p. 41 (1985). Copyright 1985, Academic Press, London.)

Crystal Defects

The electrical conductivity of CuInSe$_2$ is apparently largely determined by native defects with a high degree of compensation between defect donors and acceptors, and can therefore be varied by the ratio Cu/In used in material fabrication by vacuum evaporation and by the pressure of Se used during growth or annealing.

Neumann (1983) considered vacancies, interstitials and antisite defects in the cation lattice as possible sources of electrically active defects, and predicted that the active defects in p-type material are an indium vacancy (V_{In}) and a Cu antisite defect (Cu_{In}). In order to achieve low-resistivity, p-type $CuInSe_2$, the Cu/In ratio must be slightly larger than required for stoichiometric material.

Photoluminescence (Abou-Elfotouh *et al.*, 1984) and cathodoluminescence data (Masse and Redjai, 1984) were interpreted to indicate the following energy levels for various defects; in Cu-rich material: acceptor Cu_{In} (30 meV above the valence band edge, VB), acceptor V_{In} (85 meV above VB) and acceptor V_{Se} (130 meV above VB), and in In-rich material: acceptor V_{Cu} (45 meV above VB), acceptor V_{Se} (130 meV above VB) and donor In_{Cu} (35 meV below the conduction band edge).

Masse (1990) reviewed the various energy levels of defects with a low energy of formation, reported for I–III–VI$_2$ compounds (e.g. for $CuInSe_2$: V_{Se}, V_{Cu}, V_{In}, and Cu_{In}), and concluded that evidence favors a hydrogenic acceptor associated with a V_{Cu} rather than an antisite defect.

Hole and electron traps have been investigated using Deep Level Transient Spectroscopy (DLTS) in single crystal Schottky junctions based on $CuInSe_2$ (Igalson and Bacewicz, 1994) and in thin film $CuInSe_2/CdS/ZnO$ photovoltaic devices (Igalson and Schock, 1996). The general effect observed is a decrease in the density of shallow electron traps and an increase in the density of hole traps after an injection of electrons, metastable below 200 K. It has been proposed (Igalson, 1993) that this phenomenon can be described appropriately in terms of a model involving two charge states of the same defect, similar in many ways to the behavior of the dangling-bond defect in a-Si:H discussed in Chapter 3. The data suggest that electron traps are converted to hole traps after the capture of two electrons, and a possible candidate for the defect is a double donor In/Cu or a complex related to it. Similar indications of metastable defect behavior in Cu(In,Ga)Se$_2$ cells have been reported (Schmitt *et al.*, 1996; Walter *et al.*, 1996a), and in CuIn(SSe)$_2$ films (Herberholz *et al.*, 1996).

A method to deduce energy distributions of defects in the band gap by measuring the complex admittance of a junction has been applied to hetero-junctions of p-Cu(In,Ga)Se$_2$ thin films with n-ZnO (Walter *et al.*, 1996b). A distribution of hole traps with maximum density corresponding to an ioniza-tion energy of about 0.3 eV was indicated, similar to the results of modulated photoconductivity (Herberholz *et al.*, 1994) and DLTS measurements (Igalson and Bacewicz, 1992).

Minority Carrier Diffusion Length

The high ionized acceptor density in p-$CuInSe_2$ does not seem to cause a large decrease in the minority carrier diffusion length. Diffusion length values of

0.5 μm from the photoelectromagnetic effect (Mora and Romeo, 1977), 2.5 μm from EBIC data on p-CuInSe$_2$/CdS heterojunctions (Piekoszewski *et al.*, 1980), and 1.0 μm from surface photovoltage measured with p-CuInSe$_2$/electrolyte diodes (Johnson *et al.*, 1985), have been reported.

Optical Absorption

Figure 6.7 shows the high values of absorption constant achieved with CuInSe$_2$ which correlate with the high values of J_{sc} found in solar cells based on CuInSe$_2$. Figure 6.9 shows the variation of absorption with photon energy for single crystal and thin film CuInSe$_2$, indicative of a direct band gap with magnitude E_G for which it would be expected that $(\alpha h\nu)^2$ would be proportional to $(h\nu - E_G)$. Differences between the single crystal and thin film curves in Fig. 6.9 are probably due to small differences in composition.

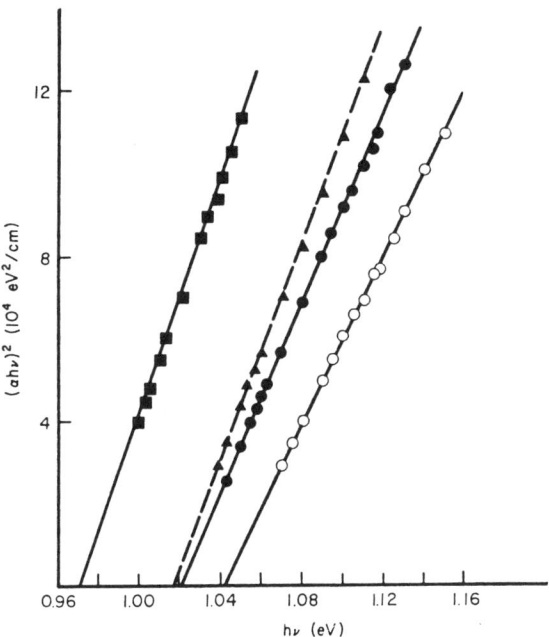

Fig. 6.9. Dependence of optical absorption of CuInSe$_2$ on photon energy plotted as $(\alpha h\nu)^2$ versus photon energy $h\nu$, which yields a linear relationship for a direct band gap. (■) single crystal at 300 K, (●) thin film at 300 K, (○) thin film at 100 K, and (▲) annealed thin film at 300 K. (Reprinted with permission from L. L. Kazmerski *et al.*, *J. Vac. Sci. Technol.* **A1**, 395 (1983). Copyright 1983, American Institute of Physics.)

The absorption edge of $CuInSe_2$ is shifted to higher energies with increasing In content. For $In/(Cu + In) < 0.65$, the band gap of the stoichiometric $CuInSe_2$ at about 1 eV is still detected. It has been suggested that for $0.50 < In/(Cu + In) < 0.66$, the material is composed of two parallel existing phases, both $CuInSe_2$ and the ordered defect compound $CuIn_2Se_{3.5}$ (Schaeffler and Schock, 1995).

Interface Properties

In Sec. 5.4 the importance of interface interactions in the CdS/CdTe heterojunction structure leading to the formation of $CdTe_{1-x}S_x$ were described. Similar investigation has been made of the possible reactions between different species at the $CdS/CuInSe_2$ interface. Analyses involving x-ray photon spectroscopy (XPS) (Kazmerski *et al.*, 1982) and electron energy loss spectroscopy (EELS) (Jamjoum *et al.*, 1982) both indicate the presence of a chemical reaction at the CdS–$CuInSe_2$ interface with the formation of Cu_2S and Cu_2Se as the reaction products.

Oxidation of CuInSe_2

Oxidation studies have been made both on p-type polycrystalline thin films fabricated by three-source deposition, and on Bridgman grown crystals (Kazmerski *et al.*, 1981, 1983a). For single crystals oxidized at 140°C XPS shows the formation of Cu_xSe at the dry oxide/$CuInSe_2$ interface. The native oxide formed on a (110) cleaved $CuInSe_2$ surface by simple exposure to dry air at room temperature is primarily In_2O_3 with less than 5% SeO_2. If the oxidation is thermally induced using either dry oxygen or dry air, the fraction of SeO_2 is approximately doubled, and evidence for Cu_xSe is seen at the interface. For anodic oxidation (exposure to electrolyte solution of sodium tartrate/H_2O/tartaric acid, pH = 6.0 at room temperature), the fraction of In_2O_3 is reduced and evidence is found for about 10% Cu_2O, 10% CuO, and 2% Se are found, with CuO at the interface. For $CdS/CuInSe_2$ cells, a heat treatment at 200–220°C in oxygen or air consistently improves the operating characteristics, as illustrated in Fig. 6.10, while heating in high vacuum, Ar, N, H, or He does not seem to degrade the cell performance.

Non-ideal current-voltage characteristics of thin film $CuInSe_2$ solar cells such as cross-over of the dark and light current-voltage curves, or strongly bias dependent current collections, have been attributed to surface oxidation occurring during several processing steps (Walter *et al.*, 1995). These include

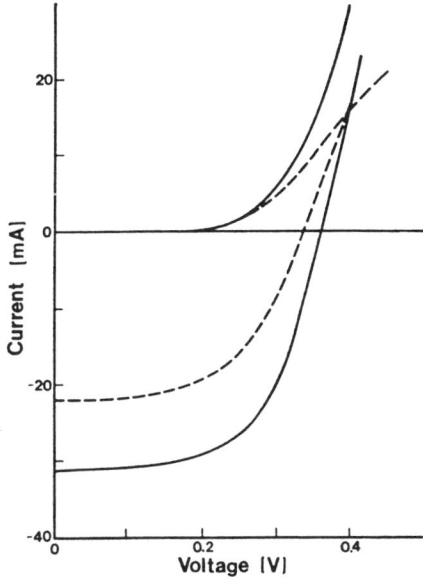

Fig. 6.10. Effect of heat-treatment in oxygen on light and dark current-voltage characteristics of CdS/CuInSe$_2$ heterojunction sólar cell. Dashed line, before O$_2$ anneal; solid line, after O$_2$ anneal. (Reprinted with permission from R. A. Mickelsen *et al.*, "Cadmium Sulfide/Copper Ternary Heterojunction Research", final Rep. XJ-9-8021-1, Boeing Aerospace Co., Seattle, WA (1982). Copyright 1982, Boeing Aerospace.)

the use of oxygen during the rf sputtering of the ZnO window layer, the air exposure of the absorber prior to the deposition of the window, and the use of an In$_x$(OH,S)$_y$ buffer layer.

Thermal Stability

Low-temperature annealing of the CdS/CuInSe$_2$ solar cell does not alter its electrical or interface composition (Kazmerski *et al.*, 1982; Kazmerski, 1983; Mickelsen and Chen, 1983; Hsiao *et al.*, 1983). Annealing at 60°C for 7500 hours causes no changes in cell performance or in the compositional appearance of the interface. Changes do begin to occur when the temperature is raised above 200°C. Two hours of annealing in Ar at 220°C caused decreases in ϕ_{oc} of 5%, J_{sc} of 17%, ff of 21%, and efficiency of 38%; these results suggest that the heat treatment has caused growth of the interfacial transition layer, and XPS identifies the existence of Cu$_x$S and Cu$_x$Se at this interface (Wager *et al.*, 1983). Heat treatment above 400°C causes a catastrophic degradation of solar cell performance due to the rapid diffusion of Cd across the interface region, an effect that is more severe in polycrystalline thin-film cells than in single crystal cells, suggesting that grain-boundary diffusion is involved.

Contacts

A basic characteristic of almost all $CuInSe_2$ solar cells is the Mo back contact. Often high-efficiency solar cells show a non-ohmic contact behavior that has been attributed to the $Mo/CuInSe_2$ back contact. Earlier measurements indicated that the nature of the Mo contact to single crystal $CuInSe_2$ depended on the defect density in the $CuInSe_2$ (Abou-Elfotouh *et al.*, 1989). The current-voltage characteristics of the $Mo/CuInSe_2$ contact on operating solar cells prepared by selenization of Cu/In precursors in H_2Se with the Mo deposited by sputtering, the CdS by chemical bath deposition, and the ZnO by rf sputtering, were measured by a technique that allowed separate evaluation of the contact properties and showed the Mo contact to be completely ohmic (Shafarman and Phillips, 1996). Simulations suggest that the $CdS/CuInSe_2$ interface is the most likely origin of the non-linear current-voltage effects.

6.3. Fabrication Techniques

A variety of techniques has been used for film and crystal growth of $CuInSe_2$, and fabrication of solar cells. These are summarized, with appropriate references, by Kazmerski and Wagner (1985). The materials requirements for the $CuInSe_2$ technology have also been reviewed by Birkmire *et al.* (1984).

Single-Crystal Growth

Methods used for single-crystal growth include (a) direct growth from elemental constituents by cooling a stoichiometric melt, zone-melting, horizontal directional solidification, and Bridgman growth (which has produced device quality p-type crystals — Bachmann 1983); (b) direct growth, not from the elemental constituents, but from binary compounds such as $Cu_2Se + In_2Se_3$, which enables the use of lower pressures in the growth ampoule; (c) chemical vapor transport, typically using iodine as the carrier for the ternary constituents; and (d) liquid-phase epitaxy on ZnSe substrates using Bi as a solvent.

Vacuum Evaporation

Three variations of vacuum evaporation have been used for the deposition of $CuInSe_2$ films. (1) Single-source deposition or flash-evaporation uses a single-phase powder as the source. Single-source deposition has not been particularly successful for CuInSe$_2$ since it often results in uncontrolled Se or Cu deficiencies. Flash evaporated films also have a very small grain size (< 100 nm). (2) Double-source deposition uses a second source of Se together with a CuInSe$_2$ source, to compensate for the loss of Se during deposition, and has been successful in producing device quality films with grain sizes of about 2.5 μm. Still the electrical properties of the films depend critically on the deposition parameters. (3) The greatest degree of control of the stoichiometry is achieved with the three-source deposition method, evaporating the three constituent elements independently (Mickelsen and Chen, 1981; Mickelsen *et al.*, 1982; Hsiao *et al.*, 1983).

Multisource vacuum evaporation has been used to analyze the growth process and the interactions of different phases (Klenk *et al.*, 1993). Growth of films has been analyzed for growth with excess Cu, CuInSe$_2$/CuGaSe$_2$ multilayers, and growth of CuIn(Se,S)$_2$ at different Cu/In ratios. It is concluded that excess Cu provides a binary copper chalcogenide that acts as a flux with a high diffusion coefficient for species involved in the growth process. It is not possible to adjust the S/Se ratio or to maintain Ga/In gradients as long as the growth is dominated by the copper chalcogenide.

Selenization of Metal Layers

A second major method for the preparation of CuInSe$_2$ films for solar cells is by selenization of previously deposited metallic Cu and In layers. Basol *et al.* (1989) and Basol and Kapur (1990b) used electron-beam evaporation to obtain Cu and In films deposited on a Mo-coated glass. Selenization was carried out in an H$_2$Se atmosphere to produce completed solar cells with an efficiency of almost 11%. Talieh and Rockett (1989) used a hybrid process consisting of sputtering Cu and In with evaporation of Se to produce final cells showing efficiencies greater than 7%. Dimmler *et al.* (1988) compared CuInSe$_2$ films produced both by thermal evaporation of the single elements and by selenization of the metal films and found many similarities between the two processes.

Process details in the selenization technique using H$_2$Se have been considered by Verma *et al.* (1991, 1992). Cu-In bilayers were selenized in H$_2$Se gas at 400°C in a tubular reactor, and results were compared with Cu-In bilayers selenized by physical vapor deposition using Se at substrate temperatures between 150°C and 400°C. It was observed that selenization in the PVD reactor with Se proceeded faster and at lower temperatures than in the tubular reactor with H$_2$Se.

High efficiency solar cells could not be made on $CuInSe_2$ films formed in a tubular reactor using only H_2Se as the reacting gas, but by mixing O_2 with the H_2Se, $CuInSe_2$ was formed from which cells with efficiencies over 10% were fabricated (Verma *et al.*, 1993). For O_2/H_2Se ratios of 0.9 to 0.2, Cu_7Se_4 and In_2O_3 were formed along with the $CuInSe_2$. But for low O_2/H_2Se ratios, 0.02 to 0.004, single phase $CuInSe_2$ films were formed suitable for photovoltaic cells. It was suggested that the presence of oxygen in a dilute H_2Se mixture stabilizes the morphology of Cu-In films when heat treated at 400°C in Ar or Ar/H_2, controls the amount of Se_x species formed in the gas phase, and causes In loss during $CuInSe_2$ film growth resulting in the formation of a copper selenide phase.

Yamanaka *et al.* (1993) evaluated the process of selenization of Cu-In layers with elemental Se at temperatures between 150° and 650°C. At temperatures above 250°C it was deduced that In in the liquid phase reacts with Se to form InSe, then reacts with Cu_7Se_4 to form $CuInSe_2$, which at these temperatures is the dominant phase in the film.

A model describing the chemical kinetics of the selenization of Cu-In layers has been developed (Russell *et al.*, 1994a,b) for both the tubular reactor growth with H_2Se and for a PVD selenization reactor.

A chemical modification of the Mo-coated substrate has been proposed as a means of obtaining better quality $CuInSe_2$ films (Basol *et al.*, 1991). Use of a Te interlayer on the Mo surface before the deposition of Cu/In precursors and subsequent selenization is reported to produce stoichiometrically, morphologically and mechanically superior $CuInSe_2$ films. It is suggested that thin interfacial layers of metals such as Cr and Ti might also effectively improve the adhesion of the $CuInSe_2$ films.

An alternate sequential deposition process involves the deposition of $InSe_x$ and Cu at a low substrate temperature followed by a high temperature selenization step, resulting in an efficiency of 14% for solar cells based on these layers (Zweigart *et al.*, 1995, 1996).

MOCVD

An extensive investigation has been made of the use of MOCVD techniques to deposit $CuInSe_2$ thin films for solar cell applications (Sagnes *et al.*, 1992; Artaud *et al.*, 1994a,b; Ouchen *et al.*, 1995a,b). It was shown that the MOCVD process performed in a horizontal reactor operating at atmospheric pressure, with H_2 as the carrier gas, was suitable to grow polycrystalline $CuInSe_2$ thin films with properties comparable to those deposited by vacuum evaporation.

The original approach involved a single step codeposition process from three precursors: TMIn (trimethylindium), (hfa)$_2$ Cu NHEt$_2$ (hexafluoroacetyl-acetonato copper mixed with diethylamine), and H$_2$Se. This produced material suitable for photovoltaic conversion, but with an efficiency below 3%. Better results involving enhanced grain size were indicated by a multistep deposition process using different substrate temperatures, and based on the growth of a ternary CuInSe$_2$ film on a starting layer of Cu(Se) or In(Se) to provide quasi-epitaxial growth.

Sputtering

Attempts to deposit CuInSe$_2$ films by radio frequency or magnetron sputtering, a technique that is promising for large-area cell production, result in films with grain size and reproducibility problems. Samaan *et al.* (1983) used rf sputtering in argon gas using *p*-type polycrystalline CuInSe$_2$ ingots or loose powders as the targets, Piekoszewski *et al.* (1980) used water-cooled, pressed powder targets of CuInSe$_2$, and Thornton *et al.* (1984) used reactive magnetron sputtering by simultaneously sputtering Cu and In in an Ar + H$_2$Se atmosphere to produce cells with an efficiency of 5–6%.

In an effort to reduce the density of defects in sputtered CuInSe$_2$ films, presumably caused by bombardment by high-energy particles from the Se target, a technique known as "hybrid sputtering" has been developed in the effort to make the voltage applied to all targets as low as possible (Nakada *et al.*, 1995). In this technique, the Se target source of high-energy ions is replaced by a Se effusion cell. Films deposited by this hybrid sputtering method show relatively large grain size and low density of defects, enabling the fabrication of an 11.3% efficient cell.

Spray Pyrolysis

The convenient, low-cost method of spray pyrolysis has been used to produce thin-film CuInSe$_2$ by spraying a freshly prepared aqueous solution of saturated CuCl, InCl$_3$ and N-dimethyl selenide in the ratio 4:1:0.5 onto a heated substrate (Pamplin and Feigelson, 1979; Gorska *et al.*, 1980). Mooney *et al.* (1980) used spray pyrolysis to produce low-resistivity *n*-type CuInSe$_2$ which was then converted to *p*-type by heat-treatment in H$_2$/H$_2$Se at 600°C, and showed a photovoltaic response in sprayed CdS/CuInSe$_2$ junctions. The technique has not yet been shown to be suitable for general solar cell applications.

Recently $CuInSe_2$/CdS heterojunction solar cells were deposited by spray pyrolysis with a maximum efficiency of 4–5% (Varner 1996). A Cu:In:Se ratio of 0.94:1:3.3 in solutions sprayed at 275°C produced the best results for $CuInSe_2$ films: $p = 10^{16}$ cm^{-3}, $\rho = 3000$ ohm cm, and $\mu = 0.2$ cm^2/V-sec. This mobility is about 25 times smaller than that obtained in evaporated $CuInSe_2$ films. A Cd:S:In ratio of 1:1.2:0.03 at 350°C with a $CdCl_2$ molarity of 10^{-2} M in solution, produced CdS films with $n = 8 \times 10^{16}$ cm^{-3}, $\rho = 40$ ohm cm and $\mu = 3$ cm^2/V-sec, a mobility that is about 1/3 that of evaporated CdS films. Both the CdS and the $CuInSe_2$ layers were subjected to a variety of analyses including electrical resistance, thermoelectric power, x-ray diffractometry, electron microprobe, and scanning electron and optical microscopy. The relatively low efficiency values were attributed to the small grain size of 0.1–0.2 μm in the sprayed $CuInSe_2$. As-deposited cells showed an efficiency of about 2%, which increased with prolonged exposure to air at room temperature, but could not be reproduced by heat treatments in air.

Electrodeposition

Pern *et al.* (1987) described the preparation of $CuInSe_2$ films by a one-step electrodeposition process. The film quality was affected by the concentration of the buffering agent ethylenediamine, which must be kept below 0.125 M to produce photovoltaically useful films.

The phases obtained under various conditions of potential and flux ratio in the application of electrodeposition techniques to the preparation of $CuInSe_2$ films have been investigated (Thouin and Vedel, 1995; Vedel *et al.*, 1996). It has been concluded that a combination of one-step electrodeposition and annealing treatment in a Se atmosphere is a promising approach for improving the quality of electrodeposited $CuInSe_2$ films for photovoltaic applications. (Guillemoles *et al.*, 1996; Lincot *et al.*, 1994). The properties of electrodeposited/selenized films were investigated by luminescence measurements, Hall effect and photoelectrochemical characterization. Films are p-type, with carrier densities of 10^{16}–10^{17} cm^{-3}, and diffusion length of about 1 μm. An efficiency of 6.5% is reported for the best $CuInSe_2$/CdS/ZnO solar cell prepared in this way to date. Interface recombination, not bulk properties, is shown to be the dominant loss mechanism.

Close-Spaced Vapor Transport

The close-spaced vapor transport method (CSVT) has been applied to the growth of $CuInSe_2$, $CuGaSe_2$ and $Cu(Ga,In)Se_2$ thin films (Masse and Djessas,

1993, 1994). The deposits were made on glass or Mo substrates below 620°C in glass tubes sealed under vacuum after the introduction of solid iodine. Films obtained above 550°C have a composition close to the stoichiometric composition. Except for low iodine concentrations (< 0.1 mg cm^{-3}) or high iodine concentrations (> 0.5 mg cm^{-3}), film growth transport is governed by diffusion.

Films deposited on SnO$_2$ show a new photovoltaic effect with the positive polarity on the SnO$_2$, presumably because of the formation of a doping gradient in the CuInSe$_2$ films due to the deposition conditions (Masse and Djessas, 1995).

Molecular Beam Epitaxy

Nearly stoichiometric epitaxial films of CuInSe$_2$ have been grown on GaAs (001) by molecular beam epitaxy, and low temperature photoluminescence measurements have been used to characterize defects involved in radiative recombination (Niki *et al.*, 1994).

6.4. Solar Cell Developments

The main developments in solar cells can be described best by distinguishing between the techniques used for film fabrication. The dominant two techniques are three-source vacuum evaporation, and selenization of pre-deposited Cu and In metal layers.

Fabricated by Three-Source Vacuum Evaporation

Mickelsen *et al.* (1987) and Devaney and Mickelsen (1988) describe the vacuum deposition processes successfully developed for the production of large area, high efficiency CuInSe$_2$ solar cells. In the fabrication of solar cells by these techniques (see Fig. 6.4), a Mo base electrode is sputter-deposited, two CuInSe$_2$ layers (one of low resistivity with Cu/In ratio slightly larger than one, and one of high resistivity with Cu/In ratio less than one) are deposited from three elemental sources by physical vapor deposition. Only Mo can be used as the contact metal because of the high reactivity of CuInSe$_2$ at these temperatures. The Zn$_x$Cd$_{1-x}$S window layer is deposited by either resistively heated or electron-beam evaporation sources with CdS, ZnS, or ZnCdS as the starting material. Replacement of the co-evaporation of ZnS and CdS sources by electron evaporation of a sintered Zn$_{0.13}$Cd$_{0.87}$S source material led to better composition uniformity. The single CdS or ZnCdS window layer was later replaced by two layers, a lower undoped layer next to the $CuInSe_2$ and an upper n-type doped layer. Heat treatment of the deposited cell for 10–40 minutes in oxygen at 225°C, needed to achieve optimum efficiency, apparently provides a gradation in composition of the two $CuInSe_2$ layers. Cells with areas up to 91 cm^2 were prepared with an efficiency of 9.5%. These cells had the following structure: soda lime glass/2–3 μm Mo/3 μm $CuInSe_2$/2 μm ZnCdS/Al/SiN/SiO. Early anti-reflection coatings of MgF_2/SiO_x were replaced by plasma enhanced chemical vapor deposition at 175°C of amorphous hydrogenated SiN and SiO films.

Several investigations were made of the effect of the specific composition of the $CuInSe_2$. Noufi *et al.* (1984) used the three-source deposition method to prepare a large number of films, varying the Cu/In ratio from 0.4 to 1.2 from film to film, while keeping the In and Se rates constant. The results indicated that the control of the electrical properties of $CuInSe_2$ films, and hence the performance of solar cells using them, required good control of stoichiometry since the electrical properties are dominated by native defects.

Noufi and Dick (1985) examined the compositional and electrical properties of the two-layer $CuInSe_2$ structure characteristic of the cells prepared by Mickelsen *et al.* (1987). They found that in the final solar cell device, the two individual $CuInSe_2$ layers had mixed and remained p-type during the cell processing steps, except for the very last few tenths of a micrometer to deposit, which was Cu-deficient and of high resistivity; the two CdS layers consisted of an almost stoichiometric layer close to the junction and a top In-doped low-resistivity layer. Rocheleau *et al.* (1987) investigated the tolerance of $CuInSe_2$ solar cells of this type to variations in film composition. They found that cell efficiencies of 10% or better could be obtained for a range of Cu from 23% to 27% and a range of indium from 25% to 28%.

The effects and mechanisms involved in the required air-bake to optimize these cells has been investigated. Noufi *et al.* (1985, 1986a) have studied the effect of the air-bake on the spectral response of $CdS/CuInSe_2$ cells, alternate methods of oxidizing or reducing the $CuInSe_2$ (Noufi *et al.*, 1986b), the dependence of the electrical conductivity of $CuInSe_2$ on composition and oxidation (Datta *et al.*, 1985), and the variation of the thermally stimulated conductivity of $CuInSe_2$ with oxidation (Datta *et al.*, 1986). Their results can be summarized as follows: (1) improvements in cell performance on air-baking consist primarily of increases in open-circuit voltage and fill factor; (2) air-baking was beneficial only if the temperature was not greater than 200°C, degradation being caused by higher temperatures; (3) the effect of the air-bake was to introduce oxygen into the CuInSe$_2$ with effects that could be duplicated by the use of chemical oxidants or reversed by the use of chemical reducing agents such as hydrazine or by exposure to an electron beam; (4) the effect of oxygen in the CuInSe$_2$ films was to increase the p-type electrical conductivity, presumably by the compensation of donors associated with interstitial In or In-on-Cu-site defects, as suggested also by a decrease in the thermally stimulated conductivity associated with these levels, or by tying up dangling bonds at the surface of selenium-deficient CIS grains. Damaskinos *et al.* (1987) showed that the resistivity of the CuInSe$_2$, which depends on its state of oxidation, is responsible in the solar cell for large reverse saturation current, low open-circuit voltage, and large cross-over between dark and light current versus voltage curves. All these phenomena appeared to be reversible by oxidation/reduction cycles.

Fabricated by Selenization of Metal Layers

Mitchell and Liu (1988) and Mitchell *et al.* (1988) summarize developments in the CuInSe$_2$ solar cell developed at ARCO Solar (now Siemens Solar) using an originally proprietary method for the details of the deposition of CuInSe$_2$, later revealed to be a two-step selenization process as described above (Love and Choudary, 1984; Kapur *et al.*, 1986; Ermer and Love, 1989; Eberspacher *et al.*, 1989). The thick ZnCdS layer is replaced by a very thin CdS layer covered by a transparent, conducting ZnO layer (Potter *et al.*, 1985). A typical cell structure for a ZnO/CdS/CuInSe$_2$ cell is glass/Mo/2 μm CuInSe$_2$/< 50 nm CdS/1 μm ZnO/grid metallization, as shown in Fig. 6.5. Unlike the cells fabricated by vacuum evaporation, no subsequent heat treatment is required. Like the cells fabricated by vacuum evaporation, the Cu/In ratio in the CIS determines the efficiency, and Cu/In = 0.92 to 0.97 is preferred (Zweibel and Barnett, 1993). The use of the very thin CdS layer and the transparent ZnO layer allow utilization of photons with wavelengths between 380 and 520 nm, which are normally absorbed in the thick CdS layer. The ZnO has a sheet resistance of less than 10 ohms/square, minimizing the series resistance losses. At 25°C the measured depletion layer width of 240 nm indicates that the thin undoped CdS is fully depleted. Single-cell efficiencies above 12% have been achieved, a 30 × 30 cm^2 module has been fabricated with a 9.7% active area efficiency, and with improved ZnO, a record short circuit current of 41.2 mA/cm^2 was achieved.

Several investigations have been directed toward the exploration of the possibility of replacing the CdS film in the ZnO/CdS/CuInSe$_2$ cell with a ZnSe film. Nouhi *et al.* (1988) reported that conducting ZnSe:In deposited by DC magnetron sputtering can be used as an alternative to CdS as an intermediate layer. In a hybrid cell of n-CdS/i-CdS/i-ZnSe/CuInSe$_2$, containing both a 1 μm thick conducting CdS film and a 1 μm thick insulating CdS film, as well as a < 50 nm thick insulating ZnSe film replacing the insulating CdS film in the ARCO ZnO/CdS/CuInSe$_2$ cell, it was shown (Yoo *et al.*, 1991) that (1) the cells with a ZnSe intermediate layer have an increase in the diode factor and a decrease in dark J_o compared to cells without a ZnSe layer, but (2) the cells with ZnSe do not show an increase in ϕ_{oc} as was hoped, because of a change in the junction transport mechanism under illumination, and (3) cells with a thicker ZnSe layer (≥ 40 nm) show strong photosuppression.

Research on ZnSe or ZnO buffer layers has shown that ZnSe/CuInSe$_2$ cells can be prepared with high voltages and efficiencies of the order of 14%, and efficient ZnO/CuInSe$_2$ cells with graded CuInSe$_2$ have been fabricated by MOCVD grown ZnO films (Olsen *et al.*, 1994, 1996). Cells with a ZnSe layer deposited by an atomic layer deposition method produced a CuIn$_{1-x}$Ga$_x$Se$_2$ thin-film solar cell with efficiency of 11.6% (Ohtake *et al.*, 1994). Cells with a chemical-bath-deposited ZnO buffer layer have yielded an efficiency of about 10% (Nii *et al.*, 1994).

Other replacements for the CdS film (or "buffer-layer") in Cu(In,Ga)(Se,S)$_2$ solar cells have also been proposed. Typical of these are In$_x$(OH,S)$_y$ and Sn(O,S)$_2$ deposited in a chemical bath process (Hariskos *et al.*, 1995, 1996). Conversion efficiencies of 15.7% were obtained with a CBD In$_x$(OH,S)$_y$ layer, but the device performance depended markedly on the illumination, correlated to light absorption in the buffer layer. Efficiencies of only 12% or less were obtained with SnO$_2$ or Sn(O,S)$_2$ buffer layers, and indications are that the absorber layer close to the heterojunction is affected by the buffer layer deposition.

Additional Solar Cell Insights

Several investigations indicate that the dominant mechanism controlling the junction current of CuInSe$_2$ solar cells at room temperature is recombination in the CuInSe$_2$ (Mitchell *et al.*, 1988; Roy *et al.*, 1988; Turner *et al.*, 1988; Yoo *et al.*, 1989; Shafarman and Phillips, 1991).

Ruberto and Rothwarf (1987) investigated the time-dependence of the open-circuit voltage with illumination time in CuInSe$_2$/CdS solar cells, and proposed a process involving the tunneling of electrons trapped in deep states in the CdS, near the junction with the CuInSe$_2$ layer, to the CuInSe$_2$ valence band.

CuInSe$_2$ junction structures have been analyzed by EBIC, optical beam-induced current (OBIC) and related measurements at room temperature (Chesarek *et al.*, 1988) and at low temperatures (Noufi *et al.*, 1988). Below 220 K a second space charge developed in the bulk of the CuInSe$_2$ in addition to the CuInSe$_2$/CdS interface, suggesting that the band diagram for the device was not that of a simple heterojunction. Minority carrier traps in CuInSe$_2$ have been detected by use of thermally stimulated capacitance spectroscopy (Ramanathan *et al.*, 1988).

A variety of models have been advanced to describe the properties of the solar cells based on CuInSe$_2$. These include a p–i–n heterojunction model (Rothwarf, 1982), a heterojunction model with the majority of the depletion region within the CuInSe$_2$ (Potter *et al.*, 1985), and a heterojunction model with charge on its interface (Rothwarf, 1986).

CuInSe$_2$/CdZnS/ITO solar cells prepared by different groups with different techniques have been compared (Shafarman and Phillips, 1993). Included were cells fabricated at the Institute of Energy Conversion (IEC) of the University of Delaware by evaporation, selenization with H$_2$Se, and selenization by Se; at Siemens Solar Industries (SSI); at Energy Photovoltaics, Inc.(EPV) by selenization by Se; and at the Institute for Physical Electronics (IPE) at the University of Stuttgart by evaporation. The cells were subjected to extensive current-voltage measurements over a wide range of temperatures. If the solar cell parameter data for all six of these quite different cell fabrication techniques are averaged, the strikingly similar results are $\phi_{oc} = 0.44 \pm 0.01$ V, $J_{sc} = 35 \pm 2$ mA/cm^2, $ff = 0.65 \pm 0.02$, and efficiency $= 10.1 \pm 0.5\%$. An analysis of the current-voltage curves shows that the basic junction mechanism is the same for all the cells with a barrier height of about 1.0 eV and a diode quality factor A of about 1.5. The different cells do show different nonlinear series resistances, which must be taken into account when analyzing the current-voltage data.

The role of the chemical-bath-deposited CdS layer on CuInSe$_2$ and Cu(In,Ga)Se$_2$ solar cells has been investigated to explore the origin of the "red kink" effect in which current-voltage curves show a "kink" around the open-circuit voltage when illuminated with red light (Hou *et al.*, 1996). The effect is associated with a low free electron density and a high trap density in the CdS layer. White light enhances the free electron density in the CdS and leads to positive charge trapping in this layer. This trapping enhances the photocarrier separating barrier in the absorber and improves the cell performance. Simulations of the effect are in agreement with the CuInSe$_2$ cell band structure suggested by Schmid *et al.* (1994).

Substrates

The commonest substrate for the $CuInSe_2$ solar cell is Mo on soda-lime glass. Effects due to Na diffusion from this glass substrate into the $CuInSe_2$ layer during cell fabrication by H_2Se selenization have been investigated (Basol *et al.*, 1994). Excessive Na diffusion through the Mo layer and its reaction with the processing atmosphere at the $Mo/CuInSe_2$ interface may limit the efficiency due to the formation of a high-resistivity layer near this contact. On the other hand, a moderate level of Na diffusion through the Mo layer dopes the $CuInSe_2$ layer *p*-type and improves the efficiency of the cells, especially those with low Cu/In ratios.

The influence of substrates, particularly alkali free substrates versus soda lime glass, on $Cu(In,Ga)Se_2$ thin films has been investigated by Ruckh *et al.* (1996b). The presence of Na in films on soda lime glass is correlated with an enhanced formation of Se–O, In–O, and Ga–O bonds at the surface after several days of exposure to air, the electrical conductivity is one order of magnitude higher, and solar cells prepared on these substrates exhibit increased open-circuit voltage. Junctions prepared on alkali free substrates show an increased space charge width. These results can be explained in terms of a higher acceptor density in the films prepared on soda lime glass.

Efforts have been directed toward developing flexible substrates for $CuInSe_2$ solar cells. The first efficient $CuInSe_2$ cell on a flexible metal substrate was described by Basol *et al.* (1993). A selenization technique was used involving H_2Se and e-beam evaporated Cu/In precursor layers to obtain $CuInSe_2$. Mo, Ti, and Al foils were evaluated as possible flexible substrate materials for solar cells. Flexible cells (capable of being bent to a radius of curvature of about 2 cm) upon Mo foil substrates had a conversion efficiency of above 8%.

Problems associated with these metal films when large-scale module fabrication was considered, led to the investigation of an insulating, light-weight and flexible polymeric substrate: a 50 μm thick KAPTON polyimide sheet supplied by Dupont (Basol *et al.*, 1996). The best 1 cm^2 area device showed a total area efficiency of 8.7%.

6.5.   Related Alloys and Materials

Chen *et al.* (1987) prepared thin-film polycrystalline solar cells based on heterojunctions of $CuGa_xIn_{1-x}Se_2/ZnCdS$ with values of x between 0.1 and 1.0. The quaternary selenide films were deposited by simultaneous elemental evaporation onto a heated Mo-coated alumina substrate, using a two-layer deposition procedure like that decribed above, and the $Zn_{0.12}Cd_{0.88}S$ window layer was deposited by vacuum evaporation in a separate chamber. Anti-reflection coatings of SiN and SiO were deposited by a low temperature, plasma enhanced, chemical vapor method. The band gap for the quaternary compounds varies from 1.0 eV for $x = 0$ to 1.7 eV for $x = 1.0$. Substitution of Ga for In causes an increase in the open-circuit voltage, but a decrease in the short-circuit current and a decrease in the fill factor. See Table 6.1 for typical values of the solar cell parameters. Effects of partial substitution of Ga for In appeared to be optimized for $x = 0.23$. $CuGa_xIn_{1-x}Se_2/ZnCdS$ heterojunctions fabricated by deposition by vacuum evaporation have also been investigated by Klenk *et al.* (1988), Dimmler *et al.* (1988) and Devaney *et al.* (1988, 1989).

Properties of solar cells involving $Cu(InGa)Se_2$ films deposited on Mo-coated glass substrates by vacuum evaporation from four elemental sources in a vacuum chamber have also been investigated (Birkmire *et al.*, 1995; Shafarman *et al.*, 1996). A simple two-step deposition process was used: (1) depositing a Cu-rich $Cu(InGa)Se_2$ film with the substrate heated to 450°C, followed by (2) increasing the substrate temperature to 600°C and shutting off the Cu flux while continuing the same In, Ga, and Se fluxes. The result is a single phase $Cu(InGa)Se_2$ film of homogeneous composition. Solar cells were completed using chemical-bath deposited CdS, sputtered ZnO window layers, evaporated Ni/Al, and MgF_2 anti-reflection layers. Figure 6.11 summarizes the dependence of (a) band gap, (b) open-circuit voltage, (c) short-circuit current, and (d) solar cell efficiency, for nine Ga/(In + Ga) ratios between 0.27 and 0.81. All the cells with Ga/(In + Ga) less than 0.5 have an efficiency of about 15%, decreasing to 8.8% for Ga/(In + Ga) = 0.81. The increase in open circuit voltage is proportional to the increase in band gap, and the short-circuit current decreases as expected with increasing band gap.

A three-stage selenization process has been used for the preparation of $CuGaInSe_2/CdS$ thin film solar cells with an efficiency greater than 15% (Romeo *et al.*, 1997). The three stages correspond to: (1) deposition of Cu, In, and Ga elemental layers on Mo-covered soda lime glass and selenization in Se vapor, (2) and (3) only Cu and Ga are used as elemental layers in order to increase the amount of Ga in the $CuGaInSe_2$ film.

The compound with composition $CuIn_3Se_5$, referred to as an ordered vacancy compound has been reported in bulk Cu-In-Se studies (Hoenle *et al.*, 1988). In spite of the apparently orderly variation of behavior with $x =$ Ga/(In + Ga) concentration indicated in Fig. 6.11, other evidence suggests that this $CuIn_3Se_5$ compound plays a significant role in solar cell performance for $x > 0.3$, replacing the standard p–i–n model of p-$Cu(InGa)Se_2/i$-$CdS/ZnO/$

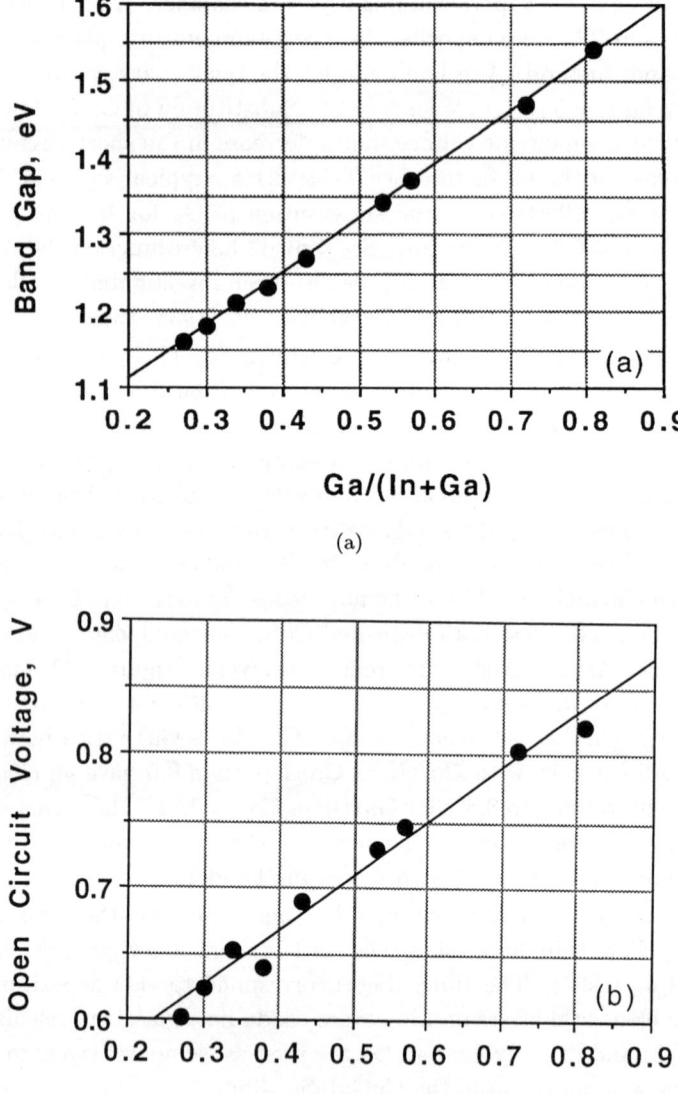

(a)

(b)

Fig. 6.11. Variation of properties of Cu(InGa)Se$_2$ solar cells deposited by four-source vacuum evaporation as a function of the Ga/(In + Ga) ratio. (a) Band gap, (b) open-circuit voltage, (c) short-circuit current, mA/cm^2, and (d) solar cell efficiency. (Data from Shafarman *et al.*, 1996.)

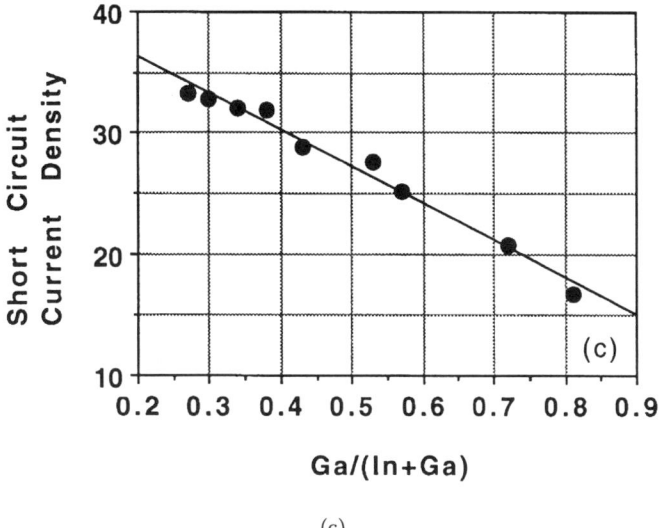

(c)

(d)

Fig. 6.11. (*Continued*)

n-ZnO) with a p–n junction model in which this n-type defect chalcopyrite material forms a p–n junction with its p-type counterpart CuInSe$_2$ (Schmid et al., 1993). Defect chalcopyrite materials in the Cu(In$_{1-x}$Ga$_x$)$_3$Se$_5$ system have been deposited in thin film form over the range of $0 < x < 1$ by coevaporation (Contreras *et al.*, 1996a,b). The band gap of these defect chalcopyrites is about 0.20 eV larger than an equivalent $Cu(In,Ga)Se_2$ material with similar Ga content. All defect materials of this type prepared to date show very high resistivity, low mobility, and low carrier concentration, but do seem to show a change in conductivity type from n-type to p-type as Ga is increased beyond $x > 0.3$.

A related investigation of secondary phases in the surface of coevaporated Cu-In-S films shows that In-rich films have an In-rich second phase at both surfaces, which has the cation ratio of $CuIn_3S_5$ (Scheer and Lewerenz, 1995). The results are interpreted as the formation of a phase of $CuIn_3S_5$ at the grain boundaries of In-rich $CuInS_2$ films. By way of contrast, Cu excess in Cu-In-S films leads to a segregation of CuS at the front surface of the films, and of MoS_2 at the back $Mo/CuInS_2$ interface.

Controlled incorporation of Na and S into the absorber layer of $Cu(In,Ga)Se_2$ solar cells has been investigated. $Cu(In,Ga)Se_2$ solar cells are reported to show appreciable improvement in performance when Na is present at the growth surface of films deposited by coevaporation (Rockett *et al.* 1996). Although it is reported that Na segregates rapidly out of the $Cu(In,Ga)Se_2$ for these deposition conditions, it leads to a strong (112) preferred orientation of the $Cu(In,Ga)Se_2$ and apparently improves the cell performance. In a related investigation of the effects of Na and S during the rapid thermal compound formation process (Karg *et al.*, 1993) involving Cu/Ga sputtering and Se evaporation, a Na-related shallow acceptor state at about 75 meV above the valence band was identified, and S incorporation apparently led to an increased open-circuit voltage because of a reduction in electrically active deep trap states (Rau *et al.*, 1996).

Nakada *et al.* (1997) investigated the effects of Na on $Cu(InGa)Se_2$ solar cells by a multistep process in which Na_2Se and Cu were evaporated simultaneously. The hole density could be held in the 10^{16} to 10^{17} cm^{-3} range for a wide range of $Cu/(In + Ga)$ ratios of 0.4 to 0.8 by the addition of Na_2Se, whereas without the Na_2Se, the hole density dropped below 10^{16} cm^{-3} for $Cu/(In + Ga)$ less than 0.8. Solar cell efficiencies of 10 to 13.5% were achieved over a wide range of $Cu/(In + Ga)$ ratios from 0.51 to 0.96. It was also found that p-type $Cu(In,Ga)_3Se_5$ phase films, which co-exist with the $Cu(In,Ga)Se_2$ phase for $Cu/(In + Ga) < 0.71$, could be obtained for the first time with a hole density high enough to allow use as an absorber layer for $Cu/(In + Ga)$ ratios in the range below 0.51, producing a 12% efficient solar cell for $Cu/(In + Ga) = 0.51$.

The technique of selenization of metal films has been applied to Cu(InGa)Se$_2$ films by Dittrich *et al.* (1988). High efficiency ($> 15\%$) Cu(InGa)Se$_2$ solar cells have been prepared by the selenization method (Birkmire *et al.*, 1995). Cu-In-Ga multilayer films were deposited by sputtering with a Cu/(In + Ga) ratio of 90% and Ga/(In + Ga) ratios from 0 to 75%. The films were reacted in H$_2$Se at 450°C for 90 minutes and then annealed at 500°C and 600°C for 90 minutes in Ar (Maraduchalam *et al.*, 1994). Solar cells were fabricated by chemical bath deposition of CdS and sputter deposition of a ZnO front contact. The as-reacted film appears to be nearly a bi-layer with CuInSe$_2$ on the surface and CuGaSe$_2$ near the substrate. Annealing at 500°C does not change this situation, but annealing at 600°C in Ar produces a single phase structure.

Several sequential deposition methods have been investigated for the fabrication of efficient Cu(In,Ga)(S,Se)$_2$ thin film solar cells (Zweigart *et al.*, 1994; Walter *et al.*, 1996). One such process involves the low-temperature deposition of In$_2$Se$_3$ and Cu precursors followed by selenization. A second process involves the indiffusion of Cu and (S,Se) into (In,Ga)$_2$(S,Se)$_3$ precursors at a high substrate temperature. These approaches lead to quite different film morphologies than does the coevaporation process, but solar cell parameters and performance are quite similar.

A two-stage Se vapor selenization of magnetron sputtered metallic precursors and Ga incorporation using a single Cu-Ga (22 at.%) alloy target, was developed for preparation of well-adherent, large, compact, well-faceted polyhedral grain CuIn$_{1-x}$Ga$_x$Se$_2$ thin films with optimum composition of Cu:In:Ga:Se of 22.95:25.03:1.40:50.63 (Dhere *et al.*, 1994).

A Zn$_{0.35}$Cd$_{0.65}$S/CuGaSe$_2$ cell corresponding to $x = 1$ was made with an efficiency of 5.8% (Dimmler *et al.*, 1987). Noufi *et al.* (1986c) investigated the optoelectronic properties and x-ray diffraction patterns of CuGaSe$_2$ thin films deposited by vacuum evaporation of the elements, as a function of composition. See Table 6.1 for solar cell parameters for ZnCdS/CuGaSe$_2$ and CdS/CuGaSe$_2$ cells.

CuInS$_2$ has a band gap of 1.5 eV that is an ideal match for the solar spectrum, and films based on CuInS$_2$ for solar cells can be prepared by a number of different processes, much like those for CuInSe$_2$. Mitchell *et al.* (1988) reported a ZnO/thin-CdS/CuInS$_2$ solar cell with efficiency of 7.3% and $\phi_{oc} = 0.59$ V. Scheer *et al.* (1993) reported a glass/Mo/p-CuInS$_2$/n-CdS/n^+-ZnO/Al cell prepared by thermal coevaporation of CuInS$_2$ and chemical bath deposition of the CdS, with an efficiency of 10.2%. The p-type CuInS$_2$ was prepared with a Cu/In ratio between 1.0 and 1.8; excess copper phases (CuS) were

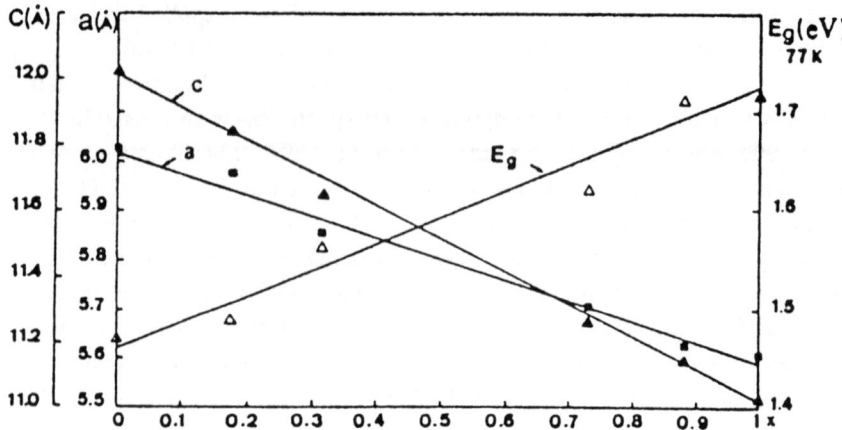

Fig. 6.12. Dependence of the lattice parameters and the energy gap on the composition x of $CuGa(Se_xTe_{1-x})_2$ alloys. (Reprinted with permission from G. Masse *et al.*, *J. Appl. Phys.* **74**, 1376 (1993). Copyright 1993 American Institute of Physics.)

removed chemically. The corresponding cell parameters were $\phi_{oc} = 0.697$ V, $J_{sc} = 21.5$ mA/cm^2, and $ff = 0.69$. Codeposition of the elements and sequential deposition processes have been compared for $CuInS_2$ with the conclusion that both types of processes lead to device efficiencies of 11 to 12% (Walter *et al.*, 1995). $In_x(OH,S)_y$ has been used to replace CdS as a buffer layer in 11.4% efficient solar cells based on $CuInS_2$ (Braunger *et al.*, 1996). An investigation of the majority and minority carriers' phototransport properties in polycrystalline $CuInS_2$ led to a density of states with four bands: shallow donors and shallow acceptors, and deep donors and deep acceptors that control recombination processes, centered between 0.3 and 0.5 eV below the conduction band and above the valence band, respectively (Lubianiker *et al.*, 1996). Properties of $CuInS_2$ thin films prepared by spray pyrolysis have been investigated (Ortega-Lopez and Morales-Acevedo, 1996), showing promise for use in solar cells.

CuGa(Se,Te)$_2$ bulk materials and thin films have been investigated by Masse *et al.* (1993). These alloys can be grown with *p*-type conductivity and with a good lattice match to CdS for a Se:Te ratio giving a band gap close to 1.5 eV. Thin films were grown using a CSVT technique, with iodine as the reagent. The dependence of band gap and lattice constants on Te/Se ratio is shown in Fig. 6.12. Good-quality CuGaSe$_2$ thin films were easily grown, but CuGa(Se$_x$Te$_{1-x}$) thin films could not be obtained when $x < 0.5$, as is expected from the CuGaTe$_2$ phase diagram. Thermal evaporation gave similar results, but flash evaporation gave thin films with a composition close to that of the source.

A general multinary system of form Cu$_{1-x}$Ag$_x$In$_{1-y-z}$Ga$_y$Al$_z$Se$_{2(1-u-w)}$ S$_{2u}$Te$_{2w}$ has been modeled by Dhingra and Rothwarf (1996), including graded band gap effects to improve the open-circuit voltage without sacrificing the short-circuit current in CuInSe$_2$/CdS solar cells. The model is based on a p–i–n structure with the graded bandgap material being the intrinsic region. Higher efficiency appears to be possible as a result of confining the incident light in the intrinsic region to achieve higher absorption efficiency, grading the conduction and valence bands to improve electron and hole collection efficiency with optimal grading being linear from 1.5 to 1.3 eV, and tailoring the recombination region to minimize recombination and achieve higher open-circuit voltage.

6.6. Multijunction Cells

A number of multijunction or stacked cells have involved a CuInSe$_2$ solar cell as one of the components. Some of these have been mentioned in earlier chapters and are summarized here for convenience. (1) A GaAs/CuInSe$_2$ multijunction gave an efficiency of 21.3% for a four-terminal device consisting of a GaAs CLEFT thin film top cell and a ZnCdS/CuInSe$_2$ thin film bottom cell (see Sec. 4.5). (2) Laminated four-terminal multijunction cells combining a-Si:H and CuInSe$_2$ cells have been developed (see Sec. 3.10). (3) Similar cells have been prepared using vacuum evaporation from elemental sources for the CuInSe$_2$, and photochemical vapor deposition for the a-Si:H cell (McCandless *et al.*, 1988). (4) Also investigated is a multijunction cell consisting of CdHgTe/CuInSe$_2$ (Meakin *et al.*, 1986).

6.7. General Materials Considerations

The fabrication and use of CuInSe$_2$ solar cells raises several questions in the areas of health and safety, and the supply of key elements (Zweibel and Barnett, 1993).

Hazardous chemicals like H$_2$Se and Cd are involved in the fabrication and structure of these cells. H$_2$Se is a highly toxic gas but it can be used safely provided that certain well-known safety precautions are used (Moskowitz *et al.*, 1986; Bottenberg and Sproull, 1988). Risks associated with it can be minimized if the gas is produced on-site and is recycled after use. Sources of Se are being sought that would eliminate the need for H$_2$Se completely.

The toxic dangers associated with Cd are certainly less in the $CdS/CuInSe_2$ cells than in the $CdS/CdTe$ cells, for which the issue has been discussed in Chapter 5, because of the smaller amount of Cd involved.

The use of Se in the cells requires proper disposal techniques to avoid contamination of groundwater. These appear to be available.

If $CuInSe_2$ cells enter into large-scale production, the availability of In might become an issue. Although it appears that there is a sufficient abundance of In to meet the growing demand of an effective solar cell market, there could be a conflict between supply and demand. Requirements for large amounts of In could be reduced by using thinner layers of $CuInSe_2$ made possible by the high absorption constant, using $CuGaInSe_2$ alloys in which about 25% of the In is replaced by Ga, and recycling cells after 30 years.

CHAPTER 7

OTHER MATERIALS OF INTEREST
FOR SOLAR CELLS

7.1. Overview

Earlier in this book we have stressed the fact that the actual number of different materials capable of being used as the light-absorbing material in stable, highly efficient, semiconductor junction, solar cells is strikingly small: Si, GaAs, CdTe, and CuInSe$_2$, with variations due to material structure and related alloys, represents the whole list. In this final chapter we take a brief look at some of the other materials and structures that either have been promising at some time in the past, or might be promising in the future, which have not been discussed previously in this book. Since 1993 a summary of the highest confirmed efficiency for a range of photovoltaic cells has been published every six months, including new entries as they qualify (Green, 1996).

7.2. Cuprous Oxide

In spite of the fact that Cu$_2$O has a direct band gap of 1.95 eV, somewhat large to match the solar spectrum, it has been thought attractive at various times in spite of its low efficiency, because of its very low production costs (Chopra and Das, 1983). Heating a Cu substrate in air at 1050°C allows the formation of fairly well oriented Cu$_2$O films, with a Schottky barrier being formed between the Cu substrate and the Cu$_2$O film. Deposition of an ohmic contact (e.g. Cu or Ni) to the Cu$_2$O completes the back-wall cell. Alternatively a metal layer can be deposited on the Cu$_2$O film to form a front-wall Schottky device.

Back-wall cells prepared by Olsen and Bohara (1975) are in effect MIS devices. Although it was suggested that efficiencies of 10% could be achieved by careful cell design, actual solar cell parameters were typically $\phi_{oc} = 0.37$ V, $J_{sc} = 7.7$ mA/cm^2, $ff = 0.57$ and efficiency = 1.6%. Just as the simple fabrication process of the Cu$_x$S/CdS cell could not make that cell practically useful because of its inherent instabilities, as described in Sec. 1.2, so the almost ideally low cost and simplicity of the fabrication process of the Cu/Cu$_2$O cell could not make that cell practically useful because of its low efficiency.

7.3. Zinc Phosphide

A careful search of the catalogue of possible materials suitable for solar cells led to the choice of Zn_3P_2 as a potentially ideal new material (Catalano *et al.*, 1978). It has a band gap of 1.5 eV and a high absorption constant ($> 10^5$ cm^{-1}), and can be readily prepared in thin-film form with p-type conductivity from inexpensive and available elements. In addition, Fe makes an excellent substrate for Zn_3P_2 films, since it has good lattice match and similar coefficient of thermal expansion, and forms a low-resistance ohmic contact. A closed-tube horizontal vapor transport system was used for the growth of large single crystals of Zn_3P_2 with and without iodine chemical transport; the as-grown resistivity of these crystals was about 50 ohm-cm (Wang *et al.*, 1981a,b). Schottky barrier Mg/Zn_3P_2 junctions on single crystal Zn_3P_2 showed solar cell efficiency of 6% (Bhushan and Catalano, 1981). Thin films can be grown by CSVT in Ar with large grains and long minority carrier diffusion lengths (0.5–4 μm). Schottky barrier cells using these polycrystalline films gave a total area efficiency of 4.3% (Bhushan, 1982).

Luminescence pair transitions (Briones *et al.*, 1981), ac-photoconductivity spectral response (Wang and Bube, 1982), electrical conductivity and Hall effect of sublimed and iodine-transport grown crystals (Wang *et al.*, 1982), and DLTS spectra corresponding to deep majority-carrier traps (Suda and Bube, 1984) have been investigated for Zn_3P_2 single crystals. It was found that vacuum evaporated contacts of Ag, Au or a Au/Ag alloy all exhibited linear current-voltage curves over the temperature range of interest. Three main deep hole traps were observed by DLTS with activation energy of 0.20, 0.36 and 0.73 eV from the top of the valence band; the densities of these levels, however, depended strongly on the thermal history of the sample.

Zn_3P_2 has not fulfilled its promise because of difficulties in obtaining good junction properties. In general n-type material could not be fabricated and thus n–p junctions were not possible. Only high-resistivity n-type Zn_3P_2 ($n < 10^{11}$ cm^{-3}), not suitable for solar cell applications, has been produced by MBE techniques (Suda *et al.*, 1996). Schottky barriers on p-type material do not follow a simple variation of barrier height with metal work function, and continuing research revealed that Mg/Zn_3P_2 junctions were the only junctions that showed the required electrical properties for solar cell fabrication. Attempts were made to prepare n–p junctions between Zn_3P_2 crystals and n-type vacuum evaporated CdS, DC-sputtered CdO, CVD ZnO, and electron-gun vacuum evaporated ITO (Wang *et al.*, 1982). All results were similar: high leakage currents in the dark and small photovoltaic effects. Similarly attempts to fabricate solar cells using n-ZnSe/p-Zn$_3$P$_2$ or n-ZnCdS/p-Zn$_3$P$_2$ heterojunctions were unsuccessful (Bhushan, 1984).

The reasons behind these materials problems with Zn$_3$P$_2$ were clarified by an investigation of the phenomena occurring at the surface of bulk polycrystalline Zn$_3$P$_2$ and at interfaces of Zn$_3$P$_2$-based devices (Casey *et al.*, 1987).

(1) Electrical measurements of Mg/Zn$_3$P$_2$ junctions formed on annealed Zn$_3$P$_2$ surfaces, and analyses of similar surfaces by Auger electron spectroscopy, indicated that loss of P from a very thin surface region is sufficient to severely degrade diode characteristics. Auger spectroscopy evidence for the surface loss of P when heated by an electron beam is shown in Fig. 7.1.

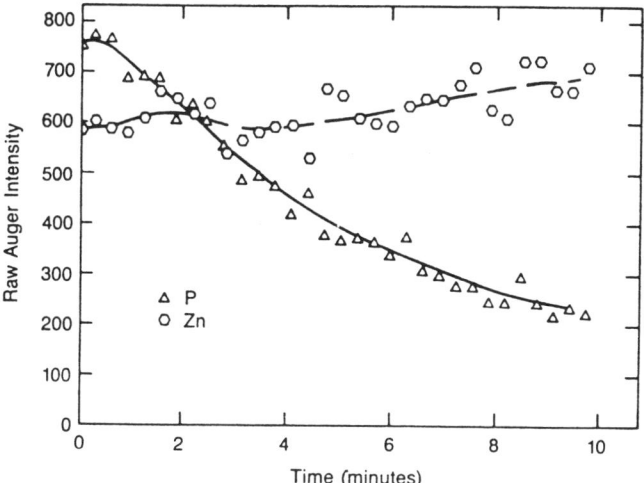

Fig. 7.1. Data from Auger spectroscopy on a Zn$_3$P$_2$ surface left under the electron beam, showing a decrease in the magnitude of the P signal with time. (Reprinted with permission from S. Casey *et al.*, *J. Appl. Phys.* **61**, 2944 (1987). Copyright 1987 American Institute of Physics).

(2) Surface photovoltage measurements showed that at the surface of as-etched or as-polished Zn$_3$P$_2$, the energy bands are nearly flat. The sign of the surface photovoltage is negative, indicating a depleted surface for p-type Zn$_3$P$_2$, but unlike the cases for CdTe and InP examined for comparison, there is no energy barrier to majority carriers for either etched or polished Zn$_3$P$_2$ surfaces. The observed barrier on p-type CdTe or InP correlated with the difficulty in producing ohmic contacts to these materials, while the

absence of a surface barrier on p-Zn_3P_2 is consistent with the observation that ohmic contacts are almost inevitable, but barriers are difficult to form.

(3) High-frequency capacitance measurements of $Al/Al_2O_3/Zn_3P_2$ structures indicated that a high interface state density ($> 10^{13}$ eV^{-1} cm^{-2}) prevents barrier formation by pinning the surface Fermi level. As shown in Fig. 7.2, these interface state densities are one order of magnitude larger than those measured on Al_2O_3 electron-beam evaporated on InP (Favennec *et al.*, 1979), two orders of magnitudes larger than those measured on Al_2O_3 electron-beam evaporated on Si (le Contellec and Morin, 1978), and almost four orders of magnitude larger than the results of similar analysis on an $Al/SiO_2/SiMOS$ device (Akinwande, 1987).

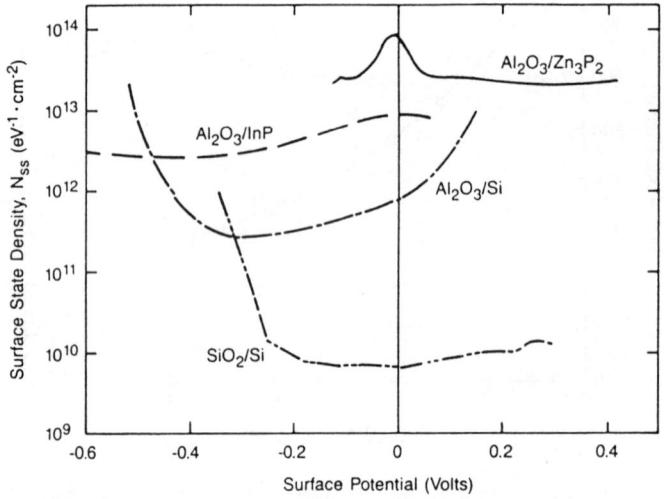

Fig. 7.2. Calculated surface state density as a function of semiconductor surface potential for an $Al/Al_2O_3/Zn_3P_2$ device. Also shown are the results of similar experiments with Al_2O_3 electron-beam evaporated on Si (le Contellec and Morin, 1978) and InP (Favennec *et al.*, 1979), and the results of a similar analysis on an $Al/SiO_2/SiMOS$ device (T. Akinwande, 1987). (Reprinted with permission from S. Casey *et al.*, *J. Appl. Phys.* **61**, 2944 (1987). Copyright 1987 American Institute of Physics).

7.4. Tungsten Diselenide

A compound with apparently promising properties is WSe_2, with a direct band gap of 1.35 eV, an extremely high optical absorption coefficient ($> 10^5$ cm^{-1}

over most of the range above its band gap energy, like CuInSe$_2$), excellent chemical stability, and the possibility of being prepared in either n- or p-type form. Schottky barrier Al/p-WSe$_2$ solar cells have been prepared with an efficiency of 5.3% (Clemen and Bucher, 1978). Difficulties were encountered with depositing films with the required structural properties for this layered material (Miller, 1981).

7.5. Polymers and Organic Materials

A fairly steady effort has been made to discover or fabricate an organic material suitable for use in solar cells. A review of organic photoconductors and photovoltaics is given by DiMarco and Giro (1994). In general organic materials have large energy gaps and low electrical conductivities, light absorption produces excitons not mobile carrier pairs, and the generation of mobile charges is strongly field dependent. The generation of photocurrent is by light absorption, exciton creation, exciton diffusion, exciton dissociation in the bulk or at the surface, field-assisted carrier separation, carrier transport, and carrier delivery to the external circuit. Models used to describe photovoltaic effects involve either Schottky-barrier type models, models based on the photoinjection of charges at the electrode interfaces, or p–n junction models formed between a p-type and an n-type organic material.

The attempt to use organic materials as the absorber layers in solar cells usually results in only low solar efficiencies because of the short minority carrier lifetimes and large series resistance characteristic of such materials. Although the monochromatic quantum efficiency of organic materials can approach 100%, the total quantum efficiency for the solar spectrum is quite low. Examples are Cr/(chlorophyll-a)Hg junctions that give $\phi_{oc} = 0.2$–0.5 V and a monochromatic conversion efficiency of 0.016% for illumination at 0.745 μm (Tang and Albrecht, 1975a,b), and a cell based on a merocyanine dye absorber layer giving $\phi_{oc} = 1.2$ V, $J_{sc} = 1.8$ mA cm^{-2} and solar efficiency of 0.7% (Morel *et al.*, 1978).

The molecular structures of several materials that have been tested in a Schottky-type cell, with an organic compound sandwiched between two metals, are given in Fig. 7.3. In the Schottky structure, illumination is through one of the two semitransparent metal electrodes. Films of these materials have been deposited by vacuum sublimation, spin coating, and electrochemical deposition. Most organic materials show p-type conductivity, so that the ohmic contact is a high-work-function metal such as Au or Ag, or ITO conducting glass, while the rectifying contact is a low-work function metal such as Al or

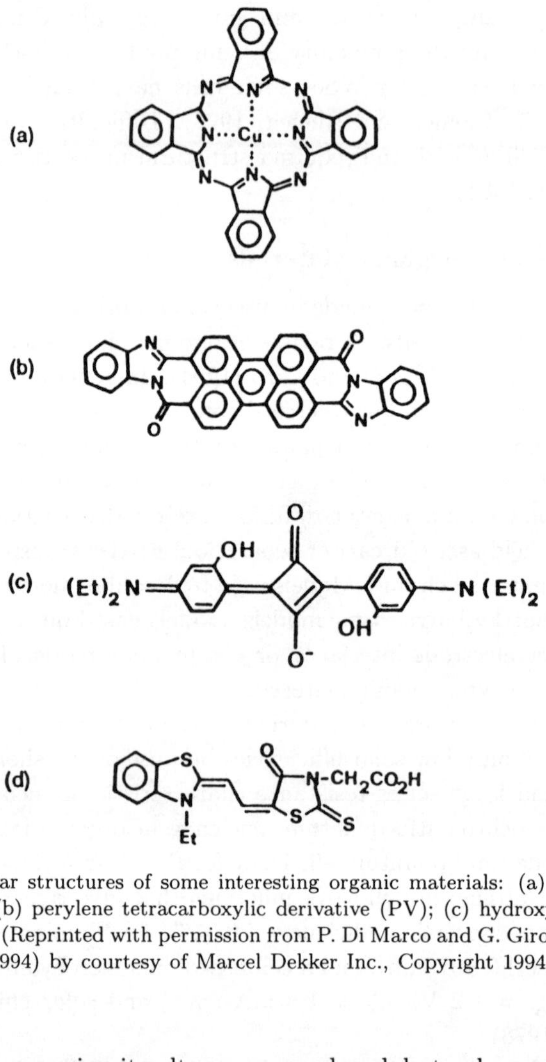

Fig. 7.3. Molecular structures of some interesting organic materials: (a) copper phthalocyanine (CuPc); (b) perylene tetracarboxylic derivative (PV); (c) hydroxysquarylium; (d) merocyanine dye. (Reprinted with permission from P. Di Marco and G. Giro in *Organic Conductors*, p. 791 (1994) by courtesy of Marcel Dekker Inc., Copyright 1994, Marcel Dekker, NY).

In. Reasonable open-circuit voltages are produced, but only small short-circuit currents. Solar efficiencies are usually less than 1%.

In p–n junction cells, illumination can be through a transparent electrode onto the n-type material, chosen to be the shorter-wavelength absorber with a fluorescence band that overlaps the absorption band of the p-type material. A notable improvement over earlier work was obtained by Tang (1986) using a cell pictured in Fig. 7.4, composed of ITO/copper-phthalocyanine (CuPc, p-type)/perylene tetracarboxylic derivative (PV, n-type)/Ag. Under AM2

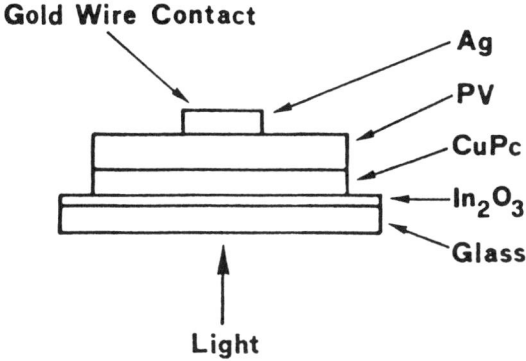

Light

Fig. 7.4. Schematic diagram of a *p–n* organic cell between *p*-type CuPc, and *n*-type PV. (Reprinted with permission from C. W. Tang, *Appl. Phys. Lett.* **48**, 183 (1986). Copyright 1986 American Institute of Physics).

illumination, a solar efficiency of 1% was achieved without correction for reflection or absorption losses.

The performance of polymer photovoltaic cells is greatly enhanced by the introduction of a network of internal donor-acceptor heterojunctions (Yu *et al.*, 1995). The semiconducting polymer is blended with C_{60} or its more soluble derivatives. Composite films of poly(2-methoxy-5-(2'-ethyl-hexyloxy)-1,4-phenylene vinylene) (MEH-PPV) and fullerenes have a collection efficiency of about 0.29 electrons per photon, and an energy efficiency of about 2.9%, values that are about two orders of magnitude larger than those obtained with devices made with pure MEH-PPV. The process depends upon the transfer of photoinduced electrons from the MEH-PPV (as donor) to C_{60} (as acceptor), via a bicontinuous network of internal donor-acceptor heterojunctions. Further research is focused on decreasing internal resistive losses, and optimization of the blend composition and network morphology.

Non-Organic Polymers

Sulfur nitride $(SN)_x$ is a material that has the possibility of behaving as a highly anisotropic semimetal with a high DC conductivity ($> 10^3$ ohm^{-1}cm^{-1}) along the polymer chains at room temperature (Chiang *et al.*, 1976). Making it of interest for Schottky barrier cells is the observation that $(SN)_x$ has a larger work function than any of the elemental metals (Scranton *et al.*, 1976; Best *et al.*, 1976). A polymer/semiconductor solar cell has been investigated using a $(SN)_x/n$-GaAs structure (Cohen and Harris, 1978). Open-circuit voltages greater than 0.70 V were obtained, compared to a value of 0.49 V for a Au/GaAs structure on the same GaAs, and without optimization a solar efficiency of 6% was obtained.

A similar metallic polymer is polyacetylene $(CH)_x$, which has a band gap of about 1.6 eV and can be doped either n-or p-type. Homojunctions, hetero-junctions with ZnS and CdS, and Schottky barriers with Ba, all show a large built-in voltage, but they appear to have low quantum efficiencies for solar illumination because of small minority carrier diffusion lengths (Waldrop *et al.*, 1981).

7.6. Semiconductor/Liquid Junctions

An introduction to photoelectrochemical (PEC) cells was given in Sec. 1.3 with descriptions in Fig. 1.12. Such cells use a junction between a semiconductor and an electrolyte to convert light to electrical and/or chemical energy. Most PEC cells are quite similar to Schottky barrier solar cells, with a depletion layer forming in the semiconductor and the metal layer replaced by a highly conductive electrolyte and a counter electrode. In the photovoltaic application, of principal interest to us here, charge transport through the electrolyte is accompanied by various redox reactions that leave the electrolyte and hopefully also the semiconductor electrode unchanged.

A review of recent developments has been given by Lewis (1995). Standard semiconductors useful in solar cells are not generally useful in photoelectro-chemical cells with an aqueous electrolyte. Silicon photoelectrodes, for example, form an insulating layer of silicon dioxide on exposure to light. GaAs photoelectrodes dissolve away when illuminated. Fujishima (1972) found that titanium dioxide was a stable electrode in the light-driven electrolysis of water into hydrogen and oxygen gases, but the band gap of TiO_2 is too large for effective conversion of solar energy and produced a solar efficiency of only 1%. Subsequent research indicated that all other materials with smaller band gaps than TiO_2 (e.g. materials obtained by replacing the oxygen in TiO_2 by sulfur, selenium or tellurium) were unstable and would eventually corrode.

It has been reported that the corrosion of silicon and other semiconductor photoelectrodes can be prevented by use of a noble metal coating (Tsubomura and Nakato, 1984). A photoelectrochemical cell was developed composed of a noble-metal-coated p–n junction Si photoanode, a counter Pt electrode, and a cation exchange membrane, by which HI is efficiently decomposed into hydrogen and iodine. A solar to chemical conversion efficiency of 8.2% was reported.

The use of nonaqueous photoelectrochemical cells, containing organically-based electrolytes, overcomes some of the difficulties and allows the use of semiconductors such as silicon that are unstable in water. In a nonaqueous photoelectrochemical cell, the carriers generated in the semiconductor by illumination are transferred to a stabilizing chemical reagent in the electrolyte solution that accepts an electron at one electrode and donates an electron at the other electrode. The stabilizer itself does not undergo a net chemical change. An example of such a cell (Lewis, 1995) is one in which the semiconductor is silicon, the nonaqueous electrolyte consists of an organic solvent (propylene carbonate), a salt (lithium perchlorate) to provide electrical conductivity, and soluble ions (an organometallic iron complex) to carry the electrical charges through the liquid. Solar efficiencies of 12 to 14% are reported for such cells. The efficiency of the cell can be optimized by selecting the chemical design of the electrolyte, the photoelectrode material, and the stabilizing agent. Stable photoelectrochemical calls in nonaqueous solutions have been made with Si, GaAs, InP, and some other semiconductors with suitable band gap.

Another approach to photoelectrochemical cells involved immersing small silicon solar cells in an electrolyte, so that the current generated by the cells is used directly to electrolyze a halogen acid, such as HBr (Johnson, 1981; Bawa and Chang, 1985). The hydrogen and bromine produced can be stored separately and then recombined in a fuel cell to produce electrical energy on demand.

Some success has also been achieved in obtaining increased quantum yields on large-band-gap materials such as TiO_2 by the use of surface dyes (O'Regan and Graetzel, 1991). Such dyes have extended the absorption range of TiO_2 into the visible portion of the spectrum. Systems with nonaqueous solvents permit greater stability and a wider choice of dye materials; solar efficiencies greater than 6% have been reported.

7.7. New Materials and Structures

A review of new materials and structures for photovoltaics has been given by Zunger *et al.* (1993). These include nontraditional alloys, materials with reduced dimensionality, spontaneously ordered alloys, interstitial semiconductors, filled tetrahedral structures, ordered vacancy compounds, and compounds based on d and f electron elements. A major motivation in the design of these new materials is the ability to achieve band gap tuning and lattice matching simultaneously.

Superlattices and multilayer structures offer the potential for producing devices with tunable properties. An a-Si:H solar cell has been developed with a p-layer with a superlattice structure (Tsuda *et al.*, 1985; Arya *et al.*, 1985); an increase in the band gap and doping of the p-layer in an otherwise normal p–i–n a-Si:H structure resulted in an increase in open-circuit voltage.

Wronski *et al.* (1987a,b,c) describe research on a-Ge:H/a-Si:H superlattices used in a Pt Schottky barrier solar cell structure. They show the existence of quantum size effects in amorphous semiconductors, a shift in the optical and carrier transport edges as a function of the thicknesses of the layers, and the possibility of efficient collection of carriers photogenerated in the barrier region.

Research on a-Si:H,F/a-SiGe:H,F multilayer superlattices by Conde *et al.* (1987, 1988a,b) indicate that optical band gap, photo-and dark conductivities, and carrier collection can be tuned within limits.

An increase in efficiency of p–i–n structures has been sought by the incorporation of quantum wells (Barnham *et al.*, 1991). The application of these ideas to InP solar cells with lattice matched $In_{0.53}Ga_{0.47}As$ multiple quantum wells makes use of the fact that strain free InGaAs wells can be incorporated in the i-region, which are deeper than for any other material used in this way (Freundlich *et al.*, 1994; Anderson and Wojtczuk, 1996; Zachariou *et al.*, 1996). Conversion efficiencies of $9 \pm 2\%$ have been achieved.

REFERENCES

A. Aberle, S. Glunz and W. Warta, *J. Appl. Phys.* **71**, 4422 (1992).

A. G. Aberle, P. P. Altermatt, G. Heiser, S. J. Robinson, A. Wang, J. Zhao, U. Krumbein and M. A. Green, *J. Appl. Phys.* **77**, 3491 (1995).

F. Abou-Elfotouh, D. J. Dunlavy and L. L. Kazmerski, *Proc. 17th IEEE Photovoltaic Specialists Conference*, IEEE Publishing, NY (1984).

F. Abou-Elfotouh, L. Kazmerski, T. Coutts, R. Matson, S. Asher, A. Nelson and A. Swartzlander-Franz, *J. Vac. Sci. Technol.* **A7**, 837 (1989).

D. Adler, in *Semiconductors and Semimetals*, Vol. 21A (J. I. Pankove, ed.) Academic Press, Orlando (1984), p. 291.

W. G. Adams and R. E. Day, *Proc. Roy. Soc.*, London, A **25**, 113 (1877).

R. K. Ahrenkiel, D. J. Dunlavy, B. M. Keyes, S. M. Vernon, S. P. Tobin and T. M. Dixon, *21st IEEE Photovoltaic Specialists Conference* (IEEE Publishing, NY), p. 432 (1990).

R. K. Ahrenkiel, *Solid State Electronics* **35**, 239 (1992).

R. K. Ahrenkiel, B. M. Keyes, S. M. Durbin and J. L. Gray, *23rd IEEE Photovoltaic Specialists Conference* (IEEE Publishing, NY), p. 42 (1993).

R. K. Ahrenkiel, B. M. Keyes, D. H. Levi, K. Emergy, T. L. Chu and S. S. Chu, *Appl. Phys. Lett.* **64**, 2879 (1994).

R. K. Ahrenkiel, D. Levi and J. Arch, *Solar Energy Materials and Solar Cells* **41**, 171 (1996).

T. Akinwande, private communication (1987).

S. K. J. Al-Ani, M. N. Makadsi, I. K. Al-Shakarchi and C. A. Hogarth, *J. Mat. Sci.* **28**, 251 (1993).

S. P. Albright, V. P. Singh and J. F. Jordan, *Solar Cells* **24**, 43 (1988).

S. P. Albright, J. F. Jordan, B. Ackerman and R. R. Chamberlin, *Photovoltaics Advanced Research and Development*, 9th Ann. Rev. Meet. SERI/CP-213-3495, Lakewood, CO (1989), p. 17.

S. Aljishi, Z. E. Smith, V. Chu, J. Kolodzey, D. Slobodin *et al.*, *International Conference on the Stability of Amorphous Silicon Alloy Materials and Devices*, Palo Alto, CA (1987).

Z. I. Alferov, U. M. Andreev, M. B. Kagan, I. I. Protasov and V. G. Trofim, *Sov. Phys. - Semicond.* **4**, 2047 (1971) [Fiz. Tekh. Poluprovodn. **4**, 2378 (1970)].

M. M. Al-Jassim, F. S. Hasoon, K. M. Joes, B. M. Keyes, R. J. Matson and H. R. Moutinho, *23rd IEEE Photovoltaic Specialists Conference*, IEEE Publishing, NY (1993), p. 459.

N. Amer and W. Jackson, in *Semiconductors and Semimetals*, Vol. 21B (J. I. Pankove, ed.) Academic Press, Orlando (1984), p. 83.

R. L. Anderson, *Solid State Electron.* **5**, 341 (1962).

W. A. Anderson, A. E. Delahoy and R. A. Milano, *J. Appl. Phys.* **45**, 3913 (1974).

W. A. Anderson, J. K. Kim and A. E. Delahoy, *IEEE Trans. Electron Devices* **ED-24**, 453 (1977).

D. A. Anderson, T. D. Moustakas and W. Paul, *7th International Conference on Amorphous and Liquid Semiconductors*, Edinburgh (1977), p. 334.

N. Anderson and S. J. Wojtczuk, *J. Appl. Phys.* **79**, 1973 (1996).

V. M. Andreev, V. P. Khvostikov, E. V. Paleeva, S. V. Sorsokina and M. Z. Shvarts, *25th IEEE Photovoltaic Specialists Conference*, IEEE Publishing, NY (1996), p. 143.

B. E. Anspaugh, *GaAs Solar Cell Radiation Handbook*, Jet Propulsion Laboratory Publication 96-9, NASA, Jet Propulsion Laboratory, California Institute of Technology, Pasadena, CA (1996).

T. C. Anthony, A. L. Fahrenbruch and R. H. Bube, *J. Electron. Mat.* **11**, 89 (1982).

T. C. Anthony, C. Fortmann, W. Huber, R. H. Bube and A. L. Fahrenbruch, *17th IEEE Photovoltaic Specialists Conference*, IEEE Publishing, NY (1984a).

T. C. Anthony, A. L. Fahrenbruch and R. H. Bube, *J. Vac. Sci. Technol.* **A2**, 1296 (1984b).

T. C. Anthony, A. L. Fahrenbruch, M. G. Peters and R. H. Bube, *J. Appl. Phys.* **57**, 400 (1985).

G. A. Antypas and R. L. Moon, *J. Electrochem. Soc.* **120**, 1574 (1974).

J. Aranovich, A. Ortiz and R. H. Bube, *J. Vac. Sci. Technol.* **16**, 994 (1979).

J. A. Aranovich, D. Golmayo, A. L. Fahrenbruch and R. H. Bube, *J. Appl. Phys.* **51**, 4260 (1980).

J. M. Arias, S. H. Shin, D. E. Cooper, M. Zandian, J. G. Pasko, E. R. Gertner, R. E. DeWames and J. Singh, *J. Vac. Sci. Technol.* **A8**, 1025 (1990).

S. Arimoto, H. Morikawa, M. Deguchi, Y. Kawama, Y. Matsuno, T. Ishihara, H. Kumabe, T. Murotani and S. Mitsui, *Technical Digest of the 7th International Photovoltaic Science and Engineering Conference*, Dept. Electrical and Computer Eng., Nagoya Inst. of Tech., Nagoya Japan (1993), p. 103.

R. A. Arndt, *11th IEEE Photovoltaic Specialists Conference*, IEEE Publishing, NY (1975), p. 40.

M. C. Artaud, F. Ouchen, S. Duchemin and J. Bougnot, *Polycrystalline semiconductors III – Solid State Phenomena* **37**, 503 (1994a).

M. C. Artaud, F. Ouchen, S. Duchemin and J. Bougnot, *12th European Community Photovoltaic Solar Energy Conference* (1994), p. 658.

R. R. Arya, A. Catalano, J. O'Dowd, J. Morris and G. Wood, *18th IEEE Photovoltaic Specialists Conference* (1985), p. 1710.

R. R. Arya, M. Bennet, L. Yang, J. Newton, Y. M. Li, N. Maley and B. Fiesel, *Technical Digest of the 7th International Photovoltaic Science and Engineering Conference*, Dept. Electrical and Computer Eng., Nagoya Inst. of Tech., Nagoya, Japan (1993), p. 41.

M. Aucouturier, *Physica B* **170**, 4369 (1991).

I. Austin, T. Nashashibi, T. Searle, P. LeComber and W. Spear, *J. Non-Cryst. Solids* **32**, 373 (1979).

K. J. Bachmann, E. Buehler, J. L. Shay and S. Wagner, *Appl. Phys. Lett.* **29**, 121 (1976).

K. J. Bachmann, H. Schreiber, Jr., W. R. Sinclair, P. H. Schmidt, F. A. Thiel, E. G. Spencer, G. Pasteur, W. L. Feldmann and K. S. Sree Harsha, *J. Appl. Phys.*

50, 3441 (1979).

K. J. Bachmann, "DOE/SERI Review of Polycrystalline Compound Solar Cells", *SERI*, Golden, Colorado (1983).

A. Baldus, A. Bett, O. V. Sulima and W. Wettling, *J. Crystal Growth* **141**, 315 (1994a).

A. Baldus, W. W. Bett, U. Blieske, T. Duong, F. Lutz, C. Schetter, W. Wettling and O. V. Sulima, *First World Conference on Photovoltaic Energy Conversion, 24th IEEE Photovoltaic Specialists Conference*, IEEE Publishing, NY (1994b), p. 1697.

A. Baldus, A. Bett, U. Blieske, O. V. Sulima and W. Wettling, *12th European Photovoltaic Solar Energy Conference*, Amsterdam (1994c), p. 1485.

A. Baldus, A. Bett, O. V. Sulima and W. Wettling, *J. Crystal Growth* **146**, 305 (1995).

M. Barbe, F. Bailly, D. Lincot and G. Cohen-Solal, *16th IEEE Photovoltaics Specialists Conference*, IEEE Publishing, NY (1982), p. 1133.

A. M. Barnett, Institute of Energy Conversion, Final Report, E(49-18)-2538 FR77, to the Energy Research and Development Administration, University of Delaware, Newark, DE (1977).

A. M. Barnett, F. A. Domian, D. H. Ford, C. L. Kendall, J. A. Rand *et al.*, *Photovoltaics Advanced Research and Development*, 9th Annu. Rev. Meet. SERI/CP-213-3495, Lakewood, CO (1989), p. 21.

A. M. Barnett, R. B. Hall, J. A. Rand, C. L. Kendall and D. H. Ford, *Solar Energy Materials* **23**, 164 (1991).

A. M. Barnett, T. B. Bledsoe, W. R. Bottenberg, D. S. Brooks, D. H. Ford, R. B. Hall, T. Huges-Lampros, E. L. Jackson, C. L. Kendall, W. P. Mulligan and K. P. Shreve, *11th EC Photovoltaic Solar Energy Conference* (L. Guimaraes, W. Palz, C. DeReyff, H. Kiess and P. Helm, eds.) Harwood Academic Publishers, Chur (1993), p. 1063.

K. W. J. Barnham and G. Duggan, *J. Appl. Phys.* **67**, 3490 (1990).

K. W. J. Barnham, B. Braun, J. Nelson and M. Paxman, *Appl. Phys. Lett.* **59**, 135 (1991).

B. M. Basol, *J. Appl. Phys.* **55**, 601 (1984).

B. M. Basol, S. S. Ou and O. M. Stafsudd, *J. Appl. Phys.* **58**, 3809 (1985).

B. M. Basol and E. S. Tseng, *Appl. Phys. Lett.* **48**, 946 (1986).

B. M. Basol, *Solar Cells* **23**, 69 (1988).

B. M. Basol, V. K. Kapur and R. C. Kullberg, *Photovoltaics Advanced Research and Development*, 9th Ann. Rev. Meet. SERI/CP-213-3495, Lakewood, CO (1989), p. 73.

B. M. Basol and V. K. Kapur, *IEEE Trans. on Electron Devices* **37**, 418 (1990b).

B. M. Basol and V. K. Kapur, U.S. Patent 4,950,615 (1990a).

B. M. Basol, V. K. Kapur and R. J. Matson, *22nd IEEE Photovoltaic Specialists Conference*, IEEE Publishing, NY (1991), p. 1179.

B. M. Basol, *J. Vac. Sci. Technol.* **A10**, 2006 (1992).

B. M. Basol, *Int. J. Solar Energy* **12**, 25 (1992).

B. M. Basol, V. K. Kapur, A. Halani and C. Leidholm, *Solar Energy Materials and Solar Cells* **29**, 163 (1993).

B. M. Basol, V. K. Kapur, C. R. Leidholm, A. Minnick and A. Halani, *First World Conference on Photovoltaic Energy Conversion, 24th IEEE Photovoltaic Specialists Conference*, IEEE Publishing, NY (1994), p. 148.

B. M. Basol, V. K. Kapur, C. R. Leidholm and A. Halani, *25th IEEE Photovoltaic Specialists Conference*, IEEE Publishing, NY (1996), p. 157.

P. A. Basore, *Progress in Photovoltaics Research and Applications* **2**, 177 (1994).

P. A. Basore, J. M. Gee, M. E. Buck, W. K. Schubert and D. S. Ruby, *Solar Energy Materials and Solar Cells* **34**, 91 (1994).

G. H. Bauer, C. E. Nebel and W. H. Bloss, *8th EC Photovoltaic Solar Energy Conf.*, Florence (1988).

G. H. Bauer, *Solid State Phenomena* **44-46**, 365 (1995).

M. S. Bawa and R. R. Chang, *J. Solar Energy Engineering, Transactions ASME*, **107** (4), 355 (1985).

A. Beck, J. Geissler, D. Helmreich and R. Wahlich, *J. Crystal Growth* **82**, 127 (1987).

N. Beck, N. Wyrsch, E. Sauvain and A. Shah, *Proc. Mater. Res. Soc. Symp.* **297**, 479 (1993).

N. Beck, J. Meier, J. Fric, Z. Remes, A. Poruba, R. Flueckiger, J. Pohl, A. Shah and M. Vanecek, *J. Non-Cryst. Solids* **198-200**, 903 (1996).

N. Beck, N. Wyrsch, Ch. Hof and A. Shah, *J. Appl. Phys.* **79**, 9361 (1996).

E. Becquerel, *Compt. Rend.* **9**, 561 (1839).

S. M. Bedair, M. F. Lamorte, J. R. Hauser and K. W. Mitchell, *Proceedings of the International Electron Devices Meeting* (1978), p. 250.

S. M. Bedair, M. F. Lamorte and J. R. Hauser, *Appl. Phys. Lett.* **34**, 38 (1979).

Bell Labs, Record, Nov. (1954) p. 436; Bell Labs, Record, May (1955), p. 166.

L. Benatar, M. Grimbergen, A. Fahrenbruch, A. Lopez-Otero, D. Redfield and R. H. Bube, *Proc. Mater. Res. Soc. Symp.* **258**, 461 (1992).

L. Benatar, Ph.D. thesis, Stanford University (1993).

M. S. Bennett, J. L. Newton, K. Rajan and A. Rothwarf, *J. Appl. Phys.* **62**, 3698 (1987).

R. G. Benz *et al.*, *J. Vac. Sci. Technol.* **A8**, 1020 (1990).

R. Bergmann, J. Kuehnle, J. H. Werner, S. Oelting, M. Albrecht, H. P. Strunk, K. Herz and M. Powalla, *First World Conference on Photovoltaic Energy Conversion, 24th IEEE Photovoltaic Specialists Conference*, IEEE Publishing, NY (1994), p. 1398.

R. Bergmann, G. Oswald, M. Albrecht and J. H. Werner, in *Polycrystalline Semiconductors IV — Physics, Chemistry and Technology* (S. Pizzini, H. P. Strunk and J. H. Werner, eds.) in Series on Solid State Phenomena, Trans. Tech. Publ., Zug, Switzerland (1995).

R. B. Bergmann, R. Brendel, M. Wolf, P. Loelgen, J. H. Werner, J. Krinke and H. P. Strunk, *25th IEEE Photovoltaic Specialists Conference*, IEEE Publishing, NY (1996), p. 365.

R. B. Bergmann, R. Brendel, M. Wolf, P. Loelgen, J. Krinke, H. P. Strunk and H. H. Werner, *Semicond. Sci. Technol.* **12**, 224 (1997).

A. K. Berry, J. L. Boone and T. P. Van Doren, *Materials Letters* **10**, 261 (1990).

K. A. Bertness, M. Ladle Ristow and H. C. Hamaker, *20th IEEE Photovoltaic Specialists Conference*, IEEE Publishing, NY (1988), p. 769.

K. A. Bertness, S. R. Kurtz. D. J. Friedman, A. E. Kibbler, C. Kramer and J. M. Olson, *24th IEEE Photovoltaic Specialists Conference*, IEEE Publishing, NY (1994), p. 1671.

J. S. Best, J. O. McCaldin, T. C. McGill, C. A. Mead and J. B. Mooney, *Appl. Phys. Lett.* **29**, 433 (1976).

A. Bett, S. Cardona, A. Ehrhardt, F. Lutz, H. Welter and W. Wettling, *22nd IEEE Photovoltaic Specialists Conference*, IEEE Publishing, NY (1991), p. 137.

A. Bett, K. Borgwarth, Ch. Schetter, O. V. Sulima and W. Wettling, *6th International Photovoltaic Science and Engineering Conference*, New Delhi (1992), p. 843.

M. Bettini, K. J. Bachmann, E. Buehler, J. L. Shay and S. Wagner, *J. Appl. Phys.* **48**, 1603 (1977).

M. Bettini, K. J. Bachmann and J. L. Shay, *J. Appl. Phys.* **49**, 865 (1978).

W. Beyer, H. Mell and H. Overhof, in *Amorphous and Liquid Semiconductors* (W. E. Spear, ed.) CICL, Edinburgh (1977), p. 328.

M. T. Bhatti, K. M. Hynes, R. W. Miles and R. Hill, *Int. J. of Solar Energy* **12**, 171 (1992).

M. Bhushan and A. Catalano, *Appl. Phys. Lett.* **38**, 39 (1981).

M. Bhushan, *Appl. Phys. Lett.* **40**, 51 (1982).

M. Bhushan, Final Report, SERI Subcontract # XE-2-02048-1 (1984).

R. N. Bicknell, N. C. Giles and J. F. Schetzina, *Appl. Phys. Lett.* **49**, 1095, 1735 (1986).

R. N. Bicknell, N. C. Giles, J. F. Schetzina and C. Hitzman, *J. Vac. Sci. Technol.* **A5**, 3059 (1987).

R. N. Bicknell-Tassius *et al.*, *Appl. Surface Sci.* **36**, 95 (1989).

R. N. Bicknell-Tassius *et al.*, *J. Cryst. Growth* **101**, 33 (1990).

R. W. Birkmire, R. B. Hall and J. E. Phillips, *Proc. 17th IEEE Photovoltaic Specialists Conference*, IEEE Publishing, NY (1984), p. 882.

R. W. Birkmire, B. E. McCandless, W. N. Shafarman and R. D. Varrin, *Proc. 19th European Photovoltaic Solar Energy Conference* (1989), p. 134.

R. W. Birkmire, B. E. McCandless and S. S. Hegedus, *Int. J. Solar Energy* **12**, 145 (1992b).

R. W. Birkmire, S. S. Hegedus, B. E. McCandless, J. E. Phillips, T. W. F. Russell, S. N. Shafarman, S. Verma and S. Yamanaka, *Photovoltaic Advanced Research and Development Project*, AIP Conference Proceedings **268**, 212 (1992).

R. W. Birkmire and P. V. Meyers, *First World Conference on Photovoltaic Energy Conversion, 24th IEEE Photovoltaic Specialists Conference*, IEEE Publishing, NY (1994), p. 76.

R. W. Birkmire, H. Hichri, R. Klenk, M. Marudachalam, B. E. McCandless, J. E. Phillips, J. M. Schultz and W. N. Shafarman, *13th NREL Photovoltaic Program Review Meeting*, Lakewood, CO (1995).

U. Blieske, S. Sterk, A. Bett, J. Schumacher, W. Wettling, A. Marti, M. J. Terron and A. Luque, *First World Conference on Photovoltaic Energy Conversion,*

24th IEEE Photovoltaic Specialists Conference, IEEE Publishing, NY (1994), p. 1902.

A. Blug, A. Baldus, A. W. Bett, U. Blieske, G. Stollwerck, O. V. Sulima and W. Wettling, *13th European Photovoltaic Solar Energy Conference*, Nice (1995), p. 910.

A. B. Bocarsly, J. M. Bolts, P. G. Cummins and M. S. Wrighton, *Appl. Phys. Lett.* **31**, 568 (1977).

K. W. Boer and J. D. Meakin, eds., *International NSF Workshop on CdS Solar Cells and Other Abrupt Heterojunctions*, Univ. of Delaware, Newark (1975).

E. C. Boes and A. Luque, in *Renewable Energy: Sources for Fuels and Electricity* (T. B. Johansson, H. Kelly, A. K. N. Reddy and R. H. Williams, eds.) Island Press, Washington, DC (1993), p. 361.

D. Bonnet, *Phys. Stat. Sol.* (a) **3**, 913 (1970).

D. Bonnet and H. Rabinhorst, *9th IEEE Photovoltaic Specialists Conference*, IEEE Publishing, NY (1972), p. 129.

P. Borden, P. Gregory, O. Moore, L. James and H. Van der Plas, *15th IEEE Photovoltaic Specialists Conference*, IEEE Publishing, NY (1981), p. 311.

E. Borne, J. P. Boyeaux and A. Laugier, *First World Conference on Photovoltaic Energy Conversion, 24th IEEE Photovoltaic Specialists Conference*, IEEE Publishing, NY (1994), p. 1637.

V. M. Botnaryuk, L. V. Gorchiak, G. M. Grigorieva, M. B. Kagan, T. A. Kozyreva, T. L. Lyubashevskaya, E. V. Russu and A. V. Simashkevich, *Solar Energy Materials* **20**, 359 (1990).

W. R. Bottenberg and R. D. Sproull, ARCO Res. and Devel. Extern. Pub. No. 88-2 (1988).

J. A. Bragagnolo, A. M. Barnett, J. E. Phillips, R. B. Hall, A. Rothwarf and J. Meakin, *IEEE Trans. Electron Devices* **ED-27**, 645 (1980).

H. Branz, R. Crandall and M. Silver, *Am. Inst. Phys. Conf. Proc.* **234**, 29 (1991).

D. Braunger, D. Hariskos, T. Walter and H. W. Schock, *Solar Energy Materials and Solar Cells* **40**, 97 (1996).

R. Brendel, *Appl. Phys. A* **60**, 523 (1995).

R. Brendel and M. Wolf, *13th European Photovoltaic Solar Energy Conference*, Nice, (1995).

R. Brendel, R. B. Bergmann, P. Loelgen, M. Wolf and J. H. Werner, *Appl. Phys. Lett.* **70**, 390 (1997).

F. Briones, F.-C. Wang and R. H. Bube, *Appl. Phys. Lett.* **39**, 805 (1981).

M. Brodsky and R. Title, *Phys. Rev. Lett.* **23**, 581 (1969).

R. H. Bube in *Critical Materials Problems in Energy Production*, Academic Press, NY (1976), p. 486.

R. H. Bube and A. L. Fahrenbruch, "Photovoltaic Effect", in *Advances in Electronics and Electron Physics* **56** (C. Marton, ed.) Academic Press, NY (1981), p. 163.

R. H. Bube, A. L. Fahrenbruch, R. Sinclair, T. C. Anthony, C. Fortmann, W. Huber, C.-T. Lee, T. Thorpe and T. Yamashita, *IEEE Trans. Electron Devices*, **ED-31**, 528 (1984).

R. H. Bube, *ISES Solar World Conference*, Hamburg (1987).

R. H. Bube and D. Redfield, *J. Appl. Phys.* **66**, 820 (1989).

R. H. Bube and D. Redfield, *J. Appl. Phys.* **66**, 3074 (1989).

R. H. Bube, L. Echeverria and D. Redfield, *Appl. Phys. Lett.* **57**, 79 (1990).

R. H. Bube, *Annual Review of Materials Science* **20**, 19 (1990).

R. H. Bube, *Photoelectronic Properties of Semiconductors*, Cambridge Univ. Press, Cambridge (1992).

R. H. Bube, L. E. Benatar, M. N. Grimbergen and D. Redfield, *J. Appl. Phys.* **72**, 5766 (1992).

R. H. Bube and D. Redfield, *J. Appl. Phys.* **71**, 5246 (1992).

R. H. Bube, "Solar Cells" in *Handbook on Semiconductors. Device Physics*, Vol. 4, (C. Hilsum, ed.) Elsevier, Amsterdam (1993).

R. H. Bube and K. W. Mitchell, *J. Electronic Mat.* **22**, 17 (1993).

R. H. Bube, L. E. Benatar, M. N. Grimbergen and D. Redfield, *J. Non-Cryst. Solids* **169**, 47 (1994).

R. H. Bube, L. E. Benatar and D. Redfield, *J. Appl. Phys.* **75**, 1571 (1994).

R. H. Bube, L. E. Benatar and K. P. Bube, *J. Appl. Phys.* **79**, 1926 (1996).

F. Buch, A. L. Fahrenbruch and R. H. Bube, *Appl. Phys. Lett.* **28**, 593 (1976).

F. Buch, A. L. Fahrenbruch and R. H. Bube, *J. Appl. Phys.* **48**, 1596 (1977).

H. C. Card and E. S. Yang, *IEEE Trans. Electron. Devices* **ED-24**, 397 (1977).

D. E. Carlson and C. R. Wronski, *Appl. Phys. Lett.* **28**, 671 (1976).

D. E. Carlson, U.S. Patent 4,064,521 (1977a).

D. E. Carlson, *IEEE Trans. Electron Devices* **ED-24**, 449 (1977b).

D. E. Carlson, *Sol. Energy Mater.* **3**, 503 (1980).

D. E. Carlson and R. W. Smith, *J. Electron. Mater.* **11**, 740 (1982).

D. E. Carlson, A. R. Moore, D. J. Szostak, B. Goldstein, R. W. Smith, P. J. Zanzucchi and W. R. French, *Solar Cells* **9**, 19 (1983).

D. E. Carlson, "Solar Cells", in *Semiconductors and Semimetals.* Vol. 21D. *Hydrogenated Amorphous Silicon: Device Applications* (J. I. Pankove, ed.) Academic Press, NY (1984), p. 7.

D. E. Carlson, "Hydrogen Motion and the Staebler-Wronski Effect in Amorphous Silicon", in *Disordered Semiconductors* (M. A. Kastner, G. A. Thomas and S. R. Ovshinsky, eds.) Plenum Press, NY (1987), p. 613.

D. E. Carlson and S. Wagner, in *Renewable Energy: Sources for Fuels and Electricity* (T. B. Johansson, H. Kelly, A. K. N. Reddy and R. H. Williams, eds.) Island Press, Washington, DC (1993), p. 403.

D. E. Carlson and K. Rajan, *13th European Photovoltaic Solar Energy Conference*, Nice (1995), p. 617.

D. E. Carlson and K. Rajan, *Appl. Phys. Lett.* **68**, 28 (1996).

H. C. Casey, Jr. and M. B. Panish, *Heterostructure Lasers. Part B*, Academic Press, NY (1978).

M. S. Casey, A. L. Fahrenbruch and R. H. Bube, *J. Appl. Phys.* **61**, 2941 (1987).

A. Catalano, V. Dalal, E. A. Fagen, R. B. Hall, J. V. Masi, J. D. Meakin, G. Warfield and A. M. Barnett, *1st European Community Photovoltaic Solar Energy Conference*, D. Reidel Pub., Dordrecht, Holland (1978), p. 644.

A. Catalano, R. V. D'Aiello, J. Dresner, B. Faughnan, A. Firester, J. Kane, H. Schade, Z. E. Smith, G. Swartz and A. Triano, *16th IEEE Photovoltaic Specialists Conference*, IEEE Publishers, NY (1982), p. 1421.

D. J. Chadi and K. Chang, *Phys. Rev. Lett.* **60**, 2187 (1988).

D. J. Chadi and K. Chang, *Appl. Phys. Lett.* **55**, 575 (1989a).

D. J. Chadi and K. Chang, *Phys. Rev. B* **39**, 10063 (1989b).

D. J. Chadi, *Ann. Rev. Mater. Sci.* **24**, 45 (1994).

Y. G. Chai, W. W. Anderson and L. B. Anderson, *IEEE Trans. Electron Devices* **ED-24**, 492 (1977).

R. R. Chamberlin and J. S. Skarman, *J. Electrochem. Soc.* **113**, 86 (1966).

D. M. Chapin, C. S. Fuller and G. L. Pearson, *J. Appl. Phys.* **25**, 676 (1954).

E. J. Charlson and J. C. Lien, *J. Appl. Phys.* **46**, 3982 (1975).

K.-F. Chien, A. L. Fahrenbruch and R. H. Bube, *J. Appl. Phys.* **64**, 2792 (1988).

L. Chen and L. Yang, *J. Non-Cryst. Solids* **137/138**, 1185 (1991).

L. Chen, J. Tauc, J. Kocka and J. Stuchlik, *Phys. Rev. B* **46**, 2050 (1992).

W. S. Chen, J. M. Stewart, B. J. Stanbery, W. E. Devaney and R. A. Mickelsen, *19th IEEE Photovoltaic Specialists Conference*, IEEE Publishing, NY (1987), p. 1445.

S. H. Chiang *et al.*, *Proc. Photovolt. Solar Energy Conf., Comm. of European Communities*, Luxembourg, Reidel, Dordrecht (1978), p. 104.

C. K. Chiang, M. J. Cohen, A. F. Garito, A. J. Heeger, C. M. Mikulski and A. G. MacDiarmid, *Solid State Commun.* **18**, 1451 (1976).

P. K. Chiang, D. D. Krut, B. T. Cavicchi, K. A. Bertness, S. R. Kurtz and J. M. Olson, *24th IEEE Photovoltaic Specialists Conference*, IEEE Publishing, NY (1994), p. 2120.

R. Chittick, J. Alexander and H. Sterling, *J. Electrochem. Soc.* **116**, 77 (1969).

S. C. Choo, *Solid-State Electronics* **11**, 1069 (1968).

K. L. Chopra and S. R. Das, *Thin Film Solar Cells*, Plenum Press, NY (1983).

M. Chu, A. L. Fahrenbruch, R. H. Bube and J. F. Gibbons, *J. Appl. Phys.* **49**, 322 (1978).

M. Chu, R. H. Bube and J. F. Gibbons, *J. Electrochem. Soc.* **127**, 483 (1980).

T. L. Chu, S. S. Chu, F. Firszt, H. A. Naseem and R. Stawski, *J. Appl. Phys.* **58**, 1349 (1985).

T. L. Chu, S. S. Chu, S. T. Ang, K. D. Han, Y. Z. Liu, K. Zweibel and H. S. Ullal, *19th IEEE Photovoltaic Specialists Conference*, IEEE Publishing, NY (1987), p. 1466.

T. L. Chu, S. S. Chu, K. D. Han and M. Mantravadi, *20th IEEE Photovoltaic Specialists Conference*, IEEE Publishing, NY (1988), p. 1422.

T. L. Chu, S. S. Chu, C. Ferekides and S. Kadamani, *Photovoltaics Advanced Research and Development*, 9th Ann. Rev. Meet. SERI/CP-213-3495, Lakewood, CO (1989), p. 75.

T. L. Chu, S. S. Chu, N. Schultz, C. Wang and C. Q. Wu, *J. Electrochem. Soc.* **139**, 2443 (1992).

V. Chu, J. P. Conde, D. S. Shen, S. Aljishi and S. Wagner, *Mater. Res. Soc. Amorphous Technology Symp. Proc.* **178** (A. Madan *et al.*, eds.) (1988), p. 167.

T. A. Chynoweth and R. H. Bube, *J. Appl. Phys.* **51**, 1844 (1980).

T. F. Ciszek, *20th IEEE Photovoltaic Specialists Conference*, IEEE Publishing, NY (1988), p. 31.

T. F. Ciszek, T. H. Wang, X. Wu, R. W. Burrows, J. Alleman, C. R. Schwerdt-feger and T. Bekkedahl, *23rd IEEE Photovoltaic Specialists Conference*, IEEE Publishing, NY (1993), p. 65.

C. Clemen and E. Bucher, *13th IEEE Photovoltaics Specialists Conference*, IEEE Publishing, NY (1978), p. 1255.

I. Clemminck, M. Burgelman, M. Casteleyn, J. DePoorter and A. Varvaet, *22nd IEEE Photovoltaic Specialists Conference* (1991), p. 1114.

M. J. Cohen and J. S. Harris, Jr., *Tech. Dig. Int. Electron Dev. Meet.* (1978), p. 247.

J. Cohen, T. Leen and F. Zhong, *J. Non-Cryst. Solids* **164/166**, 327 (1993).

C. Cohen-Solal, *J. Cryst. Growth* **72**, 512 (1985).

A. D. Compaan *et al.*, *23rd IEEE Photovoltaic Specialists Conference*, IEEE Publishing, NY (1993), p. 394.

A. D. Compaan, M. Shao, C. N. Tabory, Z. Feng, A. Fischer, I. Matulionis and R. G. Bohn, *First World Conference on Photovoltaic Energy Conversion, 24th IEEE Photovoltaic Specialists Conference*, IEEE Publishing, NY (1994), p. 111.

J. P. Conde, V. Chu, D. S. Shen, S. Aljishi, S. Tanaka and S. Wagner, *Tech. Digest 3rd Photovoltaic Sci. Eng. Conf.*, Tokyo (1987).

J. P. Conde, V. Chu, S. Tanaka, D. S. Shen and S. Wagner, *20th IEEE Photovoltaic Specialists Conference*, IEEE Publishing, NY (1988a), p. 235.

J. P. Conde, V. Chu, D. S. Shen, M. Angell and S. Wagner, *Mater. Res. Soc. Symp. Proc.*, Reno (1988b).

J. P. Conde, V. Chu, D. S. Shen and S. Wagner, *J. Appl. Phys.* **75**, 1638 (1994).

G. A. N. Connell and J. R. Pawlik, *Phys. Rev. B* **13**, 787 (1976).

M. le Contellec and F. Morin, *Thin Solid Films* **52**, 63 (1978).

M. A. Contreras, H. Wiesner, R. Matson, J. Tuttle, K. Ramanathan and R. Noufi, *Spring Meeting of the Materials Research Society*, San Francisco (1996a).

M. A. Contreras, H. Wiesner, D. Niles, K. Ramanathan, R. Matson, J. Tuttle, J. Keane and R. Noufi, *25th IEEE Photovoltaic Specialists Conference*, IEEE Publishing, NY (1996b), p. 809.

R. Corkish, *Solar Cells* **31**, 537 (1991).

R. Cortes, M. Froment, B. Mokili and D. Lincot, *Phil. Mag. Lett.* **73**, 209 (1996).

F. G. Courreges, A. L. Fahrenbruch and R. H. Bube, *J. Appl. Phys.* **51**, 2175 (1980).

R. S. Crandall, *J. Appl. Phys.* **53**, 3350 (1982).

R. S. Crandall, *Phys. Rev. B.* **36**, 2645 (1987).

R. S. Crandall, *Phys. Rev. B.* **43**, 4057 (1991).

J. Dabrowski and M. Scheffler, *Phys. Rev. B* **40**, 10391 (1989).

V. L. Dalal, *16th IEEE Photovoltaic Specialists Conference*, IEEE Publishing, NY (1982), p. 1384.

V. L. Dalal, E. Ping, S. Kaushal, M. K. Bhan and M. Leonard, *Appl. Phys. Lett.* **64**, 1862 (1994).

V. Dalal, S. Kaushal, R. Girvan, S. Hariasra and L. Sipahi, *25th IEEE Photovoltaic Specialists Conference*, IEEE Publishing, NY (1996), p. 1069.

S. Damaskinos, J. D. Meakin and J. E. Phillips, *19th IEEE Photovoltaic Specialists Conference*, IEEE Publishing, NY (1987), p. 1299.

P. D. Dapkus, R. D. Dupuis, R. D. Yingling, J. J. Yang, W. I. Simpson, L. A. Moudy, R. E. Johnson, A. G. Campbell, H. M. Manasevit and R. P. Ruth, *13th IEEE Photovoltaic Specialists Conference*, IEEE Publishing, NY (1978), p. 960.

S. K. Das and G. C. Morris, *J. Appl. Phys.* **72**, 4940 (1992).

T. Datta, R. Noufi and S. K. Deb, *Appl. Phys. Lett.* **47**, 1102 (1985).

T. Datta, R. Noufi and S. K. Deb, *J. Appl. Phys.* **59**, 1548 (1986).

G. C. Datum and S. A. Billets, *22nd IEEE Photovoltaic Specialists Conference*, IEEE Publishing, NY (1991), p. 1422.

E. Daub, P. Klopp, S. Kugler, P. Wuerfel, R. Schindler and W. Warta, *12th European Photovoltaic Solar Energy Conference*, Amsterdam (1994), p. 1772.

J. R. Davis, A. Rohatgi, R. H. Hopkins, P. D. Blais, P. Rai-Choudhury, J. R. McCormick and H. C. Mollenkopf, *IEEE Transactions on Electron Devices*, **ED-27**, 677 (1980).

L. R. Dawson, in *Progress in Solid-State Chemistry* (H. Reiss and J. O. McCaldin, eds.) Pergamon, Oxford (1972), p. 117.

S. Deane and M. Powell, *Phys. Rev. Lett.* **70**, 1654 (1993).

J. A. del Alamo, S. Swirhun and R. M. Swanson, *Solid-State Electronics* **28**, 47 (1985).

J. A. del Alamo and R. M. Swanson, *Solid State Electron.* **30**, 1127 (1987a).

J. A. del Alamo and R. M. Swanson, *Jpn. J. Appl. Phys.* **26**, 1860 (1987b).

J. A. del Alamo and R. M. Swanson, *IEEE Transactions Electron Devices* **ED-34**, 1580 (1987c).

E. Demesmaeker, S. Glunz, J. Knobloch and W. Wettling, *12th European Photovoltaic Solar Energy Conference*, Amsterdam (1994).

F. Demichelis and C. F. Pirri, *Solid State Phenomena* **44-46**, 385 (1995).

W. E. Devaney, R. A. Mickelsen and W. S. Chen, *18th IEEE Photovoltaic Specialists Conference*, IEEE Publishing, NY (1985), p. 1733.

W. E. Devaney and R. A. Mickelsen, *Solar Cells* **24**, 19 (1988).

W. E. Devaney, W. S. Chen, J. M. Stewart and R. A. Mickelsen, *Photovoltaics Advanced Research and Development*, 9th Ann. Rev. Meet. SERI/CP-213-3495, Lakewood, CO (1989), p. 77.

N. G. Dhere, S. Kuttath, K. W. Lynn, R. W. Birkmire and W. N. Shafarman, *First World Conference on Photovoltaic Energy Conversion, 24th IEEE Photovoltaic Specialists Conference*, IEEE Publishing, NY (1994), p. 190.

A. Dhingra and A. Rothwarf, *IEEE Transactions on Electron Devices* **43**, 613 (1996).

P. Di Marco and G. Giro, "Organic Photoconductors and Photovoltaics", in *Organic Conductors: Fundamentals and Applications* (J.-P. Farges, ed.) Marcel Dekker, Inc. NY (1994), p. 791.

B. Dimmler, H. Dittrich, R. Menner and H. W. Schock, *19th IEEE Photovoltaic Specialists Conference*, IEEE Publishing, NY (1987), p. 1454.

B. Dimmler, H. Dittrich and W. H. Schock, *Proc. 20th IEEE Photovoltaic Specialists Conference*, IEEE Publishing, NY (1988), p. 1416.

B. Dimmler, H. Dittrich, R. Klenk, R. H. Mauch, R. Menner and H. W. Schock, *8th EC Photovoltaic Solar Energy Conf.* (1988), p. 1583.

H. Dittrich, B. Dimmler, R. Menner and H. W. Schock, *8th EC Photovoltaic Solar Energy Conf.* (1988), p. 1102.

J. M. Dona and J. Herrero, *J. Electrochem. Soc.* **141**, 205 (1994).

M. Doty and P. Meyers, *AIP Conf. Proc.* **166**, NY (1988), p. 10.

J. Dresner, *J. Non-Cryst. Solids* **58**, 353 (1983).

R. D. Dupuis, P. D. Dapkus, R. D. Yingling and L. A. Moudy, *Appl. Phys. Lett.* **31**, 201 (1977).

C. Eberspacher, A. L. Fahrenbruch, and R. H. Bube, *17th IEEE Photovoltaic Specialists Conference*, IEEE Publishing, NY (1984), p. 459.

C. Eberspacher, J. Ermer and K. W. Mitchell, European patent application 0 318 315 A2 (filed 25 November 1988), assignee: Atlantic Richfield Company (1989).

A. Ehrhardt, W. Wettling and A. Bett, *Appl. Phys. A* **53**, 123 (1991).

H. E. Elgamel, M. Y. Ghannam, C. Vinckier, J. Nijs, R. Mertens and R. Van Overstraeten, *Solar Energy Materials and Solar Cells* **34**, 237 (1994a).

H. E. Elgamel, A. Rohatgi, Z. Chen, C. Vinckier, J. Nijs and R. Mertens, *First World Conference on Photovoltaic Energy Conversion, 24th IEEE Photovoltaic Specialists Conference*, IEEE Publishing, NY (1994b), p. 1323.

H. E. Elgamel, C. Vinckier, M. Caymax, M. Ghannam, J. Poortmans, P. De Schepper, J. Nijs and R. Mertens, *12th European Photovoltaic Solar Energy Conference*, Amsterdam (1994c), p. 724.

H. E. Elgamel, A. M. Barnett, A. Rohatgi, Z. Chen, C. Vinckier, J. Nijs and R. Mertens, *J. Appl. Phys.* **78**, 3457 (1995).

J. H. Ermer and R. B. Love. U.S. Patent number 4,798,660 (filed 22 December 1986), assignee: Atlantic Richfield Company (1989).

S. Espevik, C. Wu and R. H. Bube, *J. Appl. Phys.* **42**, 3513 (1971).

A. Eyer, R. Schindler, I. Reis, N. Schillinger, J. G. Grabmaier, *19th IEEE Photovoltaic Specialists Conference*, IEEE Publishing, NY (1987), p. 951.

A. L. Fahrenbruch and R. H. Bube, *J. Appl. Phys.* **45**, 1264 (1974).

A. L. Fahrenbruch, V. Vasilchenko, F. Buch, K. Mitchell and R. H. Bube, *Appl. Phys. Lett.* **25**, 605 (1974).

A. L. Fahrenbruch and R. H. Bube, *Fundamentals of Solar Cells: Photovoltaic Solar Energy Conversion*, Academic Press, NY (1983).

A. L. Fahrenbruch, *Solar Cells* **21**, 399 (1987).

A. L. Fahrenbruch, R. H. Bube, D. Kim and A. Lopez-Otero, *Int. J. Solar Energy* **12**, 197 (1992).

R. Falckenberg, G. Hoyler and J. G. Grabmaier, *19th IEEE Photovoltaic Specialists Conference*, IEEE Publishing, NY (1987), p. 369.

B. W. Faughnan and J. J. Hanak, *Appl. Phys. Lett.* **42**, 722 (1983).

B. W. Faughnan and R. S. Crandall, *Appl. Phys. Lett.* **44**, 537 (1984).

P. N. Favennec, M. le Contellec, H. L'Haridon, G. P. Pelous and J. Richard, *Appl. Phys. Lett.* **34**, 807 (1979).

R. S. Feigelson, A. N'Diaye, S.-Y. Yin and R. H. Bube, *J. Appl. Phys.* **48**, 3162 (1977).

C. S. Ferekides, J. Britt, Y. Ma and L. Killian, *23rd IEEE Photovoltaic Specialists Conference* (1993), p. 389.

C. S. Ferekides, K. Dugan, V. Ceekala, J. Killian, D. Oman, R. Swaminathan and D. L. Morel, *First World Conference on Photovoltaic Energy Conversion, 24th IEEE Photovoltaic Specialists Conference*, IEEE Publishing, NY (1994), p. 99.

C. S. Ferekides, D. Marinskiy, S. Marinskaya, B. Tetali, D. Oman, and D. L. Morel, *25th Photovoltaic Specialists Conference*, IEEE Publishing, NY (1996), p. 751.

F. Finger, U. Kroll, V. Viret, A. Shah, W. Beyer, X.-M. Tang, J. Weber, A. Howling and Ch. Hollenstein, *J. Appl. Phys.* **71**, 5665 (1992).

F. Finger, R. Carius, P. Hapke, K. Prasad and R. Flueckiger, *Mat. Res. Soc. Symp. Proc.* **283**, 471 (1993).

A. Fischer, C. Narayanswami, D. Grecu, E. Bykov, S. A. Nance, U. N. Jayamaha, G. Contreras-Puente, A. D. Compaan, M. A. Stan and A. R. Mason, *25th IEEE Photovoltaic Specialists Conference*, IEEE Publishing, NY (1996), p. 921.

H. Fischer and W. Pschunder, *11th IEEE Photovoltaic Specialists Conference*, IEEE Publishing, NY (1975), p. 25.

D. Fischer, S. Dubail, J. A. Anna Selvan, N. P. Vaucher, R. Platz, C. H. U. Kroll, J. Meier, P. Torres, H. Keppner, N. Wyrsch, M. Goetz, A. Shah and K.-D. Ufert, *25th IEEE Photovoltaic Specialists Conference*, IEEE Publishing, NY (1996), p. 1053.

M. Florez, W. de la Cruz, P. Teheran and G. Gordillo, *25th IEEE Photovoltaic Specialists Conference*, IEEE Publishing, NY (1996), p. 937.

R. Flueckiger, J. Meier, H. Keppner, M. Goetz and A. Shah, *23rd IEEE Photovoltaic Specialists Conference*, IEEE Publishing, NY (1993), p. 839.

R. Flueckiger, J. Meier, M. Goetz and A. Shah, *J. Appl. Phys.* **77**, 712 (1995a).

R. Flueckiger, J. Meier, A. Shah, J. Pohl, M. Tzolov and R. Carius, *Mat. Res. Soc. Symp. Proc.* **358**, 793 (1995b).

C. M. Fortmann, A. L. Fahrenbruch and R. H. Bube, *J. Appl. Phys.* **61**, 2038 (1987).

C. M. Fortmann, S. S. Hegedus, T. X. Zhou and B. N. Baron, *Solar Cells* **30**, 255 (1991).

L. M. Fraas, in *Current Topics in Photovoltaics* (T. J. Coutts and J. D. Meakin, eds.) Academic Press, NY (1985), p. 170.

L. Fraas, J. Avery, V. Sundaram, V. Dinh, T. Davenport and J. Yerkes, *21st IEEE Photovoltaic Specialists Conference*, Orlando, IEEE Publishing, NY (1990).

L. M. Fraas, J. E. Avery, P. E. Gruenbaum, R. J. Ballantyne and E. Malocsay, *Solar Cells* **30**, 355 (1991).

W. Frammelsberger, H. Ruebel, P. Lechner, R. Geyer and N. Kniffler, *Appl. Phys. Lett.* **58**, 2660 (1991).

R. J. Frank and J. L. Goodrich, *14th IEEE Photovoltaic Specialists Conference*, IEEE Publishing, NY (1980), p. 423.

A. Freundlich, V. Rossignol, M. F. Vilela, P. Renaud, A. Bensaoula and N. Medelci, *First World Conference on Photovoltaic Energy Conversion, 24th IEEE Photovoltaic Specialists Conference*, IEEE Publishing, NY (1994), p. 1886.

T. M. Friedlmeier, D. Braunger, D. Hariskos, M. Kaiser, H. N. Wanka and H. W. Schock, *25th IEEE Photovoltaic Specialists Conference*, IEEE Publishing, NY (1996), p. 845.

H. Fritzsche, C. C. Tsai and P. Persans, *Solid State Technol.* **21**, 55 (1978).

H. Fritzsche, *Amorphous Silicon and Related Materials*, 2 vols. World Scientific, Singapore (1989).

M. Froment and D. Lincot, *Electrochimica Acta* **40**, 1293 (1995).

M. Froment, M. C. Bernard, R. Cortes, B. Mokili and D. Lincot, *J. Electrochem. Soc.* **142**, 2642 (1995).

K. Fujimoto, F. Nakabeppu, Y. Sogawa, Y. Okayasu and K. Kumagai, *23rd IEEE Photovoltaic Specialists Conference*, IEEE Publishing, NY (1993), p. 83.

A. Fujishima and K. Honda, *Nature* **238**, 37 (1972).

G. Fulop, M. Doty, P. Meyers, J. Betz and C. H. Liu, *Appl. Phys. Lett.* **40**, 327 (1982).

V. V. Galavanov, R. M. Kundukhov and D. N. Nasledov, *Sov. Phys. — Solid State* **8**, 2723 (1967).

R. P. Gale, R. W. McClelland, B. D. King and J. V. Gormley, *20th IEEE Photovoltaic Specialists Conference*, IEEE Publishing, NY, Las Vegas (1988).

R. P. Gale, R. W. McCLelland, B. D. King and J. C. C. Fan, *9th Annual Review Meeting SERI/CP-213-3495*, Lakewood, CO (1989), p. 23.

G. Ganguly and A. Matsuda, *Phys. Rev. B* **47**, 3661 (1993).

J. M. Gee and G. F. Virshup, *20th IEEE Photovoltaic Specialists Conference*, IEEE Publishing, NY (1988), p. 754.

J. M. Gee, *22nd IEEE Photovoltaic Specialists Conference*, IEEE Publishing, NY (1991), p. 118.

H. Gerischer, *Electroanal. Chem. & Interfacial Chem.* **58**, 263 (1975).

T. A. Gessert, M. W. Wanlass, T. J. Coutts, X. Li and G. S. Horner, *9th Annual Review Meeting SERI/CP-213-3495*, Lakewood, CO (1989), p. 79.

S. K. Ghandhi, N. R. Taskar and I. B. Bhat, *Appl. Phys. Lett.* **50**, 900 (1987).

M. Y. Ghannam, W. Beyer, H. E. Elgamel, J. Nijs and R. Mertens, *12th European Photovoltaic Solar Energy Conference*, Amsterdam (1994), p. 1000.

B. Ghosh, S. Purakayastha, P. K. Datta, R. W. Miles, M. J. Carter and R. Hill, *Semicond. Sci. and Technol.* **10**, 71 (1995).

W. D. Gill and R. H. Bube, *J. Appl. Phys.* **41**, 3731 (1970).

H. Gleskova, P. Morin and S. Wagner, *Appl. Phys. Lett.* **62**, 2063 (1993a).

H. Gleskova, P. Morin and S. Wagner, *Proc. Mater. Res. Soc. Symp.* **297**, 589 (1993b).

S. W. Glunz, A. B. Sproul, W. Warta and W. Wettling, *J. Appl. Phys.* **75**, 1611 (1994a).

S. W. Glunz, A. B. Sproul, S. Sterk and W. Warta, *12th European Photovoltaic Solar Energy Conference*, Amsterdam (1994b), p. 492.

S. W. Glunz, C. Hebling, W. Warta and W. Wettling, *First World Conference on Photovoltaic Energy Conversion, 24th IEEE Photovoltaic Specialists Conference*, IEEE Publishing, NY (1994c), p. 1625.

S. W. Glunz and W. Warta, *J. Appl. Phys.* **77**, 3247 (1995).

S. W. Glunz, S. Sterk, R. Steeman, W. Warta, J. Knobloch and W. Wettling, *13th European Photovoltaic Solar Energy Conference*, Nice (1995), p. 409.

A. R. Gobat, M. F. Lamorte and G. W. McIver, *IRE Trans. Mil. Electron.* **6**, 20 (1962).

S. Goda, T. Moritani, Y. Hatanaka, H. Smimizu and I. Hide, *First World Conference on Photovoltaic Energy Conversion, 24th IEEE Photovoltaic Specialists Conference*, IEEE Publishing, NY (1994), p. 1227.

A. Goetzberger and A. Rauber, *20th IEEE Photovoltaic Specialists Conference*, IEEE Publishing, NY (1988), p. 1371.

C. Goradia and M. G. Goradia, *12th IEEE Photovoltaic Specialists Conference*, IEEE Publishing, NY (1976), p. 789.

M. Gorska, R. Beaulieu, J. J. Loferski, B. Roessler and J. Beall, *Solar Energy Mater.* **2**, 343 (1980).

J. E. Granata, J. R. Sites, G. Contreras-Puente and A. D. Compaan, *25th IEEE Photovoltaic Specialists Conference*, IEEE Publishing, NY (1996), p. 853.

M. A. Green and A. W. Blakers, *Solar Cells* **8**, 3 (1983).

M. A. Green, A. W. Blakers, J. Shi, E. M. Keller and S. R. Wenham, *IEEE Trans. on Electron Devices* **ED-31**, 671 (1984).

M. A. Green, *IEEE Trans. Electron Devices* **31**, 671 (1984).

M. A. Green, A. W. Blakers, S. R. Wenham, S. Narayanan, M. R. Willison, M. Taouk and T. Szpitalak, *18th IEEE Photovoltaic Specialists Conference*, IEEE Publishing, NY (1985), p. 39.

M. A. Green, *High Efficiency Silicon Solar Cells*, Trans. Tech. Publications, Aedermannsdorf (1987).

M. A. Green, *Proc. 10th European Communities Photovoltaic Solar Energy Conference* Lisbon (1991), p. 250.

M. A. Green, J. Zhao, A. Wang and S. R. Wenham, *IEEE Electron Device Lett.* **13**, 317 (1992).

M. A. Green, S. R. Wenham, J. Zhao, A. Wang, X. Dai, A. Milne, M. Taouk, J. Shi, F. Yun, B. Chan, A. B. Sproul and A. Stevens, Final Report, Sandia Contract 66-5863, (1992).

M. A. Green, "Silicon Solar Cells: Evolution, High-Efficiency Design and Efficiency Enhancements", *Semicond. Sci. Technol.* **8**, 1 (1993a).

M. A. Green, "Crystalline- and Polycrystalline-Silicon Solar Cells", in *Renewable Energy: Sources for Fuels and Electricity* (T. B. Johansson, H. Kelly, A. K. N. Reddy and R. H. Williams, eds.; L. Burnham, ex.ed.) Island Press, Washington, DC (1993b), p. 337.

M. A. Green and S. R. Wenham, *Appl. Phys. Lett.* **65**, 2907 (1994).

M. A. Green, *Silicon Solar Cells: Advanced Principles & Practice*, Centre for Photovoltaic Devices and Systems, Univ. of New South Wales, Sydney (1995a).

M. A. Green with an introduction by D. Lovejoy, "Multilayer Thin Film Silicon Solar Cells", *Natural Resources Forum* **19**(4) 269 (1995c).

M. A. Green, K. Emery, K. Buecher and D. L. King, *Progress in Photovoltaics* **3**, 51 (1995d).

M. A. Green, *Progress in Photovoltaics: Research and Applications* **4** (1), 59 (1996).

J. E. Greene, S. A. Barnett, A. Rockett and G. Bajor *Appl. Surf. Sci.* **22/23**, 520 (1985).

M. Grimbergen, R. McConville, D. Redfield and R. H. Bube, *Proc. Mat. Res. Soc. Symp.* **297**, 655 (1993).

J. Gu, T. Kitahara, K. Kawakami and T. Sakaguchi, *J. Appl. Phys.* **46**, 1184 (1975).

S. Guha, *9th Annual Review Meeting SERI/CP-213-3495*, Lakewood, CO (1989), p. 9.

S. Guha, J. Yang, R. Banerjee, T. Glattfelter, K. Hoffman and X. Xu, *Tech. Digest of the 7th International Photovoltaic Science and Engineering Conference*, Dept. Electr. and Computer Eng., Nagoya Inst. of Tech., Nagoya, Japan (1993), p. 43.

J.-F. Guillemoles, P. Cowache, A. Lusson, K. Fezzaa, F. Boisivon, J. Vedel and D. Lincot, *J. Appl. Phys.* **79**, 7293 (1996).

M. Gunes, R. M. Dawson, S. Lee, C. R. Wronski, N. Maley and Y. M. Li, *22nd IEEE Photovoltaic Specialists Conference*, IEEE Publishing, NY (1991), p. 1242.

T. Haage, S. Bauer, B. Schroeder and H. Oechsner, *First World Conference on Photovoltaic Energy Conversion, 24th IEEE Photovoltaic Specialists Conference*, IEEE Publishing, NY (1994), p. 429.

G. Habermann, A. Bett, F. Lutz, C. Schetter, O. V. Sulima and W. Wettling, *11th European Community Photovoltaic Solar Energy Conference*, Montreux (1992), p. 217.

M. Hack and M. Shur, *16th IEEE Photovoltaic Specialists Conference*, IEEE Publishing, NY (1982), p. 1429.

J.-P. Haering, J. G. Werthen, R. H. Bube, L. Gulbrandsen, W. Jansen and P. Luscher, *J. Vac. Sci. Technol. A* **1**, 1469 (1983).

P. R. Hageman, G. J. Bauhuis, A. van Geelen, P. C. van Rijsingen, J. J. Schermer and L. J. Giling, *25th IEEE Photovoltaic Specialists Conference*, IEEE Publishing, NY (1996), p. 57.

P. Hahne, W. Wettling and R. Schindler, *12th European Photovoltaic Solar Energy Conference*, Amsterdam (1994), p. 1011.

W. G. Haines and R. H. Bube, *J. Appl. Phys.* **49**, 304 (1978).

H. Haku, K. Sayama, E. Maruyama, H. Dohjoh, N. Nakamura, S. Tsuda, S. Nakano, Y. Kishi and Y. Kuwano, *Jpn. J. Appl. Phys.* **30**, 2700 (1991).

R. B. Hall, R. W. Birkmire, J. E. Phillips and J. D. Meakin, *Appl. Phys. Lett.* **38**, 925 (1981).

Y. Hamakawa, Chap. 3 in *Current Topics in Photovoltaics* (T. J. Coutts and J. D. Meakin, eds.) Academic Press, Orlando, FL (1985).

J. J. Hanak and V. Korsun, *16th IEEE Photovoltaic Specialists Conference*, IEEE Publishing, NY (1982), p. 1381.

D. Hariskos, R. Herberholz, M. Ruckh, U. Ruehle, R. Schaeffler and H. W. Schock, *13th European Photovoltaic Solar Energy Conference*, Nice, France (1995).

D. Hariskos, M. Ruckh, U. Ruehle, T. Walter, H. W. Schock, J. Hedstroem and L. Stolt, *Solar Energy Materials and Solar Cells*, **41/42**, 345 (1996).

R. L. Harper, Jr., S. Hwang, N. C. Giles, J. F. Schetzina, D. L. Dreifus and T. H. Myers, *et al.*, *Appl. Phys Lett.* **54**, 170 (1989).

K. A. Harris *et al.*, *J. Vac. Sci. Technol.* **A8**, 1013 (1990).

K. A. Harris *et al.*, *J. Vac. Sci. Technol.* **B9**, 1752 (1991).

B. Hartiti, A. Slaoui, J. C. Muller, P. Siffert, R. Schindler, I. Reis, B. Wagner and A. Eyer, *23rd IEEE Photovoltaic Specialists Conference*, IEEE Publishing, NY (1993), p. 224.

B. Hartiti, S. Sivoththaman, R. Schindler, J. Nijs, J. C. Muller and P. Siffert, *First World Conference on Photovoltaic Energy Conversion, 24th IEEE Photovoltaic Specialists Conference*, IEEE Publishing, NY (1994), p. 1519.

N. Hata and S. Wagner, *J. Appl. Phys.* **72**, 2857 (1992).

J. Haynos, J. Allison, R. Arndt and A. Meulenberg, *International Conference on Photovoltaic Power Generation*, Hamburg (1974), p. 487.

C. Hebling, R. Gaffke, P. Lanyi, H. Lautenschlager, C. Schetter, B. Wagner and F. Lutz, *25th IEEE Photovoltaic Specialists Conference*, IEEE Publishing, NY (1996), p. 649.

S. S. Hegedus, R. E. Rocheleau, W. Buchanan and B. N. Baron, *J. Appl. Phys.* **61**, 381 (1987).

S. S. Hegedus, R. E. Rocheleau, R. M. Tullman, D. E. Albright, N. Saxena, *et al.*, *20th IEEE Photovoltaic Specialists Conference*, IEEE Publishing, NY (1988), p. 129.

S. S. Hegedus and E. A. Fagen, *J. Appl. Phys.* **71**, 5941 (1992).

S. S. Hegedus, H. Liang and R. G. Gordon, *13th NREL Program Review Meeting*, Lakewood, CO (1995).

A. Hepburn, J. Marshall, C. Main, M. Powell and C. van Berkel, *Phys. Rev. Lett.* **56**, 2215 (1986).

R. Herberholz, T. Walter and H. W. Schock, *J. Appl. Phys.* **76**, 2909 (1994).

R. Herberholz, T. Walter and H. W. Schock, *Cryst. Res. Technol.* **31**, 449 (1996).

W. Herbst, A. Scholz, B. Schroeder and H. Oechsner, *23rd IEEE Photovoltaic Specialists Conf.*, Louisville, IEEE Publishing, NY, (1993).

R. Hezel and R. Ziegler, *Proc. 23rd IEEE Photovoltaic Specialists Conference*, IEEE Publishing, NY (1993), p. 260.

R. Hill and D. Richardson, *Thin Solid Films* **18**, 25 (1973).

D. E. Hill, H. W. Gutsche, M. S. Wang, K. P. Gupta, W. F. Tucker, J. D. Dowdy and R. J. Crepin, *12th IEEE Phtovoltaic Specialists Conference*, IEEE Publishing, NY (1976), p. 112.

R. Hill and J. D. Meakin, "Cadmium Sulphide-Copper Sulphide Solar Cells", in *Current Topics in Photovoltaics* (T. J. Coutts and J. D. Meakin, eds.) Academic Press, London (1985), p. 223.

Y. Hishikawa, K. Ninomiya, E. Maruyama, S. Kuroda, A. Terakawa, K. Sayama, H. Tarui, M. Sasaki, S. Tsuda and S. Nakano, *First World Conference on Photovoltaic Energy Conversion, 24th IEEE Photovoltaic Specialists Conference*, IEEE Publishing, NY (1994), p. 386.

W. Hoenle, G. Kuehn and U. Boehnke, *Cryst. Res. Technol.* **23**, 1347 (1988).

R. H. Hopkins, J. Easoz, J. P. McHugh, P. Piotrowski, R. Hundal *et al.*, *J. Crystal Growth* **82**, 142 (1987).

J. Hou, S. J. Fonash and J. Kessler, *25th Photovoltaic Specialists Conference*, IEEE Publishing, NY (1996), p. 961.

H. J. Hovel, J. M. Woodall and W. E. Howard, *Symposium on GaAs*, Boulder, CO, Institute of Physics and Physical Society, London (1972), p. 205.

Y. R. Hsiao, W. S. Chen, R. A. Mickelsen, J. M. Stewart, V. Lowe, L. C. Olsen and A. Rothwarf, "CuInSe$_2$/CdS Thin-Film Solar Cell Development", Final Rep.

XE-2-0238-1. Boeing Aerospace Co., Seattle, Washington (1983).

H. M. Hubbard, *Science* **244**, 297 (1989).

H. M. Hubbard and G. Cook, *Phys. Soc.* **18**, 9 (1989).

W. Huber, A. L. Fahrenbruch, C. Fortmann and R. H. Bube, *J. Appl. Phys.* **54**, 4038 (1983).

J. Hubin, A. V. Shah, E. Sauvain and P. Pipoz, *J. Appl. Phys.* **78**, 6050 (1995).

J. A. Hutchby and R. L. Fudurich, *J. Appl. Phys.* **47**, 3140, 3152 (1976).

S. Hwang, R. L. Harper, K. A. Harris, N. C. Giles, R. N. Bicknell, J. F. Schetzina, D. L. Dreifus, R. M. Kolbas and M. Chu, *J. Vac. Sci. Technol.* **B6**, 777 (1988).

Y. Ichikawa, S. Fujikake, T. Yoshida, T. Hama and H. Sakai, *21st IEEE Photovoltaic Specialists Conference*, IEEE Publishing, NY (1990), p. 1475.

Y. Ichikawa, K. Tabuchi, S. Kato, A. Takano, T. Sasaki, M. Tanda, S. Saito, H. Sato, S. Fujikake, T. Yoshida and H. Sakai, *First World Conference on Photovoltaic Energy Conversion, 24th IEEE Photovoltaic Specialists Conference*, IEEE Publishing, NY (1994), p. 441.

M. Igalson and R. Bacewicz, *Proc. 11th European Photovoltaic Solar Energy Conference*, Montreux, Switzerland (1992), p. 874.

M. Igalson, *Phys. Stat. Sol.* (a) **139**, 481 (1993).

M. Igalson and R. Bacewicz, *12th European Community Photovoltaic Solar Energy Conference* (1994), p. 1584.

M. Igalson and H. W. Schock, *J. Appl. Phys.* **80**, 5765 (1996).

S. Ikegami, *Solar Cells* **23**, 89 (1988).

T. Ishihara, S. Arimoto, H. Morikawa, H. Kumabe, T. Murotani and S. Mitsui, *Appl. Phys. Lett.* **63**, 3604 (1993).

M. Isomura and S. Wagner, *Proc. Mater. Res. Soc. Symp.* **258**, 473 (1992).

M. Izu and S. R. Ovshinsky, *SPIE Proc.* **407**, 42 (1983).

M. Izu, S. R. Ovshinsky, X. Deng, A. Krisko, H. C. Ovshinsky, K. L. Narasimhan and R. Young, *AIP Conference Proceedings* **306**, 12th NREL Photovoltaic Program Rev. (1993), p. 198.

M. Izu, H. C. Ovshinsky, X. Deng, A. J. Krisko, K. L. Narasimhan, R. Crucet, T. Laarman, A. Myatt and S. R. Ovshinsky, *First World Conference on Photovoltaic Energy Conversion, 24th IEEE Photovoltaic Specialists Conference*, IEEE Publishing, NY (1994), p. 820.

W. B. Jackson, N. M. Johnson and D. K. Biegelsen, *Appl. Phys. Lett.* **43**, 195 (1983).

W. B. Jackson and S. B. Zhang, in *Transport, Correlation and Structural Defects* (H. Fritzsche, ed.) World Scientific Publishing (1990), p. 63.

B. Jagannathan and W. A. Anderson, *25th IEEE Photovoltaic Specialists Conference*, IEEE Publishing, NY (1996), p. 533.

H. Jager and E. Seipp, *J. Electron. Mater.* **10**, 605 (1981).

L. W. James and R. L. Moon, *Appl. Phys. Lett.* **26**, 476 (1975).

Z. I. Jan, R. H. Bube and J. C. Knights, *J. Appl. Phys.* **51**, 3278 (1980).

L. Jastrzebski, *J. Electrochem. Soc.* **142**, 3869 (1995).

D. A. Jenny, J. J. Loferski and P. Rappaport, *Phys. Rev.* **101**, 1208 (1956).

D. G. Jensen, B. E. McCandless and R. W. Birkmire, in *Thin Films for Photovoltaic and Related Device Applications*, eds. D. Ginley, A. Catalano, H. W. Schock,

C. Eberspacher, T. M. Peterson and T. Wada, Materials Research Society Proc. **426**, Pittsburgh, PA (1996a), p. 327.

D. G. Jensen, B. E. McCandless and R. W. Birkmire, *25th IEEE Photovoltaic Specialists Conference*, IEEE Publishing, NY (1996b), p. 773.

E. L. Johnson, *Proc. of the 16th Intersociety Energy Conversion Engineering Conference*, ASME, NY (1981), p. 798.

N. M. Johnson, D. K. Biegelsen and M. D. Moyer, *Appl. Phys. Lett.* **40**, 882 (1982).

C. Johnson, S. Wagner, and K. J. Bachmann (to be published – 1985).

W. D. Johnston and W. M. Callahan, *Appl. Phys. Lett.* **28**, 150 (1976).

J. F. Jordan and S. P. Albright, *Solar Cells* **23**, 107 (1988).

J. Kakalios, R. A. Street and W. B. Jackson, *Phys. Rev. Lett.* **59**, 1037 (1987).

T. Kamei, N. Hata, A. Matsuda, T. Uchiyama, S. Amano, K. Tsukamoto, Y. Yoshioka and T. Hirao, *Appl. Phys. Lett.* **68**, 2380 (1996).

K. Kamimura, T. Suzuki and A. Kunioka, *Appl. Phys. Lett.* **38**, 259 (1981).

T. I. Kamina and P. J. Marcoux, *IEEE Electron. Dev. Lett.* **1**, 159 (1980).

N. R. Kaminar, D. D. Liu, H. F. MacMillan, L. D. Partain, M. Ladle Ristow *et al.*, *20th IEEE Photovoltaic Specialists Conference*, IEEE Publishing, NY (1988), p. 766.

A. Kampmann, P. Cowache, J. Vedel and D. Lincot, *J. Electroanalytical Chemistry* **387**, 53 (1995).

K. Kaneko, R. Kawamura and T. Misawa, *First World Conference on Photovoltaic Energy Conversion, 24th IEEE Photovoltaic Specialists Conference*, IEEE Publishing, NY (1994), p. 30.

S. Kanev, V. Stoyanov and M. Lakova, *Dok. Bolg. Akad. Nauk.* **22**, 863 (1969).

S. K. Kanev, A. L. Fahrenbruch and R. H. Bube, *Appl. Phys. Lett.* **19**, 459 (1971).

V. K. Kapur, U. V. Choudary and A. K. P. Chu, US patent 4,581,108, assignee: Atlantic Richfield (1986).

R. Kawamura, K. Sasatani, T. Onizuka and K. Kaneko, *First World Conference on Photovoltaic Energy Conversion, 24th IEEE Photovoltaic Specialists Conference*, IEEE Publishing, NY (1994), p. 1652.

L. L. Kazmerski, M. S. Ayyagari and G. A. Sanborn, *J. Appl. Phys.* **46**, 4865 (1975).

L. L. Kazmerski, F. R. White and G. K. Morgan, *Appl. Phys. Lett.* **29**, 268 (1976).

L. L. Kazmerski, Univ. Maine, Orono, Second Quarter Report to NSF-RANN and ERDA, NSF/AER 75-19576/PR/77/2 (1977).

L. L. Kazmerski and G. A. Sanborn, *J. Appl. Phys.* **48**, 3178 (1977).

L. L. Kazmerski, O. Jamjoum, P. J. Ireland, S. Deb, R. A. Mickelsen and W. S. Chen, *J. Vac. Sci. Technol.* **19**, 467 (1981).

L. L. Kazmerski, O. Jamjoum, P. J. Ireland, R. A. Mickelsen and W. S. Chen, *J. Vac. Sci. Technol.* **21**, 486 (1982).

L. L. Kazmerski, *Rev. Bras. Apl. Vacuo* **3**, 171 (1983).

L. L. Kazmerski, O. Jamjoum, J. F. Wager, P. J. Ireland and K. J. Bachmann, *J. Vac. Sci. Technol.* **A1**, 668 (1983a).

L. L. Kazmerski, M. Hallerdt, P. J. Ireland, R. A. Mickelsen and W. S. Chen, *J. Vac. Sci. Technol.* **A1**, 395 (1983).

L. Kazmerski and S. Wagner, in *Current Topics in Photovoltaics* (T. J. Coutts and J. D. Meakin, eds.) Academic Press, London (1985), p. 41.

C. J. Keavney, V. E. Haven and S. M. Vernon, *21st IEEE Photovoltaic Specialists Conference*, IEEE Publishing, NY (1990), p. 141.

M. J. Keevers and M. A. Green, *Appl. Phys. Lett.* **66**, 174 (1995).

C. P. Khattak and F. Schmid, *14th IEEE Photovoltaics Specialists Conference*, IEEE Publishing, NY (1980), p. 484.

F. Kicinski, *Chem. Industry* **17**, 54 (1948).

D. Kim, A. L. Fahrenbruch and R. H. Bube, *20th IEEE Photovoltaic Specialists Conference*, IEEE Publishing, NY (1988), p. 1487.

D. Kim, A. L. Fahrenbruch, A. Lopez-Otero and R. H. Bube, *J. Appl. Phys.* **75**, 2673 (1994).

D. Kim, S. Pozder, Y. Zhu and J. U. Trefny, *First World Conference on Photovoltaic Energy Conversion, 24th Photovoltaic Specialists Conference*, IEEE Publishing, NY (1994a), p. 334.

D. Kim, B. Qi, D. L. Williamson and J. U. Trefny, *First World Conference on Photovoltaic Energy Conversion, 24th Photovoltaic Specialists Conference*, IEEE Publishing, NY (1994b), p. 338.

H. D. Kim, D. S. Kim, J. S. Song and B. T. Ahn, *First World Conference on Photovoltaic Energy Conversion, 24th Photovoltaic Specialists Conference*, IEEE Publishing, NY (1994), p. 315.

N. P. Kim, R. M. Burgess, B. J. Stanbery, R. A. Mickelsen, J. E. Avery *et al., 20th IEEE Photovolt. Spec. Conf.*, IEEE Publishing, NY (1988), p. 457.

W. Y. Kim, M. Konagai and K. Takahashi, *20th IEEE Photovoltaic Specialists Conference*, IEEE Publishing, NY (1988a), p. 277.

W. Y. Kim, M. Konagai and K. Takahashi, *Jpn. J. Appl. Phys.* **20**, L948 (1988b).

E. F. Kingsbury and R. S. Ohl, *Bell Syst. Tech. J.* **31**, 8092 (1952).

R. Klenk, R. H. Mauch, R. Menner and H. W. Schock, *20th IEEE Photovoltaic Specialists Conference*, IEEE Publishing, NY (1988), p. 1443.

R. Klenk, T. Walter, D. Schmid and H. W. Schock, *Jpn. J. Appl. Phys.* **32**, Suppl. 32–33, 57 (1993).

J. C. Knights, G. Lucovsky and R. J. Nemanich, *J. Non-Cryst. Solids* **32**, 393 (1979).

J. Knobloch, A. Aberle and B. Voss, *9th E.C. Photovoltaic Solar Energy Conference*, Freiburg, Germany, 777 (1989).

J. Knobloch, A. Noel, E. Schaeffer, U. Schubert, F. J. Kamerewerd, S. Klussmann and W. Wettling, *23rd IEEE Photovoltaic Specialists Conference*, IEEE Publishing, NY (1993), p. 271.

J. Knobloch, S. W. Glunz, V. Henninger, W. Warta, W. Wettling, F. Schomann, W. Schmidt, A. Endroes and K. A. Muenzer, *13th European Photovoltaic Solar Energy Conference*, Nice (1995), p. 9.

J. Kocka, J. Stuchlik, M. Stutzmann, L. Chen and J. Tauc, *Phys. Rev. B* **47**, 13283 (1993).

R. J. Koestner *et al., J. Vac. Sci. Technol.* **A7**, 517 (1989).

K. S. Kolodinski, H. J. Werner and H.-J. Queisser, *Appl. Phys. A* **61**, 535 (1995).

J. Kopp, J. Knobloch and W. Wettling, *11th E.C. Photovoltaic Solar Energy Conference*, Montreux, Switzerland, 49 (1992).

P. Kordos, R. A. Powell, W. E. Spicer, G. L. Pearson and M. B. Panish, *Appl. Phys. Lett.* **34**, 366 (1979).

P. Kordos and G. L. Pearson, *Solid-State Electronics* **23**, 399 (1980).

A. V. Koval, V. Z. Nikorich, A. V. Simashkevich, R. L. Sobolevskaya and K. D. Sushkevich, *J. Cryst. Growth* **110**, 915 (1991).

G. Kroetz, J. Wind, H. Stitzl, G. Mueller, S. Kalbitzer and H. Mannsperger, *Phil Mag. B* **63**, 101 (1991).

U. Kroll, F. Finger, J. Dutta, H. Keppner, A. Shah, A. Howling, J.-L. Dorier and Ch. Hollenstein, *Mats. Res. Soc. Symp. Proc.* **258**, 135 (1992).

L.-C. Kuo, Y.-T. Tsai, W.-Y. Hsu and N.-H. Lu, *First World Conf. on Photovoltaic Energy Conversion, 24th IEEE Photovoltaic Specialists Conference*, IEEE Publishing, NY (1994), p. 690.

H. Kurita, T. Takamoto, E. Ikeda and M. Ohmori, *7th International Conference on Indium Phosphide and Related Materials* (1995), p. 516.

S. R. Kurtz, J. M. Olson and A. Kibbler, *21st IEEE Photovoltaic Specialists Conference*, IEEE Publishing, NY (1990), p. 138.

W. Kusian, E. Guenzel and R. D. Plaettner, *Solar Energy Materials* **23**, 303 (1991).

W. Kusian, K.-D. Ufert and H. Pfleiderer, *Solid State Phenomena* **44-46**, 823 (1995).

A. Kylner, J. Lindgren and L. Stolt, *J. Electrochem. Soc.* **143**, 2662 (1996).

A. Kylner and M. Wirde, *Jpn. J. Appl. Phys.* **36**, 2167 (1997a).

A. Kylner and E. Niemi, *14th European Photovoltaic Solar Energy Conference*, Spain, June–July (1997b).

Y. Kuwano, S. Tsuda, N. Nakamura, M. Nishikuni, K. Yoshida, T. Takahama, M. Isomura and S. Nakano, *AIP Conf. Proc.* **157**, 126 (1987).

G. Langguth, M. Schoefthaler, T. Sameshima and J. H. Werner, *13th European Photovoltaic Solar Energy Conference*, Nice, (1995).

P. G. LeComber and W. E. Spear, in *Amorphous Semiconductors*, **36**, of *Topics in Applied Physics* (M. H. Brodsky, ed.) Springer, NY (1986), p. 251.

C.-T. Lee and R. H. Bube, *J. Appl. Phys.* **54**, 7041 (1983).

C.-T. Lee and R. H. Bube, *J. Appl. Phys.* **58**, 880 (1985).

X. Y. Lee, A. K. Verma, C. Q. Wu, M. Goertemiller, E. Yablonovitch, J. Eldredge and D. Lillington, *25th IEEE Photovoltaic Specialists Conference*, IEEE Publishing, NY (1996), p. 53.

C. Leguijt, P. Loelgen, A. R. Burgers, J. A. Eikelboom, R. A. Steeman, W. C. Sinke, P. F. A. Alkemade and P. M. Sarro, *J. Appl. Phys.* **78**, 6596 (1995).

D. H. Levi, H. R. Moutinho, F. A. Hasoon, B. M. Keyes, R. K. Ahrenkiel, M. Al-Jassim, L. L. Kazmerski and R. W. Birkmire, *First World Conference on Photovoltaic Energy Conversion, 24th IEEE Photovoltaic Specialists Conference*, IEEE Publishing, NY (1994), p. 127.

J. D. Levine, G. B. Hotchkiss and M. D. Hammerbacher, *22nd IEEE Photovoltaic Specialists Conference*, IEEE Publishing, NY (1991), p. 1045.

J. D. Levine, *1992 Photovoltaic Advanced Research and Development Project Conference*, Denver, American Institute of Physics Press, NY (1992), p. 46.

C. R. Lewis, H. F. MacMillan, B. C. Chung, G. F. Virshup, D. D. Liu *et al., Solar Cells* **24**, 171 (1988).

N. S. Lewis, *American Scientist* **83**, 534 (1995).

X. R. Li, S. Wagner, M. Bennett, J. Y. Hou, F. Rubinelli and S. J. Fonash, *11th European Photovoltaic Solar Energy Conf.* (1992), p. 703.

X. Li, P. Sheldon, H. Moutinho and R. Matson, *25th IEEE Photovoltaic Specialists Conference*, IEEE Publishing, NY (1996), p. 933.

D. Lincot and J. Vedel, *10th European Photovoltaic Solar Energy Conference* (1991), p. 931.

D. Lincot and R. Ortega-Borges, *J. Electrochem. Soc.* **139**, 1880 (1992).

D. Lincot, R. Ortega-Borges, J. Vedel, M. Ruckh, J. Kessler, K. O. Velthaus, D. Hariskos and H. W. Schock, *11th European Photovoltaic Solar Energy Conference* (1992), p. 870.

D. Lincot, R. Ortega-Borges and M. Froment, *Phil. Mag. B*, **68**, 185 (1993).

D. Lincot, R. Ortega-Borges and M. Froment, *Appl. Phys. Lett.* **64**, 569 (1994).

D. Lincot, J.-F. Guillemoles, P. Cowache, S. Massaccesi, L. Thouin, K. Fezzaa, F. Boisivon and J. Vedel, *First World Conference on Photovoltaic Energy Conversion, 24th IEEE Photovoltaic Specialists Conference*, IEEE Publishing, NY (1994), p. 136.

D. Lincot, A. Kampmann, B. Mokili, J. Vedel, R. Cortes and M. Froment, *Appl. Phys. Lett.* **67**, 2355 (1995).

F. A. Lindholm, A. Neugroshel, M. Arienzo and P. A. Iles, *IEEE Electron Device Lett.* **6**, 363 (1985).

J. Lindmayer, *COMSAT Technical Review* **2**, 105 (1972).

J. Lindmayer and J. F. Allison, *COMSAT Tech. Rev.* **3**, 1-22; reprinted in: *11th IEEE Photovoltaic Specialists Conference*, IEEE Publishing, NY (1975), p. 209.

J. Lindmayer and C. Wrigley, *12th IEEE Photovoltaic Specialists Conference*, IEEE Publishing, NY (1976), p. 53.

P. F. Lindquist and R. H. Bube, *J. Appl. Phys.* **43**, 2839 (1972a).

P. F. Lindquist and R. H. Bube, *J. Electrochem. Soc.* **119**, 936 (1972b).

P. Loelgen, C. Leguijt, J. A. Eikelboom, R. A. Steeman, W. C. Sinke, L. A. Verhoef, P. F. A. Alkemade and E. Algra, *23rd IEEE Photovoltaic Specialists Conference*, Louisville, IEEE Publishing, NY (1993).

P. Loelgen, A. Rohatgi, P. Sana, W. C. Sinke, A. W. Weeber, C. Leguijt, R. A. Steeman and J. A. Eikelboom, *12th European Photovoltaic Solar Energy Conference*, Netherlands (1994a).

P. Loelgen, W. C. Sinke, C. Leguijt, A. W. Weeber, P. F. A. Alkemade and L. A. Verhoef, *Appl. Phys. Lett.* **65**, 2792 (1994b).

J. J. Loferski, *J. Appl. Phys.* **27**, 777 (1956).

J. J. Loferski, M. Kwietniak, J. Piekoszewski, M. Spitzer, R. Arya, B. Roessler, R. Beaulieu, E. Vera, J. Shewchun and L. L. Kazmerski, *Proc. 15th IEEE Photovoltaic Specialists Conference*, IEEE Publishing, NY (1981), p. 1056.

T. C. Lommasson, H. Talieh, J. D. Meakin and J. A. Thornton, *19th IEEE Photovoltaic Specialists Conference*, IEEE Publishing, NY (1987), p. 1285.

T. C. Lommasson, A. F. Burnett, M. Kim, L. H. Chou and J. A. Thornton, *Proc. 7th International Conference on Ternary and Multinary Compounds* (1987).

B. R. Losada, A. Moelhecke, R. Lagos and A. Luque, *Appl. Phys. Lett.* **67**, 1894 (1995).

R. J. Loveland, W. E. Spear and A. Al-Sharboty, *J. Non-Cryst. Solids* **13**, 55 (1973–1974).

Y. Y. Loginov, K. Durose, H. M. Al-Allak, S. A. Galloway, S. Oktik, A. W. Brinkman, H. Richter and D. Bonnet, *J. Cryst. Growth* **161**, 159 (1996).

A. Lopez-Otero, *Thin Solid Films* **49**, 1 (1978).

A. Lopez-Otero and W. Huber, *Surface Sci.* **86**, 167 (1979).

R. B. Love and U. V. Choudary, U.S. Patent 4,465,575 (filed 28 February 1983), assignee: Atlantic Richfield Company (1984).

Y. Lubianiker, G. Biton, I. Balberg, T. Walter, H. W. Schock, O. Resto and S. Z. Weisz, *J. Appl. Phys.* **79**, 876 (1996).

W. B. Luft, B. von Roedern, B. Stafford, D. Waddington and L. Mrig, *22nd IEEE Photovoltaic Specialists Conf.*, IEEE Publishing, NY (1991), p. 1393.

W. Luft and Y. S. Tsuo, *Hydrogenated Amorphous Silicon Alloy Deposition Processes*, Marcel Dekker, NY (1993).

A. Luque, in *Advances in Solar Energy* (M. Prince, ed.) Vol. 8, American Solar Energy Society, Boulder, CO (1993), p. 201.

W. Ma, T. Saida, C. C. Lim, S. Aoyama, H. Okamoto and Y. Hamakawa, *First World Conference on Photovoltaic Energy Conversion, 24th IEEE Photovoltaic Specialists Conference*, IEEE Publishing, NY (1994), p. 4217.

W. Ma, C. C. Lim, T. Saida, H. Okamoto and Y. Hamakawa, *Solar Energy Materials and Solar Cells* **34**, 401 (1994b).

Y. Y. Ma, Ph.D. Dissertation, Stanford University (1977).

Y. Y. Ma and R. H. Bube, *J. Electrochem. Soc.* **124**, 1430 (1977b).

Y. Y. Ma, A. L. Fahrenbruch and R. H. Bube, *Appl. Phys. Lett.* **30**, 423 (1977a).

H. F. MacMillan, H. C. Hamaker, N. R. Kaminar, M. S. Kuryla, M. Ladle Ristow *et al.*, *20th IEEE Photovoltaic Specialists Conference*, IEEE Publishing, NY (1988a), p. 462.

H. F. MacMillan, H. C. Hamaker, G. F. Virshup and J. G. Werthen, *20th IEEE Photovoltaic Specialists Conference*, IEEE Publishing, NY (1988b), p. 48.

H. F. MacMillan, B.-C. Chung, L. Partain, J. C. Schultz, G. F. Virshup and J. G. Werthen, *9th Annual Review Meeting SERI/CP-213-3495*, Lakewood, CO (1989), p. 53.

A. Madan and S. R. Ovshinky, *8th International Conference on Amorphous and Liquid Semiconductors*, Cambridge (1979).

Y. Madea and I. Hide, *Technical Digest of the 3rd Photovoltaic Science and Engineering Conference*, Tokyo (1987), p. 87.

A. H. Mahan, E. Iwaniczko, B. P. Nelson, R. C. Reedy Jr., R. S. Crandall, S. Guha and J. Yang, *25th IEEE Photovoltaic Specialists Conference*, IEEE Publishing, NY (1996), p. 1065.

C. Malone, J. L. Nicque, S. J. Fonash, C. R. Wronski and M. Bennett, *22nd IEEE Photovoltaic Specialists Conference Proc.*, IEEE Publishing, NY (1991),

p. 1219.

H. M. Manasevit, K. L. Hess, P. D. Dapkus, R. P. Ruth, J. J. Yang, A. G. Campbell, R. E. Johnson, L. A. Moudy, R. H. Bube, L. B. Fabick, A. L. Fahrenbruch and M.-J. Tsai, *13th IEEE Photovoltaic Specialists Conference*, IEEE Publishing, NY (1978), p. 165.

J. Mandelkorn, C. McAfee, J. Kesperis, L. Schwartz and W. Pharo, *J. Electrochem. Soc.*, 313 (1962).

D. Mao, L. H. Feng, Y. Zhu, J. Tang, W. Song, R. Collins, D. L. Williamson and J. U. Trefny, *13th NREL Photovoltaics Program Review, AIP Conference Proceedings* **353**, 352 (1995).

M. Marudachalam, R. W. Birkmire, J. M. Schultz and T. Yokimcus, *First World Conference on Photovoltaic Energy, 24th IEEE Photovoltaic Specialists Conference*, IEEE Publishing, NY (1994), p. 234.

E. Maruyama, A. Terakawa, K. Sayama, K. Ninomiya, Y. Hishikawa, H. Tarui, S. Tsuda, S. Nakano and Y. Kuwano, *23rd IEEE Photovoltaic Specialists Conference*, Louisville, IEEE Publishing, NY (1993a).

E. Maruyama, Y. Hoshimine, A. Terakawa, K. Sayama, K. Ninomiya, Y. Hishikawa, H. Tarui, S. Tsuda, S. Nakano and Y. Kuwano, Materials Research Society, San Francisco (1993b).

E. Maruyama, S. Tsuda and S. Nakano, *Solid State Phenomena* **44-46**, 863 Scitec Publications (1995).

E. Maruyama, Y. Hishikawa, M. Tanaka, S. Kiyama and S. Tsuda, Materials Research Society, San Francisco (1996).

G. Masse and E. Redjai, *J. Appl. Phys.* **56**, 1154 (1984).

G. Masse, *J. Appl. Phys.* **68**, 2206 (1990).

G. Masse and K. Djessas, *Thin Solid Films* **226**, 254 (1993).

G. Masse and K. Djessas, *Thin Solid Films* **237**, 129 (1994).

G. Masse and K. Djessas, *Thin Solid Films* **257**, 137 (1995).

G. Masse, K. Djessas and L. Yarzhou, *J. Appl. Phys.* **74**, 1376 (1993).

J. H. Matlock, *Semiconductor International* **2**, 33 (1979).

A. Matsuda, S. Yamasaki, K. Nakagawa, H. Okushi, K. Tanaka, S. Iizima, M. Matsumura and H. Yamanioto, *Jpn. J. Appl. Phys.* **19**, L305 (1980).

H. Matsumoto, K. Kuribayashi, H. Uda, Y. Komatou, A. Nakano and S. Ikegami, *Solar Cells* **11**, 367 (1984).

R. H. Mauch, J. Hedstrom, D. Lincot, M. Ruckh, J. Kessler, R. Klinger, L. Stolt, J. Vedel and H. Schock, *Proc. 22nd IEEE Photovoltaic Specialists Conference*, IEEE Publishing, NY (1991a).

R. H. Mauch, M. Ruckh, J. Hedstrom, D. Lincot, J. Kessler, R. Klinger, L. Stolt, J. Vedel and H. W. Schock, *10th European Photovoltaic Solar Energy Conference* (1991b), p. 1415.

B. E. McCandless, R. W. Birkmire, W. A. Buchanan, J. E. Phillips and R. E. Rocheleau, *20th IEEE Photovoltaic Specialists Conference*, IEEE Publishing, NY (1988), p. 381.

B. E. McCandless and S. S. Hegedus, *22nd IEEE Photovoltaic Specialists Conference*, IEEE Publishing, NY (1991).

B. E. McCandless and R. W. Birkmire, *Solar Cells* **31**, 527 (1991).

B. E. McCandless, Y. Qu and R. W. Birkmire, *First World Conference on Photovoltaic Energy Conversion, 24th IEEE Photovoltaic Specialists Conference,* IEEE Publishing, NY (1994), p. 107.

B. E. McCandless, A. Mondal and R. W. Birkmire, *Solar Energy Materials and Solar Cells* **36**, 369 (1995).

B. E. McCandless, H. Hichri, G. Hanket and R. W. Birkmire, *25th Photovoltaic Specialists Conference,* IEEE Publishing, NY (1996), p. 781.

P. J. McElheny, H. Okushi, S. Yamasaki and A. Matsuda, *J. Non-Cryst. Solids* **137&138**, 243 (1991).

T. J. McMahon, *Proc. Mater. Res. Soc. Symp.* **258**, 325 (1992).

J. D. Meakin, *Society of Photo-Optical Instrumentation Engineers — Photovoltaics* **543**, 108 (1985).

J. D. Meakin, R. W. Birkmire, L. D. Di Netta, P. G. Lasswell and J. E. Phillips, *Solar Cells* **16**, 447 (1986).

M. Meaudre, P. Jensen and R. Meaudre, *Phil. Mag. B* **63**, 815 (1991).

M. Meaudre and R. Meaudre, *Phys. Rev.* **45**, 4524 (1992a).

M. Meaudre and R. Meaudre, *Phys. Rev.* **45**, 12134 (1992b).

J. Meier, R. Flueckiger, H. Keppner and A. Shah, *Appl. Phys. Lett.* **65**, 860 (1994a).

J. Meier, S. Dubail, R. Flueckiger, S. Fischer, H. Heppner and A. Shah, *First World Conference on Photovoltaic Energy Conversion, 24th IEEE Photovoltaic Specialists Conference,* IEEE Publishing, NY (1994), p. 409.

J. Meier, P. Torres, R. Platz, S. Dubail, U. Kroll, J. A. Anna Selvan, N. Pellaton Vaucher, Ch. Hof, D. Fischer, H. Keppner, A. Shah, K.-D. Ufert, P. Giannoules and J. Koehler, *Mat. Res. Soc. Symp. Proc.,* San Francisco (1996).

O. de Melo, M. Melendez-Lira, I. Hernandez-Calderon, L. Banos and A. Morales-Acevedo, *First World Conference on Photovoltaic Energy Conversion, 24th Photovoltaic Specialists Conference,* IEEE Publishing, NY (1994), p. 369.

A. Mettler, N. Wyrsch, M. Goetz and A. Shah, *Solar Energy Materials and Solar Cells* **34**, 533 (1994).

P. V. Meyers, Proc. Polycrystalline Thin Film Program Meet., *Solar Energy Research Institute,* Golden, CO (1987), p. 9.

P. V. Meyers, *Solar Cells* **23**, 59 (1988).

P. V. Meyers, *Solar Cells* **27**, 91 (1989).

P. V. Meyers, *Photovoltaics Advanced Research and Development,* 9th Ann. Rev. Meet. SERI/CP-213-3495, Lakewood, CO (1989).

P. V. Meyers, C. H. Liu and M. E. Doty, U.S. Patent 4,873,198, assignee: Ametek, Inc. (1989).

P. V. Meyers and J. E. Phillips, *25th IEEE Photovoltaic Specialists Conference,* IEEE Publishing, NY (1996), p. 789.

P. P. Michiels, L. A. Verkhoef, J. C. Stroom, W. C. Sinke, R. J. C. van Zolingen, C. M. M. Denisse and M. Hendriks, *21st IEEE Photovoltaic Specialists Conference,* IEEE Publishing, NY (1990), p. 638.

R. A. Mickelsen and W. S. Chen, *Proc. 15th IEEE Photovoltaic Specialists Conference,* IEEE Publishing, NY (1981), p. 800.

R. A. Mickelsen, W. S. Chen and L. F. Buldhaupt, "CadmiumSulfide/Copper Ternary Heterojunction Research", Final Rep. XJ-9-8021-1. Boeing Aerospace Co., Seattle, Washington (1982).

R. A. Mickelsen and W. S. Chen, *Proc. 16th IEEE Photovoltaic Specialists Conference*, IEEE Publishing, NY (1982), p. 781.

R. A. Mickelsen and W. S. Chen, *Proc. 16th IEEE Photovoltaic Specialists Conference*, IEEE Publishing, NY (1983), p. 781.

R. A. Mickelsen and W. S. Chen, *Proc. 7th International Conference on Ternary and Multinary Compounds*, Snowmass, CO (1986).

R. A. Mickelsen, B. J. Stanbery, J. D. Avery, W. S. Chen and W. E. Devaney, *Proc. 19th IEEE Photovoltaic Specialists Conference*, IEEE Publishing, NY (1987), p. 744.

D. Miller, *SERI Emerging Materials Contract Review Meeting*, Golden, CO (1981).

A. G. Milnes and D. L. Feucht, *Heterojunctions and Metal-Semiconductor Junctions*, Academic Press, NY (1972), p. 9.

J. Mimila-Arroyo, A. Bouazzi and G. Cohen-Solal, *Rev. Phys. Appl.* **12**, 423 (1977).

J. Mimila-Arroyo, Y. Marfaing, G. Cohen-Solal and R. Triboulet, *Sol. Energy Mater.* **1**, 171 (1979).

S.-K. Min, H. Y. Cho, W. C. Choi, M. Yamaguchi and T. Takamoto, *23rd IEEE Photovoltaic Specialists Conference*, IEEE Publishing, NY (1993).

K. W. Mitchell, A. L. Fahrenbruch and R. H. Bube, *Solid-State Electronics* **20**, 559 (1977a).

K. W. Mitchell, A. L. Fahrenbruch and R. H. Bube, *J. Appl. Phys.* **48**, 4365 (1977b).

K. W. Mitchell, D. Tanner, S. Vasquez, D. Willett and S. Lewis, *18th IEEE Photovoltaic Specialists Conference*, IEEE Publishing, NY (1985a), p. 914.

K. W. Mitchell, C. Eberspacher, F. Cohen, J. Avery, G. Duran and W. Bottenberg, *18th IEEE Photovoltaic Specialists Conference*, IEEE Publishing, NY (1985b), p. 1359.

K. W. Mitchell, C. Eberspacher, J. Ermer and D. Pier, *20th IEEE Photovoltaic Specialists Conference*, IEEE Publishing, NY (1988), p. 1384.

K. W. Mitchell and H. I. Liu, *Proc. 20th IEEE Photovoltaic Specialists Conference*, IEEE Publishing, NY (1988), p. 1461.

K. W. Mitchell, G. A. Pollock and A. V. Mason, *20th IEEE Photovoltaic Specialists Conference*, IEEE Publishing, NY (1988), p. 1542.

K. W. Mitchell, D. Willett, C. Eberspacher, J. Ermer, D. Pier and K. Pauls, *Proc. 21st IEEE Photovoltaic Specialists Conference*, IEEE Publishing, NY (1990), p. 1481.

K. W. Mitchell, *Progress in Photovoltaics* **2**, 115 (1994).

K. W. Mitchell, R. R. King, T. L. Jester and M. McGraw, *First World Conference on Photovoltaic Energy Conversion, 24th IEEE Photovoltaic Specialists Conference*, IEEE Publishing, NY (1994), p. 1266.

K. Miyachi, N. Ishiguro, T. Miyashita, N. Yanagawa, H. Tanaka, M. Koyama, Y. Ashida and N. Fukuda, *11th E.C. Photovoltaic Solar Energy Conference*, Montreux, Switzerland (1992), p. 88.

T. Mizrah and D. Adler, *IEEE Trans. Electron Devices* **ED-24**, 458 (1977).

J. Moesslein, A. Lopez-Otero, A. L. Fahrenbruch, D. Kim and R. H. Bube, *J. Appl. Phys.* **73**, 8359 (1993).

H.-D. Mohring, M. B. Schubert, G. H. Bauer and W. H. Bloss, *ISES Solar World Congress*, Hamburg (1987).

B. Mokili, M. Froment and D. Lincot, *Journal de Physique IV. Colloque C3*, supplement au *Journal de Physique III*, **5**, C3-261 (1995).

A. Mondal, R. W. Birkmire and B. E. McCandless, *22nd IEEE Photovoltaic Specialists Conference*, IEEE Publishing, NY (1991).

A. Mondal, B. E. McCandless and R. W. Birkmire, *Solar Energy Materials and Solar Cells* **26**, 181 (1992).

R. Monna, D. Angermeier, A. Slaoui and J. C. Muller, *25th IEEE Photovoltaic Specialists Conference*, IEEE Publishing, NY (1996), p. 701.

J. B. Mooney, R. H. Lamoreaux and C. W. Bates, Rep. SERI/PR-8104-4-T1. SERI, Golden, CO (1980).

S. Mora and N. Romeo, *J. Appl. Phys.* **48**, 4826 (1977).

D. L. Morel, A. K. Ghosh, T. Feng, E. L. Stogryn, P. E. Purwin, R. F. Shaw and C. Fishman, *Appl. Phys. Lett.* **32**, 495 (1978).

D. L. Morel, *Solar Cells* **24**, 157 (1988).

A. Morimoto, T. Miura, M. Kumeda and T. Shimizu, *J. Appl. Phys.* **53**, 7299 (1982).

P. D. Moskowitz, P. K. Fowler, D. G. Dobryn, C. M. Lee and V. M. Fthenakis, "Control of toxic gas release during the production of copper indium diselenide PV cells", draft report, MIT/BNL-86-4 (1986).

P. D. Moskowitz, K. Zweibel and V. M. Fthenakis, SERI/TR-211-3621 (DE900-00310), Solar Energy Research Institute, Golden, CO (1990).

P. D. Moskowitz and V. M. Fthenakis, *Solar Cells* **29**, 63 (1990).

P. D. Moskowitz and K. Zweibel, *20th IEEE Photovoltaic Specialists Conference*, IEEE Publishing, NY (1990), p. 1040.

P. D. Moskowitz, *First World Conference on Photovoltaic Energy Conversion, 24th IEEE Photovoltaic Specialists Conference*, IEEE Publishing, NY (1994), p. 115.

T. D. Moustakas, D. A. Anderson and W. Paul, *Solid State Commun.* **23**, 155 (1977).

H. R. Moutinho, R. G. Dhere, K. Ramanathan, P. Sheldon and L. L. Kazmerski, *25th Photovoltaic Specialists Conference*, IEEE Publishing, NY (1996), p. 945.

T. H. Myers, *J. Vac. Sci. Technol.* **A7**, 300 (1989).

T. Nakada, K. Migita, S. Niki and A. Kunioka, *Jpn. J. Appl. Phys.* **34**, 4715 (1995).

T. Nakada, D. Iga, H. Ohbo and A. Kunioka, *Jpn. J. Appl. Phys.* **36**, 732 (1997).

G. Nakamura, K. Sato, H. Kondo, Y. Yukimoto and K. Shirahata, *4th Eur. Community Photovoltaic Solar Energy Conference*, Stressa, Italy, D. Reidel, Dordrecht, Holland (1982), p. 616.

G. Nakamura, K. Sato, T. Ishihara, M. Usui, H. Sasaki, K. Okaniwa and Y. Yukimoo, *Proc. Amorphous Silicon Subcontractors Annual Review Meeting*, Washington, D.C. (1985), p. 149.

M. Nakata, S. Wagner and T. Peterson, *J. Non-Cryst. Solids* **164/166**, 179 (1993).

N. Nakayama, H. Matsumoto, K. Yamaguchi, S. Ikegami and Y. Hioki, *Jpn. J. Appl. Phys.* **15**, 2281 (1976).

N. Nakayama, H. Matsumoto, A. Nakano, S. Ikegami, H. Uda and T. Yamashita, *Jpn. J. Appl. Phys.* **19**, 703 (1980).

T. Nakazawa, K. Takamizawa and K. Ito, *Appl. Phys. Lett.* **50**, 279 (1987).

T. R. Nash and R. L. Anderson, *IEEE Trans. Electron Devices* **ED-24**, 468 (1977).

P. Nath, K. Hoffman, J. Call, G. DiDio, C. Vogeli and S. R. Ovshinsky, *20th IEEE Photovoltaic Specialists Conference*, IEEE Publishing, NY (1988), p. 293.

C. E. Nebel, M. Schubert, H. C. Weller, G. H. Bauer and W. H. Bloss, *8th EC Photovoltaic Solar Energy Conference*, Florence (1988a).

C. E. Nebel, H. C. Weller and G. H. Bauer, *20th IEEE Photovoltaic Specialists Conference*, IEEE Publishing, NY (1988), p. 229.

H. Neber-Aeschbacher, ed., *Hydrogenated Amorphous Silicon, Solid State Phenomena* **44-46**, 1-1048, Scitec Publications (1995).

B. P. Nelson, E. Iwaniczko, R. E. I. Schropp, H. Mahan, E. C. Molenbroek, S. Salamon and R. S. Crandall, *12th European Photovoltaic Solar Energy Conference*, Amsterdam, the Netherlands (1994), p. 679.

G. F. Neumark, in *Wide-Gap II–VI Compounds for Opto-electronic Applications* (H. Ruda, ed.) Chapman & Hall, London (1992).

N. H. Nickel, W. B. Jackson and N. M. Johnson, *Phys. Rev. Lett.* **71**, 2733 (1993).

F. H. Nicoll, *J. Electrochem. Soc.* **110**, 1165 (1963).

T. Nii, I. Sugiyama, T. Kase, M. Sato, Y. Kaniyama, S. Kuriyagawa, K. Kushiya and H. Takeshita, *First World Conference on Photovoltaic Energy Conversion, 24th IEEE Photovoltaic Specialists Conference*, IEEE Publishing, NY (1994), p. 254.

S. Niki, Y. Makita, A. Yamada, H. Shibata, P. J. Fons, A. Obara, T. Kurafuji, S. Chichibu and H. Nakanishi, *First World Conference on Photovoltaic Energy Conversion, 24th IEEE Photovoltaic Specialists Conference*, IEEE Publishing, NY (1994), p. 132.

K. Nishimura and R. H. Bube, *J. Appl. Phys.* **58**, 420 (1985).

T. Nishio, K. Omura, A. Hanafusa, T. Arita, H. Higuchi, T. Aramoto, S. Shibutani, S. Kumazawa, M. Murozono, Y. Yabuuchi and H. Takakura, *25th Photovoltaic Specialists Conference*, IEEE Publishing, NY (1996), p. 953.

S. Noel, R. Schindler, S. W. Glunz, W. Warta and W. Wettling, *13th European Photovoltaic Solar Energy Conference*, Nice (1995), p. 1406.

K. Nomoto, H. Saitoh, A. Chida, H. Sannomiya, M. Itoh and Y. Yamamoto, *Tech. Digest of the 7th International Photovoltaic Science and Engineering Conference*, Dept. Electr. and Computer Eng., Nagoya Inst. of Tech., Nagoya, Japan (1993), p. 275.

R. Noufi, R. Axton, C. Herrington and S. K. Deb, *Appl. Phys. Lett.* **45**, 668 (1984).

R. Noufi and J. Dick, *J. Appl. Phys.* **58**, 3884 (1985).

R. Noufi, P. Souza and C. Osterwald, *Solar Cells* **15**, 87 (1985).

R. Noufi, R. C. Powell and R. J. Matson, *Proc. 7th International Conference on Ternary and Multinary Compounds*, Snowmass, CO (1986a).

R. Noufi, R. J. Matson, R. C. Powell and C. Herrington, *Solar Cells* **16**, 479 (1986b).

R. Noufi, R. Powell, C. Herrington and T. Coutts, *Solar Cells* **17**, 303 (1986c).

A. Nouhi and R. J. Stirn, *J. Vac. Sci. Technol.* **A4**, 403 (1986).

A. Nouhi, R. J. Stirn and A. Hermann, *Proc. 19th IEEE Photovoltaic Specialists Conference*, IEEE Publishing, NY (1987), p. 1461.

A. Nouhi, R. J. Stirn, P. V. Meyers and C. H. Liu, *J. Vac. Sci. Technol.* (1989).

T. Nunoi, S. Okamoto, K. Nakajima, S. Tanaka, N. Shibuya, K. Okamoto, T. Nammori and H. Itoh, *21st IEEE Photovoltaic Specialists Conference*, IEEE Publishing, NY (1990), p. 664.

S. Oelting, D. Martini and D. Bonnet, *11th E.C. Photovoltaic Solar Energy Conference*, (L. Guimaraes, W. Palz, C. de Reyff, H. Kiess and P. Helm, eds.) Harwood Academic, Chur (1993), p. 491.

K. Ohata, J. Saraie and T. Tanaka, *Japan. J. Appl. Phys.* **12**, 1641 (1973).

R. S. Ohl, US Patent 240252 (1941); US Patent 2443542 (1941).

Y. Ohtake, K. Kushiya, A. Yamada and M. Konagai, *First World Conference on Photovoltaic Energy Conversion, 24th IEEE Photovoltaic Specialists Conference*, IEEE Publishing, NY (1994), p. 218.

H. Okazaki, T. Takamoto, H. Takamura, T. Kamei, M. Ura, A. Yamamoto and M. Yamaguchi, *12th IEEE Photovoltaic Specialists Conference*, IEEE Publishing, NY (1988), p. 886.

J. M. Olson, S. R. Kurtz, A. E. Kibbler and E. Beck, *9th Annual Review Meeting* SERI/CP-213-3495, Lakewood, CO (1989), p. 51.

L. C. Olsen and R. C. Bohara, *11th IEEE Photovoltaic Specialists Conference*, IEEE Publishing, NY (1975), p. 381.

L. C. Olsen, F. W. Addis, D. Greer, W. Lei and F. Abulfotah, *First World Conference on Photovoltaic Energy Conversion, 24th IEEE Photovoltaic Specialists Conference*, IEEE Publishing, NY (1994), p. 194.

L. C. Olsen, X. Deng, W. Lei, F. W. Addis and J. Li, *25th IEEE Photovoltaic Specialists Conference*, IEEE Publishing, NY (1996), p. 61.

L. C. Olsen, H. Aguilar, F. W. Addis, W. Lei and J. Li, *25th IEEE Photovoltaic Specialists Conference*, IEEE Publishing, NY (1996), p. 997.

B. O'Regan and M. Graetzel, *Nature* **353**, 737 (1991).

R. Ortega Borges, D. Lincot and J. Vedel, *11th European Photovoltaic Solar Energy Conference*, Montreux, Switzerland (1992), p. 862.

R. Ortega-Borges and D. Lincot, *J. Electrochem. Soc.* **140**, 3464 (1993).

M. Ortega-Lopez and A. Morales-Acevedo, *25th IEEE Photovoltaic Specialists Conference*, IEEE Publishing, NY (1996), p. 1009.

H.-C. Ostendorf, W. Kusian, W. Kruehler and R. Schwarz, *J. Non-Cryst. Solids* **164-166**, 659 (1993).

F. Ouchen, P. Gallon, M. C. Artaud, P. Cowache, C. Llinares and S. Duchemin, *13th European Community Photovoltaic Solar Energy Conference* (1995a), p. 1957.

F. Ouchen, P. Gallon, M. C. Artaud, J. Bougnot and S. Duchemin, *10th International Conference on Ternary and Multinary Compounds*, Stuttgart (1995b).

J. Daey Ouwens, R. E. I. Schropp, W. van der Weg and D. L. Williamson, *12th European Photovoltaic Solar Energy Conference*, Amsterdam, the Netherlands (1994), p. 1292.

H. Overhof and W. Beyer, *Philos. Mag. B* **43**, 433 (1981).

S. R. Ovshinsky, *Proc. International Photovoltaic Solar Energy Conversion* (1988), p. 577.

M. E. Ozsan, D. R. Johnson, M. Sadeghi, D. Sivapathasundaram, L. M. Peter, M. J. Furlong, G. Goodlet, A. Shingleton, D. Lincot, B. Mokili and J. Vedel, *First World Conference on Photovoltaic Energy Conversion, 24th IEEE Photovoltaic Specialists Conference*, IEEE Publishing, NY (1994), p. 327.

M. E. Ozsan, D. R. Johnson, M. Sadeghi, D. Sivapathasundaram, D. Lincot, B. Mokili, M. Froment, J. Vedel, L. M. Peter, G.Goodlet and R. C. Walker, *13th European Photovoltaic Solar Energy Conference* (1995), p. 2115.

F. A. Padovani and R. Stratton, *Solid State Electron.* **9**, 695 (1966).

B. Pamplin, *Prog. Cryst. Growth Charact.* **1**, 395 (1979).

B. Pamplin and R. S. Feigelson, *Thin Solid Films* **60**, 141 (1979).

R. Pandya and E. A. Schiff, *J. Non-Cryst. Solids* **77&78**, 623 (1985).

M. P. R. Panicker, M. Knaster and F. A. Kroger, *J. Electrochem. Soc.* **125**, 566 (1978).

J. I. Pankove, ed., *Hydrogenated Amorphous Silicon*, **21** A, B, C and D, *Semiconductors and Semimetals*, Academic Press, Inc. (1984).

H. Park, J. Liu and S. Wagner, *Appl. Phys. Lett.* **55**, 2658 (1989).

J. Parkes, R. D. Tomlinson and M. J. Hampshire, *J. Appl. Cryst.* **6**, 414 (1973).

M. H. Patterson and R. H. Williams, *J. Phys. D* **11**, L83 (1978).

M. S. Patterson, A. K. Turner, M. Sadeghi and R. J. Marshall, *12th European Photovoltaic Solar Energy Conference* (1994), p. 950.

W. Paul, R. A. Street and S. Wagner, *J. Electronic Materials* **22**, 39 (1993).

K. L. Pauls, K. W. Mitchell and W. Chesarek, *23rd IEEE Photovoltaic Specialists Conference*, IEEE Publishing, NY (1993), p. 209.

P. D. Paulson, V. Dutta and C. Singh, *First World Conference on Photovoltaic Energy Conversion, 24th Photovoltaic Specialists Conference*, IEEE Publishing, NY (1994), p. 331.

A. Pawlikiewicz, A. Banerjee and S. Guha, in *Amorphous Silicon Technology, Proc. Mat. Res. Soc. Symp.* **219**, 141 (1991).

G. L. Pearson, *18th IEEE Photovoltaic Specialists Conference*, IEEE Publishing, NY (1985).

I. Perichaud, F. Floret and S. Martinuzzi, *23rd IEEE Photovoltaic Specialists Conference*, IEEE Publishing, NY (1993), p. 243.

F. J. Pern, J. Goral, R. J. Matson, T. A. Gessert and R. Noufi, *8th Photovoltaic Advanced Research and Development Project Review Meet.*, Denver CO (1987).

F. J. Pern, R. Noufi, A. Mason and A. Swartzlander, *19th IEEE Photovoltaic Specialists Conference*, IEEE Publishing, NY (1987), p. 1295.

M. G. Peters, A. L. Fahrenbruch and R. H. Bube, *J. Appl. Phys.* **64**, 3106 (1988a).

M. G. Peters, A. L. Fahrenbruch and R. H. Bube, *J. Vac. Sci. Technol.* **A6**, 3098 (1988b).

S. Peulon and D. Lincot, *13th European Photovoltaic Solar Energy Conference*, Nice, France (1995), p. 1750.

S. Peulon and D. Lincot, *Advanced Materials* **8**, 166 (1996).

J. Piekoszewski, J. J. Loferski, R. Beualieu, J. Beall, B. Roessler and J. Shewchun, *Solar Energy Mater.* **2**, 363 (1980).

J. Piekoszewski, L. Castaner, J. J. Loferski, J. Beall and W. Giriat, *J. Appl. Phys.* **51**, 5375 (1980).

G. E. Pike and C. H. Seager, *J. Appl. Phys.* **50**, 3414 (1979).

M. Pinarbasi, N. Maley, L. H. Chou, A. Myers, M. J. Kushner *et al., Proc. Am. Vacuum Soc.* (1988).

F. A. Ponce, R. Sinclair and R. H. Bube, *Appl. Phys. Lett.* **39**, 951 (1981).

J. Ponpon, *Solid-State Electron.* **28**, 689 (1985).

W. M. Pontuschka, W. E. Carlos, P. C. Taylor and R. W. Griffith, *Phys. Rev. B* **25**, 4362 (1982).

O. Porre, M. Pasquinelli, S. Martinuzzi and I. Perichaud, *11th E.C. Photovoltaic Solar Energy Conf.* (L. Guimaraes, W. Palz, C. de Reyff, H. Kiess and P. Helm, eds.) Harwood Academic, Chur (1993), p. 1053.

R. R. Potter, C. Eberspacher and L. B. Fabick, *18th IEEE Photovoltaic Specialists Conference*, IEEE Publishing, NY (1985), p. 1659.

M. J. Powell, C. van Berkel and S. Deane, *J. Non-Cryst. Solids* **137/138**, 1215 (1991).

Y. Qu, P. V. Meyers and B. E. McCandless, *25th Photovoltaic Specialists Conference*, IEEE Publishing, NY (1996), p. 1013.

J. A. Rand, J. E. Cotter, C. J. Thomas, A. E. Ingram, Y. B. Bai, T. R. Ruffins and A. M. Barnett, *First World Conference on Photovoltaic Energy Conversion, 24th IEEE Photovoltaic Specialists Conference*, IEEE Publishing, NY (1994), p. 1262.

U. Rau, M. Schmitt, D. Hilburger, F. Engelhardt, O. Seifert, J. Parisi, W. Riedl, J. Rimmasch and F. Karg, *25th IEEE Photovoltaic Specialists Conference*, IEEE Publishing, NY (1996), p. 1005.

P. K. Raychaudhuri, *J. Appl. Phys.* **62**, 3025 (1987).

D. Redfield, *Appl. Phys. Lett.* **25**, 647 (1974).

D. Redfield, *Appl. Phys. Lett.* **33**, 531 (1978).

D. Redfield, *Appl. Phys. Lett.* **35**, 182 (1979).

D. Redfield, *IEEE Transactions on Electron Devices* **ED-27**, 766 (1980).

D. Redfield, *Appl. Phys. Lett.* **52**, 492 (1988).

D. Redfield, *Appl. Phys. Lett.* **54**, 398 (1989).

D. Redfield, *Mod. Phys. Lett. B* **5**, 933 (1991).

D. Redfield, *Proc. Mater. Res. Soc. Symp.* **258**, 341 (1992).

D. Redfield and R. H. Bube, *Phys. Rev. Lett.* **65**, 464 (1990).

D. Redfield and R. H. Bube, *22nd IEEE Photovoltaic Specialists Conference Proc.*, IEEE Publishing, NY (1991), p. 1319.

D. Redfield and R. H. Bube, *Phil. Mag. B* **74**, 309 (1996).

D. Redfield and R. H. Bube, *Photoinduced Defects in Semiconductors*, Cambridge Univ. Press, Cambridge (1996).

I. Reis, A. Eyer and A. Rauber, *20th IEEE Photovoltaic Specialists Conference*, IEEE Publishing, NY (1988), p. 1405.

D. C. Reynolds, G. Leies, L. L. Antes and R. E. Marburger, *Phys. Rev.* **96**, 533 (1954).

D. R. Rhiger and R. E. Kvaas, *J. Vac. Sci. Technol.* **A1**, 1712 (1983).

A. R. Riben and D. L. Feucht, *Solid State Electron.* **9**, 1055 (1966); *Int. J. Electron.* **20**, 583 (1966).

E. S. Rittner and R. A. Arndt, *J. Appl. Phys.* **47**, 2999 (1976).

A. Rockett and R. W. Birkmire, *J. Appl. Phys.* **70**, R81 (1991).

A. Rockett, M. Bodegard, K. Granath and L. Stolt, *25th IEEE Photovoltaic Specialists Conference*, IEEE Publishing, NY (1996), p. 985.

M. Rodot, M. Barbe, J. E. Bouree, V. Perraki, G. Revel, R. Kishore, J. L. Pastol, R. Mertens, M. Caymax and M. Eyckmans, *Revue Phys. Appl.* **22**, 687 (1987).

A. Rohatgi, C. J. Summers, A. Erbil, R. Sudharsanan and S. Ringel, SERI Annu. Rep. XL-7-06031-1 (1988).

A. Rohatgi, S. A. Ringel, R. Sudharsanan, P. V. Meyers, C. H. Liu and V. Ramanathan, *Photovoltaics Advanced Research and Development*, 9th Ann. Rev. Meet. SERI/CP-213-3495, Lakewood, CO (1989), p. 55.

A. Rohatgi, Z. Chen, P. Sana, R. Ramanachalam, J. Crotty and J. Salami, in *Techn. Digest of the 7th International Photovoltaic Science and Engineering Conference*, Dept. Elect. and Computer Eng., Nagoya Inst. of Tech., Nagoya, Japan (1993), p. 93.

A. Rohatgi, S. Narasimha, S. Kamra, P. Doshi, C. P. Khattak, K. Emery and H. Field, *25th IEEE Photvoltaic Specialists Conference*, IEEE Publishing, NY (1996), p. 741.

N. Romeo, A. Bosio, R. Tedeschi, A. Romeo, V. Canevari and D. Leone, *14th European Photovoltaic Solar Energy Conference*, Barcelona, Spain, June-July (1997).

N. Romeo, A. Bosio, R. Tedeschi, A. Romeo, V. Canevari and F. Fermi, *14th European Photovoltaic Solar Energy Conference*, Barcelona, Spain, June-July (1997).

D. H. Rose, D. S. Albin, R. J. Matson, A. B. Swartzlander, X. S. Li, R. G. Dhere, S. Asher, F. S. Hasoon and P. Sheldon, *Proc. of the 1996 Spring MRS Conference*, April (1996a).

D. H. Rose, D. H. Levi, R. J. Matson, D. S. Albin, R. G. Dhere and P. Sheldon, *25th IEEE Photovoltaic Specialists Conference*, IEEE Publishing, NY (1996b), p. 777.

M. Rosmeulen, H. E. Elgamel, J. Poortmans, M.-A. Trauwaert, J. Vanhellemont and J. Nijs, *First World Conference on Photovoltaic Energy Conversion, 24th IEEE Photovoltiac Specialists Conference*, IEEE Publishing, NY (1994), p. 1621.

M. Rosmeulen, S. W. Glunz, W. Warta, J. Knobloch and W. Wettling, *13th European Photovoltaic Solar Energy Conference*, Nice (1995), p. 403.

A. Rothwarf, *J. Vac. Sci. Technol.* **20**, 282 (1982).

A. Rothwarf, *IEEE Trans. on Electron Devices* **ED-29**, 1513 (1982).

A. Rothwarf, *Solar Cells* **16**, 567 (1986).

A. Rothwarf, *Solar Cells* **21**, 1 (1987).

A. Rothwarf and A. M. Barnett, *IEEE Trans. Electron Devices* **ED-24**, 381 (1977).

M. Ruckh, D. Hariskos, U. Ruehle, H. W. Schock, R. Menner and B. Dimmler, *25th IEEE Photovoltaic Specialists Conference*, IEEE Publishing, NY (1996a), p. 825.

M. Ruckh, D. Schmid, M. Kaiser, R. Schaeffler, T. Walter and H. W. Schock, *Solar Energy Materials and Solar Cells* **41/42**, 335 (1996b).

J. M. Ruiz, J. Casado and A. Luque, *12th European Photovoltaic Solar Energy Conference* (1994), p. 572.

T. W. F. Russell, S. Verma and R. W. Birkmire, *12th European Community Photovoltaic Solar Energy Conference* (1994a).

T. W. F. Russell, N. Orbey and R. W. Birkmire, *First World Conference on Photovoltaic Energy Conversion, 24th IEEE Photovoltaic Specialists Conference*, IEEE Publishing, NY (1994), p. 238.

E. S. Sabisky and J. L. Stone, *20th IEEE Photovoltaic Specialists Conference*, IEEE Publishing, NY (1988), p. 39.

B. Sagnes, A. Salesse, M. C. Artaud, S. Duchemin, J. Bougnot and G. Bougnot, *J. Crystal Growth* **124**, 620 (1992).

C.-T. Sah, R. N. Noyce and W. Shockley, *Proc. IRE* **45**, 1228 (1957).

R. Sahai, D. D. Edwall and J. S. Harris, Jr., *13th IEEE Photovoltaic Specialists Conference*, IEEE Publishing, NY (1978), p. 946.

I. Sakata, M. Yamanaka, S. Numase and Y. Hayashi, *J. Appl. Phys.* **71**, 4344 (1992).

Z. M. Saleh, H. Tarui, N. Nakamura, M. Nishikuni, S. Tsuda, S. Nakano and Y. Kuwano, *Jpn. J. Appl. Phys.* **12**, 3801 (1992).

Z. M. Saleh, H. Taruji, S. Tsuda, S. Nakano and Y. Kuwano, *Proc. Mater. Res. Soc. Symp.* **297**, 501 (1993).

A. N. Y. Samaan, I. S. Al-Saffar, S. M. Wasim, A. E. Hill, D. G. Armour and R. D. Tomlinson, *Nuovo Cimento Soc. Ital. Fis.* **2D**, 1784 (1983).

H. Sannomiya, K. Nomoto, A. Chida, Y. Nakata and Y. Mamamoto, *First World Conference on Photovoltaic Energy Conversion, 24th IEEE Photovoltaic Specialists Conference*, IEEE Publishing, NY (1994), p. 405.

J. Saraie, M. Akiyama and T. Tanaka, *Jpn. J. Appl. Phys.* **11**, 1758 (1972).

D. K. Sardana, *Semiconductor International*, 362 (1985).

R. A. Sasala, T. Zhou and W. M. Kocher, *First World Conference on Photovoltaic Energy Conversion, 24th IEEE Photovoltaic Specialists Conference*, IEEE Publishing, NY (1994), p. 311.

E. Sauvain, P. Pipoz, A. Shah and J. Hubin, *J. Appl. Phys.* **75**, 1722 (1994).

K. Sayama, A. Terakawa, M. Shima, E. Maruyama, K. Ninomiya, H. Tarui, S. Tsuda and S. Nakano, *Solar Energy Materials and Solar Cells* **34**, 423 (1994).

H. Schade, in *Semiconductor and Semimetals*, Vol. 21B (J. I. Pankove, ed.) Academic Press, Orlando (1984), p. 359.

C. H. Seager, *J. Appl. Phys.* **52**, 3960 (1981).

C. H. Seager, D. J. Sharp, J. K. G. Panitz and J. I. Hanoka, *J. de Physique* (Paris) Colloque **C1**, 103 (1982).

R. Schaeffler and H. W. Schock, *POLYSE 95*, Gargano (Lago di Garda), Italy (1995).

P. Schaetzle, T. Zoellner, R. Schindler and A. Eyer, *23rd IEEE Photovoltaic Specialists Conference*, IEEE Publishing, NY (1993), p. 78.

R. Scheer, T. Walter, H. W. Schock, M. L. Fearheiley and H. J. Lewerenz, *Appl. Phys. Lett.* **63**, 3294 (1993).

R. Scheer and H. J. Lewerenz, *J. Vac. Sci. Technol.* **13**, 1924 (1995).

C. Schetter, H. Lautenschlager and F. Lutz, *13th Euopean Photovoltaic Solar Energy Conference*, Nice (1995), p. 407.

R. Schindler, I. Reis, B. Wagner, A. Eyer, H. Lautenschlager, C. Schetter, W. Warta, B. Hartiti, A. Slaoui, J. C. Muller and P. Siffert, *23rd IEEE Photovoltaic Specialists Conference*, IEEE Publishing, NY (1993), p. 162.

D. Schmid, R. Ruckh, F. Grunwald and H. W. Schock, *J. Appl. Phys.* **73**, 2902 (1993).

D. Schmid, M. Ruckh and H. W. Schock, *First World Conference on Photovoltaic Energy Conversion, 24th IEEE Photovoltaic Specialists Conference*, IEEE Publishing, NY (1994), p. 198.

M. Schmitt, U. Rau, J. Parisi, W. Riedl, J. Rimmasch and F. Karg, *25th IEEE Photovoltaic Specialists Conference*, IEEE Publishing, NY (1996), p. 909.

U. Schneider, B. Schroeder and F. Finger, *J. Non-Cryst. Solids* **96/98**, 795 (1987).

H. Schoch *et al.*, *Proc. Materials Research Society Meeting*, San Francisco (1996).

A. Scholz and B. Schroeder, *Am. Inst. Phys. Conf. Proc.* **234**, 178 (1990).

A. Scholz and B. Schroeder, *J. Non-Cryst. Sol.* **137&138**, 259 (1991).

R. E. I. Schropp, A. Sluiter, M. B. von der Linden and J. Daey Ouwens, *J. Non-Cryst. Solids* **164-166**, 709 (1993a).

R. E. I. Schropp, J. Daey Ouwens, M. B. von der Linden, C. H. M. van der Werf, W. F. van der Weg and P. F. A. Alkemade, *Proc. Materials Research Society* **297**, 797 (1993b).

R. E. I. Schropp, M. B. von der Linden, J. Daey Ouwens and H. de Gooijer, *Solar Energy Materials and Solar Cells* **34**, 455 (1994).

R. E. I. Schropp, *Solid State Phenomena* **44-46**, 853 (1995).

D. L. Schulz, M. Pehnt, E. Urgiles, D. W. Niles, K. M. Jones, C. J. Curtis and D. S. Ginley, *25th IEEE Photovoltaic Specialists Conference*, IEEE Publishing, NY (1996), p. 929.

G. Schumm, *Phys. Rev. B* **49**, 2427 (1994).

B. A. Scott, R. M. Pleenik and E. E. Simony, *Appl Phys. Lett.* **39**, 73 (1981).

R. A. Scranton, J. B. Mooney, J. O. McCaldin, T. C. McGill and C. A. Mead, *Appl. Phys. Lett.* **29**, 47 (1976).

F. A. Selim and F. A. Kroger, *J. Electrochem. Soc.* **124**, 401 (1977).

R. F. Service, *Science* **272**, 1744 (1996).

J. Y. Seto, *J. Appl. Phys.* **46**, 5247 (1975).

W. N. Shafarman and J. E. Phillips, *22nd IEEE Photovoltaic Specialists Conference*, IEEE Publishing, NY (1991).

W. N. Shafarman and J. E. Phillips, *23rd IEEE Photovoltaic Specialists Conference*, IEEE Publishing, NY (1993).

W. N. Shafarman, R. Klenk and B. E. McCandless, *25th IEEE Photovoltaic Specialists Conference*, IEEE Publishing, NY (1996), p. 763.

W. N. Shafarman and J. E. Phillips, *25th IEEE Photovoltaic Specialists Conference*, IEEE Publishing, NY (1996), p. 917.

A. Shah, E. Sauvain, N. Wyrsch, H. Curtins, B. Leutz *et al.*, *20th IEEE Photovoltaic Specialists Conference*, IEEE Publishing, NY (1988), p. 282.

M. Shao, C. N. Tabory and A. D. Compaan, *25th IEEE Photovoltaic Specialists Conference*, IEEE Publishing, NY (1996), p. 869.

P. R. Sharps, A. L. Fahrenbruch, A. Lopez-Otero and R. H. Bube, *J. Appl. Phys.* **68**, 6406 (1990).

P. R. Sharps, M. L. Timmons, R. Venkatasubramanian, R. Pickett, J. S. Hills, J. Hancock, J. Hutchby, P. Iles, C. L. Chu, M. Wanlass and J. S. Ward, *23rd IEEE Photovoltaic Specialists Conference*, IEEE Publishing, NY (1993), p. 633.

P. R. Sharps, M. L. Timmons, Y. C. M. Yeh and C. L. Chu, *24th IEEE Photovoltaic Specialists Conference*, IEEE Publishing, NY (1994), p. 1725.

J. L. Shay, S. Wagner and H. M. Kasper, *Appl. Phys. Lett.* **27**, 89 (1975).

J. L. Shay, S. Wagner, K. J. Bachmann and E. Buehler, *J. Appl. Phys.* **47**, 614 (1976).

J. L. Shay, S. Wagner, M. Bettini, K. J. Bachmann and E. Buehler, *IEEE Trans. Electron Devices* **ED-24**, 483 (1977).

I. Shih and C. X. Qiu, *19th IEEE Photovoltaic Specialists Conference*, IEEE Publishing, NY (1987), p. 1291.

Z. Shi, W. Zhang, G. F. Zheng, V. L. Chin, A. Stephens, M. A. Green and R. Bergmann, *First World Conference on Photovoltaic Energy Conversion, 24th IEEE Photovoltaic Specialists Conference*, IEEE Publishing, NY (1994), p. 1339.

A. Shibata, Y. Kazama, K. Seki, W. Y. Kim, S. Yamanaka *et al.*, *20th IEEE Photovoltaic Specialists Conference*, IEEE Publishing, NY (1988), p. 317.

S. M. Shibli, *et al.*, *J. Vac. Sci. Technol.* **B8**, 187 (1990).

F. Shimura, *Semiconductor Silicon Crystal Technology*, Academic Press, NY (1989).

F. Shimura, ed., *Oxygen in Silicon*, Semiconductors and Semimetals **42**, Academic Press, San Diego (1994).

H. Shinohara, M. Abe, K. Nishi and Y. Arai, *First World Conference on Photovoltaic Energy Conversion, 24th IEEE Photovoltaic Specialists Conference*, IEEE Publishing, NY (1994), p. 682.

J. A. Silberman, D. Laser, I. Lindau, W. E. Spicer and J. A. Wilson, *J. Vac. Sci. Technol.* **A1**, 1706 (1983).

J. Singh and J. Arias, *J. Vac. Sci. Technol.* **A7**, 2562 (1989).

R. A. Sinton, Y. Kwark, P. Gruenbaum and R. M. Swanson, *18th IEEE Photovoltaic Specialists Conference*, IEEE Publishing, NY (1985), p. 61.

R. A. Sinton, Y. Kwark, J. Y. Gan and R. M. Swanson, *IEEE Electron Device Lett.* **EDL-7**, 567 (1986a).

R. A. Sinton, Y. Kwark and R. M. Swanson, *14th Project Integration Meeting, Photovoltaic Concentrator Technology Project* (1986b), p. 117.

R. A. Sinton and R. M. Swanson, *IEEE Electron Device Lett.* **EDL-8**, 547 (1987a).

R. A. Sinton and R. M. Swanson, *IEEE Trans. Electron Devices* **ED-34**, 1380 (1987b).

R. A. Sinton and R. M. Swanson, *IEEE Trans. Electron Devices* **ED-34**, 2116 (1987c).

R. A. Sinton and R. M. Swanson, *IEEE Transactions on Electron Devices* **37**, 348 (1990).

R. A. Sinton, P. J. Verlinden, R. M. Swanson, R. A. Crane, K. Wickham and J. Perkins, *13th European Photovoltaic Solar Energy Conference*, Nice (1995).

S. Sivoththaman, W. Laureys, J. Nijs and R. Mertens, *Appl. Phys. Lett.* **67**, 2335 (1995).

A. Skumanich, N. Amer and W. Jackson, *Phys. Rev. B* **31**, 2263 (1985).

K. D. Smith, H. K. Gummel, J. D. Bode, D. B. Cuttriss, R. J. Nielson and W. Rosenzweig, *Bell Syst. Tech. J.* **41**, 1765 (1963).

Z Smith and S. Wagner, *Phys. Rev. Lett.* **59**, 688 (1987).

Z Smith, S. Aljishi, V. Chu, J. Conde and S. Wagner, *Tech. Digest 3rd Photovoltaic Sci. Eng. Conf.*, Tokyo (1987).

Z Smith and S. Wagner, in *Amorphous Silicon and Related Materials* (H. Fritzsche, ed.) World Scientific, Singapore (1989), p. 409.

H. Somberg, *21st IEEE Photovoltaic Specialists Conference*, IEEE Publishing, NY (1990), p. 608.

W. Song, D. Mao, Y. Zhu, J. Tang and J. U. Trefny, *25th IEEE Photovoltaic Specialists Conference*, IEEE Publishing, NY (1996), p. 873.

B. L. Sopori, K. M. Jones, X. Deng. R. Matson, M. Al-Jassim, S. Tsuo, A. Doolittle and A. Rohatgi, *21st IEEE Photovoltaic Specialists Conference*, IEEE Publishing, NY (1991), p. 833.

B. L. Sopori, X. Dang, J. P. Benner, A. Rohatgi, P. Sana, S. K. Estreicher, Y. K. Park and M. S. Roberson, *First World Conference on Photovoltaic Energy Conversion, 24th IEEE Photovoltaic Specialists Conference*, IEEE Publishing, NY (1994), p. 1615.

B. Sopori, L. Jastrzebski and T. Tan, *25th IEEE Photovoltaic Specialistists Conference*, IEEE Publishing, NY (1996), p. 625.

W. Spear and P. LeComber, *Phil. Mag.* **33**, 935 (1976).

W. E. Spear, *Adv. Phys.* **26**, 312 (1977).

A. B. Sproul and M. A. Green, *J. Appl. Phys.* **70**, 846 (1991).

A. B. Sproul, Z. Shi, J. Zh'ao, A. Wang, Y. H. Tang, F. Yun, T. Young, Y. Hang, S. Edmiston, S. R. Wenham and M. A. Green, *First World Conference on Photovoltaic Energy Conversion, 24th IEEE Photovoltaic Specialists Conference*, IEEE Publishing, NY (1994), p. 1410.

K. S. Sree Harsha, K. J. Bachmann, P. H. Schmidt, E. G. Spencer and F. A. Thiel, *Appl. Phys. Lett.* **30**, 645 (1977).

D. L. Staebler and C. R. Wronski, *Appl. Phys. Lett.* **31**, 292 (1977).

D. L. Staebler and C. R. Wronski, *J. Appl. Phys.* **51**, 3262 (1980).

D. L. Staebler, R. Crandall and R. Williams, *Appl. Phys. Lett.* **39**, 733 (1981).

B. Stafford and E. Sabisky, eds., *Stability of Amorphous Silicon Alloy Materials and Devices*, Am. Inst. Phys. Conf. Proc. 157, AIP, New York (1987).

B. Stafford, ed., *Amorphous Silicon Materials and Solar Cells*, Am. Inst. Phys. Conf. Proc. 234, AIP, New York (1991).

B. J. Stanbery, W. S. Chen and R. A. Mickelsen, *166th Society Meeting of the Electrochemical Society*, October (1984).

B. J. Stanbery, J. E. Avery, R. M. Burgess, W. S. Chen, W. E. Devaney, D. H. Doyle, R. A. Mickelsen, R. W. McClelland, B. D. King, R. P. Gale and John C. C. Fan *et al.*, *19th IEEE Photovoltaic Specialists Conference*, IEEE Publishing, NY (1987), p. 280.

A. G. Stanley, *Appl. Solid State Sci.* **5**, 251 (1975).

S. Sterk, S. W. Glunz, J. Knobloch and W. Wettling, *First World Conference on Photovoltaic Energy Conversion, 24th IEEE Photovoltiac Specialists Conference*, IEEE Publishing, NY (1994), p. 1303.

R. J. Stirn and Y. C. M. Yeh, *Appl. Phys. Lett.* **27**, 95 (1975).

R. J. Stirn and Y. C. M. Yeh, *IEEE Trans. Electron Devices* **ED-24**, 476 (1977).

R. J. Stirn, Y. C. M. Yeh, E. Y. Wang, F. P. Ernest and C. J. Wu, *Tech. Dig. Int. Electron Devices Meet.* (1977), p. 48.

R. J. Stirn and A. Nouhi, *Appl. Phys. Lett.* **48**, 1790 (1986).

P. Stradins and H. Fritzsche, *Proc. Materials Research Society Symp.*, San Francisco (1993).

R. A. Street, D. K. Biegelsen and J. Stuke, *Philos. Mag. B* **40**, 451 (1979).

R. A. Street, *Appl. Phys. Lett.* **41**, 1060 (1982a).

R. A. Street, *Phys. Rev. Lett.* **49**, 1187 (1982b).

R. A. Street, *J. Non-Cryst. Solids* **77&78**, 1 (1985).

R. A. Street, J. Kakalios and T. Hayes, *Phys. Rev. B* **34**, 3030 (1986).

R. A. Street, *Appl. Phys. Lett.* **59**, 1084 (1991a).

R. A. Street, *Hydrogenated Amorphous Silicon*, Cambridge University Press (1991b).

R. A. Street, W. B. Jackson and M. Hack, *Materials Research Society Symposium Proc.*, San Francisco (1993).

C. Stuerke, *13th IEEE Photovoltaic Specialists Conference*, IEEE Publishing, NY (1978), p. 551.

T. Suda and R. H. Bube, *Appl. Phys. Lett.* **45**, 775 (1984).

T. Suda, K. Kakishita, H. Sato and K. Sasaki, *Appl. Phys. Lett.* **69**, 2426 (1996).

N. Suyama, N. Ueno, K. Omura, H. Takada, S. Kitamura, T. Hibino and M. Murozono, *19th IEEE Photovoltaic Specialists Conference*, IEEE Publishing, NY (1987), p. 1470.

R. M. Swanson, S. K. Beckwith, R. A. Crane, W. D. Eades, Y. H. Kwark, R. A. Sinton and S. E. Swirhun, *IEEE Transactions on Electron Devices*, **ED-31**, 661 (1984).

S. E. Swirhun, J. A. del Alamo and R. M. Swanson, *IEEE Electron Device Letters* **EDL-7**, 168 (1986).

S. M. Sze, *Physics of Semiconductor Devices*, Wiley, New York (1969), p. 382.

J. Szlufcik, K. De Clercq, P. De Schepper, J. Poortmans, A. Buczkowski, J. Nijs and R. Mertens, *13th European Photovoltaic Solar Energy Conference*, Amsterdam (1994), p. 1018.

K. Takahashi, M. Kohbata, A. Ohnishi, T. Hayashi, A. Ushirokawa, M. Yamaguchi, S. Ikegami, K. Hashimoto, H. Arai, T. Orii, T. Takamoto, H. Okazaki, H. Takamura, M. Ura and M. Ohmori, *Proceedings of the European Space Power Conference* (1991), p. 501.

T. Takamoto, H. Okazaki, H. Takamura, M. Ohmori, M. Ura and M. Yamaguchi, *Second International Conference on Indium Phosphide and Related Materials*, Denver (1990).

T. Takamoto, E. Ikeda, H. Kurita and M. Ohmori, *First World Conference on Photovoltaic Energy Conversion, 24th IEEE Photovoltaic Specialists Conference*, IEEE Publishing, NY (1994), p. 1729.

H. Talieh and A. Rockett, *Photovoltaics Advanced Research and Development*, 9th Annual Rev. Meet. SERI/CP-213-3495, Lakewood, CO (1989), p. 81.

C. W. Tang and A. C. Albrecht, *Nature* (London) **254**, 507 (1975a).

C. W. Tang and A. C. Albrecht, *J. Chem. Phys.* **63**, 953 (1975b).

C. W. Tang, *Appl. Phys. Lett.* **48**, 183 (1986).

J. Tang, L. Feng, D. Mao, W. Song and J. U. Trefny, *25th IEEE Photovoltaic Specialists Conference*, IEEE Publishing, NY (1996), p. 925.

Y. Tang, S. Dong, R. Braunstein and B. von Roedern, *Appl. Phys. Lett.* **68**, 640 (1996).

Y. Tang and R. Braunstein, *J. Appl. Phys.* **79**, 850 (1996).

D. E. Tarrant and R. R. Gay, Final Tech. Rept., NREL No. ZN-1-19019-5 (1995).

J. Tauc, *Rev. Mod. Phys.* **29**, 308 (1957).

P. C. Taylor, in *Semiconductors and Semimetals*, Vol. 21C (J. I. Pankove, ed.) Academic Press, Orlando (1984), p. 99.

Y. Tawada, M. Kondo, H. Okamoto and Y. Hamakawa, *15th IEEE Photovoltaic Specialists Conference Proc.*, IEEE Publishers, NY (1981), p. 245.

Y. Tawada, K. Tsuge, M. Kondo, H. Okamoto and Y. Hamakawa, *J. Appl. Phys.* **53**, 5273 (1982).

A. Terakawa, M. Shima, K. Sayama, H. Tarui, H. Nishiwaki and S. Tsuda, *Mater. Res. Soc. Symp. Proc.* **336**, 487 (1994).

C. Thero and P. Singh, *25th IEEE Photovoltaic Specialists Conference*, IEEE Publishing, NY (1996), p. 941.

W. G. Thompson, S. L. Franz, R. L. Anderson and O. H. Winn, *IEEE Trans. Electron Devices* **ED-24**, 463 (1977).

J. A. Thornton, D. G. Cornog, R. B. Hall, S. P. Shea and J. D. Meakin, *Proc. 17th IEEE Photovoltaic Specialists Conference*, IEEE Publishing, NY (1984), p. 781.

J. A. Thornton, T. C. Lommasson and H. Talieh, Annual Tech. Progress Rept., 1/1/86-12/31/86, Contract No. XL-5-04131-1, SERI, Golden, CO (1987).

T. P. Thorpe, Jr., A. L. Fahrenbruch and R. H. Bube, *J. Appl. Phys.* **60**, 3622 (1986).

L. Thouin and J. Vedel, *J. Electrochem. Soc.* **142**, 2996 (1995).

A. N. Tiwari *et al., J. Cryst. Growth* **111**, 730 (1991).

S. P. Tobin, C. Bajgar, S. M. Vernon, C. J. Keavney, M. A. Chung and D. S. Ruby, *20th IEEE Photovoltaic Specialists Conference*, IEEE Publishing, NY (1988), p. 469.

M. Tran, H. Fritzsche and P. Stradins, *Proc. Mater. Res. Soc. Symp.* **297**, 195 (1993).

A. Triska, D. Dennison and H. Fritzsche, *Bull. Am. Phys. Soc.* **20**, 392 (1975).

C. Y. Tsai, W. H. Bloss, K. Zieger, F. Scholz, V. Frese and U. Blieske, *13th European Photovoltaic Solar Energy Conference* (1995), p. 918.

M.-J. Tsai and R. H. Bube, *J. Appl. Phys.* **49**, 3397 (1978).

M.-J. Tsai, A. L. Fahrenbruch and R. H. Bube, *J. Appl. Phys.* **51**, 2696 (1980).

H. Tsubomura and Y. H. Nakato, *35th Meeting — International Society of Electrochemistry*, Int. Soc. of Electrochemistry, Graz, Austria (1984), p. 136.

S. Tsuda, N. Nakamura, M. Nishiwaki, K. Watanabe, T. Takahama, Y. Hishikawa, M. Ohnishi, S. Kishi, S. Nakano and Y. Kuwano, *J. Non-Cryst. Solids* **77&78**, 1465 (1985).

S. Tsuda, T. Takahama, H. Tarui, K. Watanabe, N. Nakamura *et al.*, *18th IEEE Photovoltaic Specialists Conference*, IEEE Publishing, NY (1985), p. 1295.

F. S. Turco-Sandroff *et al.*, *Appl. Phys. Lett.* **59**, 688 (1991).

A. K. Turner, J. M. Woodcock, M. E. Ozsan and J. G. Summers, in *Proc. of 10th European Photovoltaic Solar Energy Conference* (1991), p. 791.

G. B. Turner, R. J. Schwartz, J. L. Gray and J. W. Park, *19th IEEE Photovoltaic Specialists Conference*, New Orleans, IEEE Publishing, NY (1987), p. 249.

G. B. Turner, R. J. Schwartz and J. L. Gray, *20th IEEE Photovoltaic Specialists Conference*, IEEE Publishing, NY (1988), p. 1457.

J. Tuttle *et al.*, *Proc. Materials Research Society*, San Francisco (1996).

Y.-S. Tyan and E. A. Perez-Albuerne, *16th IEEE Photovoltaic Specialists Conference*, IEEE Publishing, NY (1982), p. 794.

Y. S. Tyan, F. Vazan and S. Barge, *17th IEEE Photovoltaic Specialist Conference* (1984), p. 804.

Y. Uchica, "DC Glow Discharge" in *Semiconductors and Semimetals*. Vol. 21A. *Hydrogenated Amorphous Silicon, Preparation and Structure* (J. I. Pankove, ed.) Academic Press, NY (1984), p. 41.

H. Uda, H. Matsumoto, Y. Komatsu, A. Nakano and S. Ikegami, *16th IEEE Photovoltaic Specialists Conference*, IEEE Publishing, NY (1982), p. 801.

T. Uematsu, K. Kanda, S. Kokunai, T. Warabisako, S. Iida and T. Saitoh, *21st IEEE Photovoltaic Specialists Conference*, IEEE Publishing, NY (1990), p. 229.

T. Unold, J. Hautala and J. D. Cohen, *Phys. Rev. B* **50**, 16985 (1994).

C. van Berkel and M. Powell, *Appl. Phys. Lett.* **51**, 1094 (1987).

M. W. M. van Cleef, J. K. Rath, F. A. Rubinelli, C. H. M. van der Werf and R. E. I. Schropp, *25th IEEE Photovoltaic Specialists Conference*, IEEE Publishing, NY (1996), p. 429.

H. Van der Plas, L. W. James, R. L. Moon and N. J. Nelson, *13th IEEE Photovoltaic Specialists Conference*, IEEE Publishing, NY (1978), p. 934.

M. Vanecek, A. H. Mahan, B. P. Nelson and R. S. Crandall, *11th European Photovoltaic Solar Energy Conf.* (1992), p. 96.

M. Vanecek, Z. Remes, J. Fric, E. Sipek, A. Fejfar, J. Kocka, U. Kroll, A. H. Mahan and R. S. Crandall, *13th European Photovoltaic Solar Energy Conference*, Nice (1995).

R. J. Van Overstraeten and R. P. Mertens, *Physics, Technology and Use of Photovoltaics*, Adam Hilger Ltd., Boston (1986).

L. Varner, Ph.D. Dissertation, Stanford University (1996).

K. Vasanth, S. Wagner, D. Caputo and M. Bennett, *First World Conference on Photovoltaic Energy Conversion, 24th IEEE Photovoltaic Specialists Conference*, IEEE Publishing, NY (1994), p. 488.

J. Vedel, L. Thouin and D. Lincot, *J. Electrochem. Soc.* **143**, 2173 (1996).

R. Venkatasubramanian, M. L. Timmons, P. R. Sharps, J. A. Hutchby, E. Beck and K. Emergy, *23rd IEEE Photovoltaic Specialists Conference*, IEEE Publishing, NY (1993), p. 691.

R. Venkatasubramanian, D. P. Malta, M. L. Timmons, J. B. Posthill, J. A. Hutchby,

R. Ahrenkiel, B. Keyes and T. Wangensteen, *24th IEEE Photovoltaic Specialists Conference*, IEEE Publishing, NY (1994), p. 1692.

R. Venkatasubramanian, B. C. O'Quinn, J. S. Hills, P. R. Sharps, M. L. Timmons, J. A. Hutchby, H. Field, R. Ahrenkiel and B. Keyes, *25th IEEE Photovoltaic Specialists Conference*, IEEE Publishing, NY (1996), p. 31.

L. Ventura, A. Slaoui, R. Schindler, M. Loghmarti, J. C. Muller, R. Stuck and P. Siffert, *12th European Photovoltaic Solar Energy Conference*, Amsterdam (1994), p. 560.

L. A. Verhoef, P.-P. Michiels, S. Roorda, W. C. Sinke, and R. J. C. Van Zolingen, *Materials Science and Engineering*, **B7**, 49 (1990).

L. A. Verhoef, P.-P. Michiels, R. J. C. van Zolingen, H. H. C. De Moor, A. R. Burgers, R. A. Steeman and W. C. Sinke, *First World Conference on Photovoltaic Energy Conversion, 24th IEEE Photovoltaic Specialists Conference*, Hawaii, IEEE Publishing, NY (1994), p. 1547.

P. Verlinden, F. van de Wiele, G. Stehelin, F. Floret and J. P. David, *19th IEEE Photovoltaic Specialists Conference*, IEEE Publishing, NY (1987), p. 405.

P. Verlinden, R. A. Sinton and R. M. Swanson, *Int. J. Sol. Energy* **6**, 347 (1988).

P. J. Verlinden, R. M. Swanson, R. A. Sinton, R. A. Crane, C. Tilford, J. Perkins and K. Garrison, *23rd IEEE Photovoltaic Specialists Conference*, IEEE Publishing, NY (1993), p. 58.

P. J. Verlinden, R. A. Crane, R. M. Swanson, T. Iwata, K. Handa, H. Ogasa and D. L. King, *12th European Photovoltaic Solar Energy Conference*, Amsterdam (1994a).

P. J. Verlinden, R. M. Swanson and R. A. Crane, *12th European Photovoltaic Solar Energy Conference*, Amsterdam (1994b).

P. J. Verlinden, R. M. Swanson, R. A. Crane, K. Wickham and J. Perkins, *13th European Photovoltaic Solar Energy Conference*, Nice (1995).

S. Verma, R. D. Varrin, Jr., R. W. Birkmire and T. W. F. Russell, *22nd IEEE Photovoltaic Specialists Conference*, IEEE Publishing, NY (1991).

S. Verma, S. Yamanaka, R. W. Birkmire, B. E. McCandless and T. W. F. Russell, *11th European Community Photovoltaic Solar Energy Conference* (1992), p. 807.

S. Verma, T. W. F. Russell and R. W. Birkmire, *23rd IEEE Photovoltaic Specialists Conference*, IEEE Publishing, NY (1993).

S. M. Vernon, S. P. Tobin, S. J. Wojtczuk, C. J. Keavney, C. Bajgar, *et al.*, *Photovoltaics Advanced Research and Development*, 9th Ann. Rev. Meet. SERI/CP-213-3495, Lakewood, CO (1989).

G. F. Virshup, B. C. Chung and J. G. Werthen, *20th IEEE Photovoltaic Specialists Conference*, IEEE Publishing, NY (1988), p. 441.

U. Vogel-Grote, W. Kuemmerle, R. Fischer and J. Stuke, *Philos. Mag. B* **41**, 127 (1980).

B. Von Roedern and A. Madan, *Phil. Mag.* **B63**, 293 (1991).

B. von Roedern, *11th E.C. Photovoltaic Solar Energy Conf.*, Montreux, Switzerland (1992), p. 295.

B. von Roedern, D. K. Paul, J. Blake, R. W. Collins, G. Moddell and W. Paul, *Phys. Rev. B* **25**, 7678 (1992).

B. von Roedern, *Appl. Phys. Lett.* **62**, 1368 (1993).

B. von Roedern, B. Kroposki, T. Strand and L. Mrig, *13th European Photovoltaic Solar Energy Conference*, Nice (1995), p. 1672.

J. F. Wager, O. Jamjoum and L. L. Kazmerski, *Solar Cells* **9**, 159 (1983).

B. F. Wagner, Ch. Schetter, O. V. Sulima and A. Bett, *Proc. 23rd IEEE Photovoltaic Specialists Conference*, IEEE Publishing, NY (1993), p. 356.

B. W. Wagner, X. Xu, X. R. Li, D. S. Shen, M. Isomura, M. Bennett, A. E. Delahoy, X. Li, J. K. Arch, J.-L. Nicque and S. J. Fonash, *22nd IEEE Photovoltaic Specialists Conference*, IEEE Publishing, NY (1991), p. 1307.

S. Wagner, J. L. Shay, P. Migliorato and H. M. Kasper, *Appl. Phys. Lett.* **25**, 434 (1974).

S. Wagner, J. L. Shay, K. J. Bachmann and E. Buehler, *Appl. Phys. Lett.* **26**, 229 (1975).

S. Wagner and P. M. Bridenbaugh, *J. Cryst. Growth* **39**, 151 (1977).

S. Wagner and Z Smith, *Int. Conf. Hydrogenated Amorphous Silicon Devices and Technology*, Yorktown Heights, NY (1988).

F. V. Wald, *19th IEEE Photovoltaic Specialists Conference*, IEEE Publishing, NY (1987), p. 514.

J. R. Waldrop, M. J. Cohen, A. J. Heeger and A. G. MacDiarmid, *Appl. Phys. Lett.* **38**, 53 (1981).

T. Walter, D. Hariskos, R. Herberholz, V. Nadenau, R. Schaeffler and H. W. Schock, *13th European Photovoltaic Solar Energy Conference*, Nice, France (1995a).

T. Walter, D. Braunger, D. Hariskos, Ch. Koeble and H. W. Schock, *13th European Photovoltaic Solar Energy Conference*, Nice, France (1995b).

T. Walter, R. Herberholz, C. Mueller and H. W. Schock, *Materials Research Society Symp. Proc.*, San Francisco (1996a).

T. Walter, R. Herberholz, C. Mueller and H. W. Schock, *J. Appl. Phys.* **80**, 4411 (1996b).

T. Walter, D. Braunger, H. Dittrich, Ch. Koeble, R. Herberholz and H. W. Schock, *Solar Energy Materials and Solar Cells* **41/42**, 355 (1996c).

F. Wang, A. Schwartzman, A. L. Fahrenbruch, R. Sinclair, R. H. Bube and C. M. Stahle, *J. Appl. Phys.* **62**, 1469 (1987).

F.-C. Wang, A. L. Fahrenbruch and R. H. Bube, *15th IEEE Photovoltaic Specialists Conference*, IEEE Publishing, NY (1981a), p. 1265.

F.-C. Wang, R. H. Bube, R. S. Feigelson and R. K. Route, *J. Cryst. Growth* **55**, 268 (1981b).

F.-C. Wang and R. H. Bube, *J. Appl. Phys.* **53**, 3335 (1982).

F.-C. Wang, A. L. Fahrenbruch and R. H. Bube, *J. of Electronic Materials* **11** (1), January (1982).

F.-C. Wang, A. L. Fahrenbruch and R. H. Bube, *J. Appl. Phys.* **53**, 8874 (1982).

F. F. Wang, A. L. Fahrenbruch and R. H. Bube, *J. Appl. Phys.* **65**, 3552 (1989).

M. Wanlass, J. Ward, K. Emergy, T. Gessert, C. Osterwald and T. Coutts, *Solar Cells* **30**, 363 (1991).

M. W. Wanlass, J. S. Ward, K. A. Emergy, A. Duda and T. J. Coutts, *24th IEEE Photovoltaic Specialists Conference*, IEEE Publishing, NY (1994), p. 1717.

H. Watanabe, K. Hirawasa, K. Masuri, K. Okada, M. Takayama, K. Fukui and H. Yamashita, *Optoelectronics* **5**, 223 (1990).

H. C. Weller, S. M. Paasche, C. E. Nebel, F. Kessler and G. H. Bauer, *19th IEEE Photovoltaic Specialists Conference*, IEEE Publishing, NY (1987), p. 872.

H. Welter, A. Bett, A. Ehrhardt and W. Wettling, *10th European Photovoltaic Solar Energy Conference* (1991), p. 537.

S. R. Wenham, M. A. Green, S. Edmiston, P. Campbell, L. Koschier, C. B. Honsberg, A. B. Sproul, D. Thorpe, Z. Shi and G. Heiser, *First World Conference on Photovoltaic Energy Conversion, 24th IEEE Photovoltaic Specialists Conference*, IEEE Publishing, NY (1994), p. 1234.

J. Werner, W. Jantsch, K. H. Froehner and H. J. Queisser, in *Grain Boundaries in Semiconductors* (H. J. Leamy, G. E. Pike and C. H. Seager, eds.), North-Holland, NY (1982), p. 101.

J. H. Werner, S. Kolodinski, U. Rau, J. K. Arch and E. Bauser, *Appl. Phys. Lett.* **62**, 2998 (1993).

J. H. Werner, R. Brendel and H. J. Queisser, *First World Conference on Photovoltaic Energy Conversion, 24th IEEE Photovoltaic Specialists Conference*, IEEE Publishing, NY (1994a), p. 1742.

J. H. Werner, S. Kolodinski and H. J. Queisser, *Phys. Rev. Letters* **72**, 3851 (1994b).

J. H. Werner, R. Bergmann and R. Brendel, in *Festkoerperprobleme/Advances in Solid State Physics* (R. Helbig, ed.) Vieweg, Braunscheig/Wiesbaden, **34**, 115 (1994c).

J. H. Werner, J. K. Arch, R. Brendel, G. Langguth, M. Konuma, E. Bauser, G. Wagner, B. Steiner and W. Appel, *12th European Community Photovoltaic Solar Energy Conference*, Harwood Academic, Chur (1994d).

J. H. Werner, R. Brendel and H.-J. Queisser, *Appl. Phys. Lett.* **67**, 1028 (1995a).

J. H. Werner, B. Winter, M. Wolf, S. Kolodinski, R. Brendel, M. Hirsch, H. J. Queisser, J. Wollweber and W. Schroeder, *13th European Community Photovoltaic Solar Energy Conference*, Stephens, Bedford (1995), p. 111.

J. G. Werthen, J.-P. Haering and R. H. Bube, *J. Appl. Phys.* **54**, 1159 (1983a).

J. G. Werthen, A. L. Fahrenbruch, R. H. Bube and J. C. Zesch, *J. Appl. Phys.* **54**, 2750 (1983b).

J. G. Werthen, J.-P. Haering, A. L. Fahrenbruch and R. H. Bube, *J. Appl. Phys.* **54**, 5982 (1983c).

J. G. Werthen, J.-P. Haering, A. L. Fahrenbruch and Richard H. Bube, *J. Phys. D: Appl. Phys.* **16**, 2391 (1983).

R. H. Williams, R. R. Varma, W. E. Spear and P. G. LeComber, *J. Phys. C* **12**, L209 (1979).

K. Winer, *Phys. Rev. B* **41**, 12150 (1990).

K. Winer, *J. Non-Cryst. Solids* **137/138**, 157 (1991).

M. Wolf, *Energy Convers.* **11**, 63 (1971) [Reprinted in "Solar Cells", (C. E. Backus, ed.) 191, IEEE Press, New York (1976)].

J. Wollweber, D. Schultz, W. Schoeder and N. V. Abrosimov, *First World Conference on Photovoltaic Energy Conversion, 24th IEEE Photovoltaic Specialists Conference*, IEEE Publishing, NY (1994), p. 1372.

J. M. Woodcock, A. K. Turner, M. E. Oezan and J. G. Summers, *22nd IEEE Photovoltaic Specialists Conference*, IEEE Publishing, NY (1993), p. 842.

J. M. Woodcock *et al.*, *Proc. 12th European PVSEC Meeting*, Amsterdam (1994).

J. M. Woodall and H. J. Hovel, *Appl. Phys. Lett.* **21**, 379 (1972).

J. M. Woodall and H. J. Hovel, *Appl. Phys. Lett.* **27**, 447 (1975).

J. M. Woodall and H. J. Hovel, *Appl. Phys. Lett.* **30**, 492 (1977).

C. R. Wronski, *Optoelectronics-Devices Technol.* **1**, 11 (1986).

C. R. Wronski, *19th IEEE Photovoltaic Specialists Conference*, IEEE Publishing, NY (1987a), p. 321.

C. R. Wronski and M. Hicks, *Jpn. J. Appl. Phys.* **26**, L105 (1987b).

C. R. Wronski, M. Hicks and S. Lee, *Tech. Digest 3rd Photovoltaic Sci. Eng. Conf.*, Tokyo (1987c), p. 721.

C. Wu and R. H. Bube, *J. Appl. Phys.* **45**, 648 (1974).

Y. S. Wu, C. R. Becker, A. Waag, R. N. Bicknell-Tassius and G. Landwehr, *J. Appl. Phys.* **69**, 268 (1991).

N. Wyrsch and A. Shah, *Solid State Communications* **80**, 807 (1991).

N. Wyrsch and A. Shah, *First World Conference on Photovoltaic Energy Conversion, 24th IEEE Photovoltaic Specialists Conference*, IEEE Publishing, NY (1994), p. 583.

N. Wyrsch, N. Beck, P. Pipoz, M. Goerlitzer, H. Beck and A. Shah, *Solid State Phenomena* **44-46**, 525 (1995).

N. Wyrsch, M. Goerlitzer, N. Beck, J. Meier and A. Shah, *Mat. Res. Soc. Symp. Proc.*, San Francisco (1996).

J. Xi, D. Shugar and H. Volltrauer, *First World Conference on Photovoltaic Energy Conversion, 24th IEEE Photovoltaic Specialists Conference*, IEEE Publishing, NY (1994), p. 401.

X. Xu, J. Yang and S. Guha, *Appl. Phys. Lett.* **62**, 1399 (1993).

E. Yablonovitch, T. Gmitter, R. W. Swanson and Y. K. Kwark, *Appl. Phys. Lett.* **47**, 1211 (1985).

K. Yamaguchi, H. Matsumoto, N. Nakayama and S. Ikegami, *Jpn. J. Appl. Phys.* **15**, 1575 (1976).

K. Yamaguchi, N. Nakayama, H. Matsumoto and S. Ikegami, *Jpn. J. Appl. Phys.* **16**, 1203 (1977).

M. Yamaguchi, *J. Appl. Phys.* **78**, 1476 (1995).

M. Yamaguchi, T. Hayashi, A. Ushirokawa, Y. Takahashi, M. Koubata, M. Hashimoto, H. Okazaki, T. Takamoto, M. Ura, M. Ohmori, S. Ikegami, H. Arai and T. Orii, *21st IEEE Photovoltaic Specialists Conference*, Kissimmee, IEEE Publishing, NY (1990).

M. Yamaguchi, T. Hayashi, A. Ushirokawa, K. Takahashi, T. Takamoto, H. Okazaki, M. Ohmori, S. Ikegami and H. Arai, *22nd IEEE Photovoltaic Specialists Conference*, Las Vegas, IEEE Publishing, NY (1991).

M. Yamaguchi, T. Takamoto, E. Ideka, H. Kurita, M. Ohmori, K. Ando and C. Vargas-Aburto, *Jpn. J. Appl. Phys.* **34**, 6222 (1995).

N. Yamaguchi, *Proc. 6th Int. Photovoltaic Science and Engineering Conf.*, New Delhi: Oxford/IBM (1992), p. 503.

S. Yamanaka, B. E. McCandless and R. W. Birkmire, *23rd IEEE Photovoltaic Specialists Conference*, IEEE Publishing, NY (1993).

S.-Y. Yin, A. L. Fahrenbruch and R. H. Bube, *J. Appl. Phys.* **49**, 1294 (1978).

J.-B. Yoo, A. L. Fahrenbruch and R. H. Bube, *Solar Cells* **31**, 171 (1991).

A. Yoshikawa and Y. Sakai, *J. Appl. Phys.* **45**, 3521 (1974).

A. Yoshikawa and Y. Sakai, *Solid State Electronics* **20**, 133 (1977).

M. Yoshimi, W. Ma, T. Horiuchi, C. C. Lim, S. C. De, K. Hattori, H. Okamoto and Y. Hamakawa, *Proc. Mat. Res. Soc. Symp.*, *Materials Research Society*, **258**, 845 (1992).

G. Yu, J. Gao, J. C. Hummelen, F. Wudl and A. J. Heeger, *Science* **270**, 1789 (1995).

A. Zachariou, K. W. J. Barnham, P. Griffin, J. Nelson, C. Button, M. Hopkinson, M. Pate and J. Epier, *25th IEEE Photovoltaic Specialists Conference*, IEEE Publishing, NY (1996), p. 113.

K. Zanio, *Semiconductors and Semimetals. Vol. 13, Cadmium Telluride*, Academic Press, NY (1978).

Q. Zhang, H. Takashima, J.-H. Zhou, M. Kumeda and T. Shimizu, *Phys. Rev. B* **50**, 1551 (1994).

J. Zhao, A. Wang, M. Taouk, S. R. Wenham, M. A. Green and D. L. King, *IEEE Electron Device Letters* **14**, 539 (1993).

J. Zhao, A. Wang and M. A. Green, *Progress in Photovoltaics* **2**, 227 (1994).

J. Zhao, A. Wang, P. Altermatt and M. A. Green, *Appl. Phys. Lett.* **66**, 3636 (1995).

G. F. Zheng, Z. Shi, R. Bergmann, X. Dai, S. Robinson, A. Wang, J. Kurianski and M. A. Green, *Solar Energy Materials and Solar Cells* **32**, 129 (1994).

T. X. Zhou, N. Reiter, R. C. Powerll, R. Sasaia and P. V. Meyers, *First World Conference on Photovoltaic Energy Conversion, 24th IEEE Photovoltaic Specialists Conference*, IEEE Publishing, NY (1994), p. 103.

A. Zunger, S. Wagner and P. M. Petroff, *J. Electronic Materials* **22**, 3 (1993).

K. Zweibel and A. M. Barnett, Chapter 10 in *Renewable Energy: Sources for Fuels and Electricity* (T. B. Johansson, H. Kelly, A. K. N. Reddy and R. H. Williams, eds.) Island Press, Washington, D.C. (1993), p. 437.

S. Zweigart, Th. Walter, Ch. Koeble, S. M. Sun, U. Ruehle and H. W. Schock, *First World Conference on Photovoltaic Energy Conversion, 24th IEEE Photovoltaic Specialists Conference*, IEEE Publishing, NY (1994), p. 60.

S. Zweigart, D. Schmid, J. Kessler, H. Dittrich and H. W. Schock, *J. of Crystal Growth* **146**, 233 (1995).

S. Zweigart, S. M. Sun, G. Bilger and H. W. Schock, *Solar Energy Materials and Solar Cells* **41/42**, 219 (1996).

BIBLIOGRAPHY

Listed Chronologically

A. G. Milnes and D. L. Feucht, *Heterojunctions and Metal-Semiconductor Junctions*, Academic Press, NY (1972), p. 9.

H. J. Hovel, *Solar Cells*, Vol. 11 of *Semiconductors and Semimetals*, Academic Press, NY (1975).

S. W. Angrist, *Direct Energy Conversion*, Allyn & Bacon, 3rd ed. (1976).

C. Stein, ed., *Critical Materials Problems in Energy Production*, Academic Press, NY (1976).

K. Zanio, *Cadmium Telluride*, Vol. 13 of *Semiconductors and Semimetals*, Academic Press, NY (1978).

L. E. Murr, ed., *Solar Materials Science*, Academic Press, NY (1980).

F. Cardon, W. P. Gomes and W. Dekeyser, eds., *Photovoltaic and Photoelectrochemical Solar Energy Conversion*, Plenum Press, NY (1981).

S. J. Fonash, *Solar Cell Device Physics*, Academic Press, NY (1981).

M. A. Green, *Solar Cells: Operating Principles, Technology, and System Applications*, Prentice-Hall, Englewood Cliffs, NJ (1982).

"Science and Technology of Non-Crystalline Semiconductors", *Solar Energy Materials*, **8**, No. 1–3 (1982).

J. A. Amick, V. K. Kapur and J. Dietl, eds., *Proceedings of the Symposium on Materials and New Processing Technologies for Photovoltaics*, Proc. Volume **83-11**, Electrochemical Society, Pennington, NJ (1983).

K. L. Chopra and S. R. Das, *Thin Film Solar Cells*, Plenum Press, NY (1983).

A. L. Fahrenbruch and R. H. Bube, *Fundamentals of Solar Cells: Photovoltaic Solar Energy Conversion*, Academic Press, NY (1983).

J. I. Pankove, ed., *Hydrogenated Amorphous Silicon*, **21** A, B, C and D, *Semiconductors and Semimetals*, Academic Press, Inc. (1984).

T. J. Coutts and J. D. Meakin, eds., *Current Topics in Photovoltaics*, Academic Press, NY (1985).

R. J. Van Overstraeten and R. P. Mertens, *Physics, Technology and Use of Photovoltaics*, Adam Hilger Ltd., Bristol (1986).

M. A. Green, *High Efficiency Silicon Solar Cells*, Trans Tech Publications Ltd., Zurich, Switzerland (1987).

A. Madan and M. P. Shaw, *The Physics and Applications of Amorphous Semiconductors*, Academic Press, NY (1988).

H. Fritzsche, *Amorphous Silicon and Related Materials*, 2 vols. World Scientific, Singapore (1989).

H. J. Moeller, H. P. Strunk and J. H. Werner, eds. *Polycrystalline Semiconductors — Grain Boundaries and Interfaces.* Springer Proc. Phys. Vol. 35, Springer, Berlin (1989).

F. Shimura, *Semiconductor Silicon Crystal Technology*, Academic Press, NY (1989).

R. A. Street, *Hydrogenated Amorphous Silicon*, Cambridge Univ. Press (1991).

J. H. Werner and H. P. Strunk, eds., *Polycrystalline Semiconductors II*, Springer Proc. Phys. Vol. 54, Springer, Berlin (1991).

R. H. Bube, *Photoelectronic Properties of Semiconductors*, Cambridge Univ. Press (1992).

W. Luft and Y. S. Tsuo, *Hydrogenated Amorphous Silicon Alloy Deposition Processes*, Marcel Dekker, NY (1993).

Renewable Energy: Sources for Fuels and Electricity, eds. T. B. Johansson, H. Kelly, A. K. N. Reddy and R. H. Williams, ex.ed. L. Burnham, Island Press, Washington, D.C. (1993).

F. Shimura, ed., *Oxygen in Silicon*, Semiconductors and Semimetals **42**, Academic Press, San Diego (1994).

Organic Conductors: Fundamentals and Applications, eds. J.-P. Farges, Marcel Dekker, NY (1994).

H. P. Strunk, J. H. Werner, B. Fortin and O. Bonnaud, eds., *Polycrystalline Semiconductors III – Physics and Technology*, Solid-State Phenomena **37-38**, Trans Tech, Aedermannsdorf (1994).

M. A. Green, *Silicon Solar Cells: Advanced Principles & Practice*, M. A. Green, Electrical Engineering, UNSW, Sydney, Australia 2052 (1995a).

M. A. Green, *Solar Cells: Operating Principles, Technology and System Applications*, M. A. Green, Electrical Engineering, UNSW, Sydney, Australia 2052 (1995b).

H. Neber-Aeschbacher, ed., *Hydrogenated Amorphous Silicon*, Solid State Phenomena **44-46**, Scitec Publications (1995), pp. 1–1048.

R. C. Nevill, *Solar Energy Conversion: The Solar Cell*, Elsevier, Amsterdam (1995).

Semiconductor Processing and Characterization with Lasers: Applications in Photovoltaics, eds. M. Briege, H. Dittrich, M. Klose, H. W. Schock and J. Werner, Trans Tech. Pub., Zuerich, Switzerland (1995).

B. E. Anspaugh, *GaAs Solar Cell Radiation Handbook*, Jet Propulsion Laboratory Publication 96-9, NASA, Jet Propulsion Laboratory, California Institute of Technology, Pasadena, CA (1996).

J. A. Mazer, *Solar Cells: An Introduction to Crystalline Photovoltaic Technology*, Kluwer Academic Pub., Norwell, MA (1996).

R. K. Pandey, S. M. Sahu and S. Chandra, *Handbook of Semiconductor Electrodeposition*, Marcel Dekker, New York (1996).

S. Pizzini, H. P. Strunk and J. H. Werner, eds., *Polycrystalline Semiconductors IV – Physics, Chemistry and Technology*, Trans Tech Publications, Zuerich-Uetikon, Switzerland (1996).

D. Redfield and R. H. Bube, *Photoinduced Defects in Semiconductors*, Cambridge Univ. Press (1996).

M. Suezawa and H. Katayama-Yoshida, eds., *Defects in Semiconductors*, ICDS-18. Trans Tech Publications, Zuerich-Uetikon, Switzerland (1996).

S. R. Wenham, M. A. Green and M. E. Watt, *Applied Photovoltaics*, M.A. Green, Electrical Engineering, UNSW, Sydney, Australia 2052 (1996).

REVIEW PUBLICATIONS

Listed Chronologically

A. L. Fahrenbruch, V. Vasilchenko, F. Buch, K. Mitchell and R. H. Bube, "II–VI Photovoltaic Heterojunctions for Solar Energy Conversion", *Appl. Phys. Lett.* **25**, 605 (1974).

R. H. Bube, "Electronic Transport in Polycrystalline Films", *Annual Review of Materials Science* **5**, 201 (1975).

R. H. Bube, "Non-Conventional Heterojunctions for Solar Energy Conversion", in *Critical Materials Problems in Energy Production*, Academic Press, NY (1976), p. 486.

R. H. Bube, F. Buch, A. L. Fahrenbruch, Y. Y. Ma and K. W. Mitchell, "Photovoltaic Energy Conversion with n-CdS-p-CdTe Heterojunctions and Other II–VI Junctions", *IEEE Trans. on Electron Devices* **ED-24**, 487 (1977).

R. H. Bube, "Photovoltaic Effects in II–VI Heterojunctions", *Proc. Soc. Photo-Opt. Instr. Eng.* **114**, 7 (1977).

A. L. Fahrenbruch, "II–VI Compounds in Solar Energy Conversion", *J. Crystal Growth* **39**, 73 (1977).

A. J. Nozik, "Photoelectrochemistry: Applications to Solar Energy Conversion", *Annual Review of Physical Chemistry* **29**, 189 (1978).

A. L. Fahrenbruch and J. Aranovich, "Heterojunction Phenomena and Interfacial Defects in Photovoltaic Convertors", in *Solar Energy Conversion — Solid State Physics Aspects*, (B. O. Seraphin, ed.), Springer Series on Topics in Applied Physics, Springer-Verlag (1979).

R. H. Bube, "Heterojunctions for Thin Film Solar Cells", Ch. 17 in *Solar Materials Science*, Academic Press, Inc., NY (1980), p. 585.

J. R. Davis, A. Rohatgi, R. H. Hopkins, P. D. Blais, P. Rai-Choudhury, J. R. Mc-Cormick and H. C. Mollenkopf, "Impurities in Silicon Solar Cells", *IEEE Transactions on Electron Devices*, **ED-27**, 677 (1980).

D. Redfield, "Unified Models of Fundamental Limitations on the Performance of Silicon Solar Cells", *IEEE Transactions on Electron Devices* **ED-27**, 766 (1980).

R. H. Bube and A. L. Fahrenbruch, "Photovoltaic Effect", in *Advances in Electronics and Electron Physics* **56** (C. Marton, ed.) Academic Press, NY (1981), p. 163.

A. Rothwarf, "Polycrystalline Thin Films for Terrestrial Solar Cells", *J. Vac. Sci. Technol.* **20**, 282 (1982).

R. H. Bube, "Cadmium Telluride Solar Cells, Proc. of Symp. on Materials and New Processing Technologies for Photovoltaics", (J. A. Amick, V. K. Kapur and J. Dietl, eds.) Proc. Vol. **83-11**, 359 Electrochem. Soc., Pennington, NJ (1983).

R. H. Bube, A. L. Fahrenbruch, R. Sinclair, T. C. Anthony, C. Fortmann, W. Huber, C.-T. Lee, T. Thorpe and T. Yamashita, "Cadmium Telluride Films and Solar Cells", *IEEE Transactions on Electron Devices* **ED-31**, 528 (1984).

E. Bucher, "New Technologies and Materials for Photovoltaic Solar Energy Conversion", *Electrochemical Society Extended Abstracts* **84-2**, Electrochemical Society, Pennington, NJ (1984), p. 450.

D. E. Carlson, "(*a*-Si:H) Solar Cells", in *Hydrogenated Amorphous Silicon* **21** A, B, C and D, *Semiconductors and Semimetals* (J. I. Pankove, ed.) Academic Press, Inc. (1984), p. 7.

S. J. Fonash and A. Rothwarf, "Heterojunction Solar Cells", in *Current Topics in Photovoltaics* (T. J. Coutts and J. D. Meakin, eds.) Academic Press, London (1985), p. 2.

L. M. Fraas, "Advanced Concentrator Solar Cells", in *Current Topics in Photovoltaics* (T. J. Coutts and J. D. Meakin, eds.) Academic Press, London (1985), p. 170.

Y. Hamakawa, "Amorphous-Silicon Solar Cells", in *Current Topics in Photovoltaics* (T. J. Coutts and J. D. Meakin, eds.) Academic Press, London (1985), p. 111.

R. Hill and J. D. Meakin, "Cadmium Sulphide-Copper Sulphide Solar Cells", in *Current Topics in Photovoltaics* (T. J. Coutts and J. D. Meakin, eds.) Academic Press, London (1985), p. 223.

G. Hodes, S. J. Fonash, A. Heller and B. Miller, "Photoelectrochemical Cells Based on Polycrystalline Semiconductors", *Advances in Electrochemistry and Electrochemical Engineering* **13**, 113 (1985).

L. Kazmerski and S. Wagner, "Cu-Ternary Chalcopyrite Solar Cells", in *Current Topics in Photovoltaics* (T. J. Coutts and J. D. Meakin, eds.) Academic Press, London (1985), p. 41.

J. D. Meakin, "Status of CuInSe$_2$ Solar Cells", *Photovoltaics*, SPIE Vol. **543** Society of Photo-Optical Instrumentation Engineers, Bellingham, WA (1985), p. 108.

K. Rajeshwar, "Materials Aspects of Photoelectrochemical Energy Conversion", *Journal of Applied Electrochemistry* **15**, 1 (1985).

T. J. Coutts, L. L. Kazmerski and S. Wagner, eds., *Copper Indium Diselenide for Photovoltaic Applications*, Elsevier, NY (1986).

R. H. Bube, "Thin-Film Polycrystalline Solar Cells", *ISES Solar World Congress*, Hamburg (1987).

A. L. Fahrenbruch, "Ohmic Contacts and Doping of CdTe", Solar Cells **21**, 399 (1987).

Y. Hamakawa, "Photovoltaic Power", *Scientific American* **256**, 76 (1987).

A. Rothwarf, "Criteria for the Design of High Efficiency Thin Film Solar Cells: Theory and Practice", *Solar Cells* **21**, 1 (1987).

B. M. Basol, "Electrodeposited CdTe and HgCdTe Solar Cells", *Solar Cells* **23**, 69 (1988).

E. C. Boes, "A Summary of Recent Photovoltaic Concentrator Technology Developments", *20th IEEE Photovoltaic Specialists Conference*, IEEE Publishing, NY (1988), p. 21.

R. H. Bube, "CdTe Junction Phenomena", *Solar Cells* **23**, 1 (1988).

T. L. Chu, "Cadmium Telluride Solar Cells", *Curr. Top. Phovotaics* **3**, 236 (1988).

H. Curtins and M. Favre, "Surface and Bulk States Determined by Photothermal Deflection Spectroscopy", Amorphous Silicon and Related Materials (H. Fritzsche, ed.) World Scientific Publishing (1988), p. 329.

K. W. Mitchell, C. Eberspacher, J. Ermer and D. Pier, "Copper Indium Diselenide Photovoltaic Technology", *20th IEEE Photovoltaic Specialists Conference*, IEEE Publishing, NY (1988), p. 1384.

A. L. Fahrenbruch, K.-F. Chien, D. Kim, A. Lopez-Otero, P. Sharps and R. H. Bube, "Ion-Assisted Doping of CdTe", *Solar Cells* **27**, 137 (1989).

Y. Hamakawa and H. Okamoto, "Amorphous Silicon Solar Cells", in *Advances in Solar Energy* **5** (K. W. Boer, ed.) Plenum Publishing Corp. (1989).

H. M. Hubbard, "Photovoltaics Today and Tomorrow", *Science* **244**, 297 (1989).

H. M. Hubbard and G. Cook, "Photovoltaics in our Energy Future", *Phys. Soc.* **18**, 9 (1989).

L. L. Kazmerski, *International Materials Reviews* **34**, 185 (1989).

R. H. Bube, "Materials for Photovoltaics", *Annual Review of Materials Science* (R. A. Huggins, G. A. Giordmaine and J. B. Wachtman, Jr., eds.) Ann. Rev., Palo Alto, CA (1990).

W. B. Jackson and S. B. Zhang, "Hydrogen Complexes in Amorphous Silicon", in *Transport, Correlation and Structural Defects* (H. Fritzsche, ed.) World Scientific Publishing (1990), p. 63.

L. A. Verhoef, P.-P. Michiels, S. Roorda, W. C. Sinke and R. J. C. Van Zolingen, "Gettering in Polycrystalline Silicon Solar Cells", *Materials Science and Engineering* **B7**, 49 (1990).

D. E. Carlson, "Multijunction Amorphous Silicon Solar Cells", *Phil. Mag. B* **63**, 305 (1991).

C. F. Gay, "The Role of Photovoltaics in the World Energy Supply", *International Workshop on Mass Production of Photovoltaics. Commercialization and Policy Options*, UN Centre for Science and Technology for Development, Sao Paulo, Brazil (1991).

Y. Hamakawa, "Recent Advances in Solar Photovoltaic Technologies in Japan", *Solar Energy Materials* **23**, 139 (1991).

A. Rockett and R. W. Birkmire, "CuInSe$_2$ for Photovoltaic Applications", *J. Appl. Phys.* **70**, R81 (1991).

B. M. Basol, "Processing High Efficiency CdTe Solar Cells", *Int. J. Solar Energy* **12**, 25 (1992).

B. M. Basol, "I–III–VI$_2$ Compound Semiconductors for Solar Cell Applications", *J. Vac. Sci. Technol. A* **10**, 2006 (1992).

A. L. Fahrenbruch, R. H. Bube, D. Kim and A. Lopez-Otero, "Ion-and Photon-assisted p-type Doping of CdTe during Physical Vapor Deposition", *Int. J. Solar Energy*, **12**, 197 (1992).

D. Redfield and R. H. Bube, "Stability of Amorphous Silicon PV Modules, Cells and Films", *11th EC Photovoltaic Specialists Conference*, Harwood Acad. Publ., Switzerland (1992), p. 665.

A. Shah, J. Dutta, N. Wyrsch, K. Prasad, H. Curtins, F. Finger, A. Howling and Ch. Hollenstein, "VHF Plasma Deposition: A Comparative Overview", *Mat. Res. Soc. Symp. Proc.* **258**, 15 (1992).

K. Takahashi, M. Yamaguchi, T. Takamoto, S. Ikegami, A. Ohnishi, T. Hayashi, A. Ushirokawa, M. Kohbata, H. Arai, K. Hashimoto, T. Orii, H. Okazaki,

H. Takamura, M. Ura and M. Ohmori, "InP Solar Cells and their Flight Experiments", The Institute of Space and Astronautical Science, Report No. 644, Kanagawa, Japan (1992).

B. M. Basol, "Preparation Techniques for Thin Film Solar Cell Materials: Processing Perspective", *Jpn. J. Appl. Phys.* **32**, Suppl. 32, 35 (1993).

E. C. Boes and A. Luque, "Photovoltaic Concentrator Technology", in *Renewable Energy: Sources for Fuels and Electricity* (T. B. Johansson, H. Kelly, A. K. N. Reddy, and R. H. Williams, eds.; L. Burnham, ex.ed.) Island Press, Washington, D.C. (1993), p. 361.

R. H. Bube and K. W. Mitchell, "Research Opportunities in Polycrystalline Compound Solar Cells", *J. Electron. Mat.* **22**, 17 (1993).

R. H. Bube, "Solar Cells", in *Handbook on Semiconductors, Vol. 4, Device Physics* (C. Hilsum, ed.) Elsevier, Amsterdam (1981); Revised edition (1993).

D. E. Carlson and S. Wagner, "Amorphous Silicon Photovoltaic Systems", in *Renewable Energy: Sources for Fuels and Electricity* (T. B. Johansson, H. Kelly, A. K. N. Reddy and R. H. Williams, eds.; L. Burnham, ex.ed.) Island Press, Washington, D.C. (1993), p. 403.

M. A. Green, "Silicon Solar Cells: Evolution, High-Efficiency Design and Efficiency Enhancements", *Semicond. Sci. Technol.* **8**, 1 (1993a).

M. A. Green, "Crystalline- and Polycrystalline-Silicon Solar Cells", in *Renewable Energy: Sources for Fuels and Electricity* (T. B. Johansson, H. Kelly, A. K. N. Reddy and R. H. Williams, eds.; L. Burnham, ex.ed.) Island Press, Washington, D.C. (1993b), p. 337.

H. Kelly, "Introduction to Photovoltaic Technology", in *Renewable Energy: Sources for Fuels and Electricity* (T. B. Johansson, H. Kelly, A. K. N. Reddy and R. H. Williams, eds.; L. Burnham, ex.ed.) Island Press, Washington, D.C. (1993), p. 297.

W. Luft, H. M. Branz, V. L. Dalal, S. S. Hegedus and E. A. Schiff, "Recent Progress in Amorphous Silicon PV Technology," *NREL 12th PV Program Review*, Denver (1993).

W. Paul, R. A. Street and S. Wagner, "Hydrogenated Amorphous Semiconductors", *J. Electronic Materials* **22**, 39 (1993).

A. Zunger, S. Wagner and P. M. Petroff, "New Materials and Structures for Photovoltaics", *J. Electronic Materials* **22**, 3 (1993).

K. Zweibel and A. M. Barnett, "Polycrystalline Thin-Film Photovoltaics", in *Renewable Energy: Sources for Fuels and Electricity* (T. B. Johansson, H. Kelly, A. K. N. Reddy and R. H. Williams, eds.; L. Burnham, ex.ed.) Island Press, Washington, D.C. (1993), p. 437.

R. R. Arya, R. S. Oswald, Y. M. Li, N. Maley, K. Jansen, L. Yang, L. F. Chen, F. Willing, M. S. Bennett, J. Morris and D. E. Carlson, "Progress in Amorphous Silicon Based Multijunction Modules", *First World Conference on Solar Energy Conversion, 24th IEEE Photovoltaic Specialists Conference*, IEEE Publishing, NY (1994), p. 394.

P. A. Basore and J. M. Gee, "Crystalline-Silicon Photovoltaics: Necessary and Sufficient", *IEEE First World Conference on Photovoltaic Energy Conversion,*

24th IEEE Photovoltaic Specialists Conference, IEEE Publishing, NY (1994), p. 2254.

R. W. Birkmire and P. V. Meyers, "Processing Issues for Thin-Film CdTe Cells and Modules", *First World Conference on Solar Energy Conversion, 24th IEEE Photovoltaic Specialists Conference*, IEEE Publishing, NY (1994), p. 76.

R. H. Bube, "Cadmium Telluride", in *Encyclopedia of Advanced Materials* (D. Bloor, R. J. Brook, M. C. Flemings, S. Mahajan and R. W. Cahn, eds.) Pergamon Press (1994).

A. Catalano, "Polycrystalline Thin-Film Technologies: Status and Prospects", *First World Conference on Solar Energy Conversion, 24th IEEE Photovoltaic Specialists Conference*, IEEE Publishing, NY (1994), p. 52.

V. L. Dalal, S. Kaushal, J. Xu and K. Han, "A Critical Review of the Growth and Properties of a-(Si,Ge):H", *First World Conference on Solar Energy Conversion, 24th IEEE Photovoltaic Specialists Conference*, IEEE Publishing, NY (1994), p. 464.

P. Di Marco and G. Giro, "Organic Photoconductors and Photovoltaics", in *Organic Conductors: Fundamentals and Applications* (J.-P. Farges, ed.) Marcel Dekker, Inc. NY (1994), p. 791.

Y. Hamakawa, "Recent Advances of Thin Film Solar Cells and Their Technologies", *First World Conference on Solar Energy Conversion, 24th IEEE Photovoltaic Specialists Conference*, IEEE Publishing, NY (1994), p. 34.

Y. Hishikawa, K. Ninomiya, E. Maruyama, S. Kuroda, A. Terakawa, K. Sayama, H. Tarui, M. Sasaki, S. Tsuda and S. Nakano, "Approaches for Stable Multi-Junction a-Si Solar Cells", *First World Conference on Solar Energy Conversion, 24th IEEE Photovoltaic Specialists Conference*, IEEE Publishing, NY (1994), p. 386.

R. Mertens, "Silicon Solar Cells", *12th EC Photovoltaic Solar Energy Conf.*, Amsterdam (1994).

R. Mertens, J. Nijs and R. Van Overstraeten, "Solar Energy, the Importance for Europe", *European Conf. on Energy and the Environment*, Antwerp (1994).

A. Rockett, F. Abou-Elfotouh, D. Albin, M. Bode, J. Ermer, R. Klenk, T. Lommasson, T. W. F. Russell, R. D. Tomlinson, J. Tuttle, L. Stolt, T. Walter and T. M. Peterson, "Structure and Chemistry of $CuInSe_2$ for Solar Cell Technology: Current Understanding and Recommendations", *Thin Solid Films* **237**, 1 (1994).

M. X. Tan, P. E. Laibinis, S. T. Nguyen, J. M. Kesselman, C. E. Stanton and N. S. Lewis, "Principles and Applications of Semiconductor Photoelectrochemistry", Progress in Inorganic Chemistry **41**, 21 (1994).

H. S. Ullal, K. Zweibel, B. G. von Roedern, R. Noufi and P. Sheldon, "Overview of the U.S. DOE/NREL Polycrystalline Thin-Film Photovoltaic Technologies", *First World Conference on Photovoltaic Energy Conversion, 24th IEEE Photovoltaic Specialists Conference*, IEEE Publishing, NY (1994), p. 266.

J. H. Werner, R. Bergmann and R. Brendel, "The Challenge of Crystalline Thin Film Silicon Solar Cells", in *Festkoerperprobleme/Advances in Solid State Physics* (R. Helbig, ed.) Vieweg, Braunschweig/Wiesbaden **34**, 115 (1994c).

C. R. Wronski, "Amorphous Silicon Technology: Coming of Age", *IEEE First World Conference on Photovoltaic Energy Conversion, 24th IEEE Photovoltaic Specialists Conference*, IEEE Publishing, NY (1994), p. 373.

G. H. Bauer, "Amorphous Hydrogenated Silicon-Germanium Alloys", *Solid State Phenomena* **44-46**, 365 (1995).

W. H. Bloss, F. Pfisterer, M. Schubert and T. Walter, "Thin-film Solar Cells", *Progress in Photovoltaics Research and Applications* **3**, 3 (1995).

F. Demichelis and C. F. Pirri, "Hydrogenated Amorphous Silicon Based Alloy: a-$Si_{1-x}C_x$:H", *Solid State Phenomena* **44-46**, 385 (1995).

M. A. Green, K. Emery, K. Buecher and D. L. King, "Solar Cell Efficiency Tables (Version 5)", *Progress in Photovoltaics Research and Applications* **3**, 51 (1995).

M. A. Green with an introduction by D. Lovejoy, "Multilayer Thin Film Silicon Solar Cells", *Natural Resources Forum*, Vol. **19**, No. 4, 269 (1995c).

Y. Hamakawa, W. Ma and H. Okamoto, "Amorphous-Silicon-Based Devices", in *Plasma Deposition of a-Si based Materials* (G. Bruno and A. Madan, eds.) Academic Press, NY (1995).

G. Hodes, "Electrodeposition of II–VI Semiconductors", in *Physical Chemistry, Methods and Applications* (I. Rubinstein, ed.) Marcel Dekker Inc., NY (1995), p. 515.

H. Keppner, U. Kroll, J. Meier and A. Shah, "Very High Frequency Glow Discharge: Plasma- and Deposition Aspects", *Solid State Phenomena* **44-46**, 97 (1995).

W. Kusian, K.-D. Ufert and H. Pfleiderer, *Solid State Phenomena* **44-46**, 823 (1995).

C. Leguijt, "Surface Passivation for Silicon Solar Cells", Univ. Utrecht (1995).

N. S. Lewis, "Artificial Photosynthesis", *American Scientist* **83**, 534 (1995).

P. Loelgen, "Surface and Volume Recombination in Silicon Solar Cells", Univ. Utrecht (1995).

E. Maruyama, S. Tsuda and S. Nakano, "Industrialization of Amorphous Silicon Solar Cells and their Future Applications", Solid State Phenomena **44-46**, 863 (1995).

P. D. Moskowitz and V. M. Fthenakis, "Bibliography", Photovoltaic Environmental, Health and Safety Project, Biomedical and Environmental Assessment Group, Brookhaven National Laboratory (1995).

V. Nadenau, D. Braunger, D. Hariskos, M. Kaiser, Ch. Koeble, A. Oberacker, M. Ruckh, U. Ruehle, R. Schaeffler, D. Schmid, T. Walter, S. Zweigart and H. W. Schock, "Solar Cells Based on CuInSe₂ and Related Compounds: Material and Device Properties and Processing", *Progress in Photovoltaics Research and Applications* **3**, 363 (1995).

J. Nijs, H. E. Elgamel, J. Szlufcik, S. Sivoththaman, O. Evrard, K. De Clercq, P. De Schepper, J. Poortmans, M. Ghannam, R. Mertens, P. Fath and G. Willeke, "Recent Results Obtained at IMEC on Multicrystalline Silicon Solar Cells", *Renewable Energy* **6**, (5–6), 573 (1995).

R. E. I. Schropp, "Efficiency Optimization Techniques for Amorphous Silicon Solar Cells", *Solid State Phenomena* **44-46**, 853 (1995).

A. V. Shah, R. Platz and H. Keppner, "Thin-film Silicon Solar Cells: A Review and Selected Trends", *Solar Energy Materials and Solar Cells* **38**, 501 (1995).

K. Shimakawa, A. Kolobov and S. R. Elliott, "Photo-induced Effects and Metastability in Amorphous Semiconductors and Insulators", *Advances in Physics* **44**, 475 (1995).

D. E. Tarrant and R. R. Gay, "Research on High Efficiency, Large Area CuInSe$_2$-based Thin Film Modules", Final Technical Report, NREL Subcontract No. ZN-1-19019-5 (1995).

J. R. Tuttle, J. R. Sites, A. Delahoy, W. Shafarman, B. Basol, S. Fonash, J. Gray, R. Menner, J. Phillips, A. Rockett, J. Scofield, F. R. Shapiro, P. Singh, V. Suntharalingam, D. Tarrant, T. Walter, S. Wiedeman and T. M. Peterson, "Characterization and Modeling of Cu(In,Ga)(S,Se)$_2$-based Photovoltaic Devices: A Laboratory and Industrial Perspective", *Progress in Photovoltaics: Research and Applications* **3**, 89 (1995).

W. Wettling, "High Efficiency Silicon Solar Cells: State of the Art and Trends", *Solar Energy Materials and Solar Cells* **38**, 487 (1995).

K. Zweibel, "Thin Films: Past, Present and Future", *Progress in Photovoltaics* **3**, 279 (1995).

R. W. Birkmire, J. E. Phillips, W. A. Buchanan, E. Eser, S. S. Hegedus, B. E. McCandless, P. V. Meyers and W. N. Shafarman, "Processing & Modeling Issues for Thin Film Solar Cell Devices", Annual Report to NREL, Subcontract No. XAV-3-13170-01 (1996).

T. J. Coutts, M. W. Wanlass, J. S. Ward and S. Johnson, "A Review of Recent Advances in Thermophotovoltaics", *25th IEEE Photovoltaic Specialists Conference Proc.*, IEEE Publishing, NY (1996), p. 25.

S. Guha, "Amorphous Silicon Alloy Solar Cells and Modules — Opportunities and Challenges", *25th IEEE Photovoltaic Specialists Conference*, IEEE Publishing, NY (1996), p. 1017.

H. Keppner, U. Kroll, P. Torres, J. Meier, D. Fischer, M. Goetz, R. Tscharner and A. Shah, "Scope of VHF Plasma Deposition for Thin-Film Silicon Solar Cells", *25th IEEE Photovoltaic Spec. Conf. Proc.*, IEEE Publishing, NY (1996).

K. W. Mitchell, R. R. King and T. L. Jester, "Cz Si Photovoltaics: Towards 100 MW", *25th IEEE Photovoltaic Specialists Conference*, IEEE Publishing, NY (1996), p. 541.

H. W. Schock, "Thin Film Photovoltaics", *Applied Surface Science* **92**, 606 (1996).

R. F. Service, "New Solar Cells Seem to Have Power at the Right Price", *Science* **272**, 1744 (1996).

K. Zweibel, H. S. Ullal and B. von Roedern, "Progress and Issues in Polycrystalline Thin-Film PV Technologies", *25th IEEE Photovoltaic Specialists Conference*, IEEE Publishing, NY (1996), p. 745.

7th International Conference on II-VI Compounds and Devices, *J. of Cryst. Growth* **159** (1996).

INDEX